Plasma Applications in Biomedicine

Plasma Applications in Biomedicine

Editor

Christoph Viktor Suschek

Basel • Beijing • Wuhan • Barcelona • Belgrade • Novi Sad • Cluj • Manchester

Editor
Christoph Viktor Suschek
Heinrich-Heine-University Dusseldorf
Düsseldorf
Germany

Editorial Office
MDPI AG
Grosspeteranlage 5
4052 Basel, Switzerland

This is a reprint of articles from the Special Issue published online in the open access journal *Biomedicines* (ISSN 2227-9059) (available at: https://www.mdpi.com/journal/biomedicines/special_issues/R2G7144VVN).

For citation purposes, cite each article independently as indicated on the article page online and as indicated below:

Lastname, A.A.; Lastname, B.B. Article Title. *Journal Name* **Year**, *Volume Number*, Page Range.

ISBN 978-3-7258-1547-0 (Hbk)
ISBN 978-3-7258-1548-7 (PDF)
doi.org/10.3390/books978-3-7258-1548-7

Cover image courtesy of Christoph Viktor Suschek

Contents

About the Editor

Christoph Viktor Suschek

Christoph Viktor Suschek, born November 1958, pursued an extensive academic and research career in the fields of physics, chemistry, and biology. From 1980 to 1989, he studied at Heinrich Heine University Düsseldorf, completing his diploma thesis in biology on June 6, 1989. Following his diploma, Suschek undertook doctoral research at the Institute of Immunobiology at Heinrich Heine University, earning his PhD in natural sciences in January 1996. Suschek continued as a scientific assistant at the Institute of Immunobiology during which he habilitated in July 2002, receiving the Venia Legendi in biochemistry and molecular biology. From February 2006 to March 2011, Suschek served as the head of research at the Clinic for Plastic Surgery, Hand, and Burn Surgery at the University Hospital RWTH Aachen. Since April 2011, Christoph Viktor Suschek has been the laboratory head at the Clinic of Orthopedics and Trauma Surgery at Heinrich Heine University Düsseldorf. His career reflects a deep commitment to advancing the understanding and application of biochemistry and molecular biology in medical sciences.

 *biomedicines*

Editorial

Plasma Applications in Biomedicine: A Groundbreaking Intersection between Physics and Life Sciences

Christoph V. Suschek

Department for Orthopedics and Trauma Surgery, Medical Faculty, Heinrich Heine University Düsseldorf, Moorenstraße 5, 40225 Düsseldorf, Germany; suschek@hhu.de

Citation: Suschek, C.V. Plasma Applications in Biomedicine: A Groundbreaking Intersection between Physics and Life Sciences. *Biomedicines* **2024**, *12*, 1029. https://doi.org/10.3390/biomedicines12051029

Received: 23 April 2024
Accepted: 26 April 2024
Published: 7 May 2024

Plasma applications in biomedicine represent a groundbreaking intersection between physics and life sciences, unveiling novel approaches to disease treatment and tissue regeneration [1]. The journey of plasma medicine begins with a profound understanding of the interactions between cold atmospheric plasmas (CAPs) and biological entities. Initially conceived as a tool for material processing, plasmas have emerged as multifaceted actors, engaging in intricate dialogues with living cells and tissues. The novelty of this field lies not only in its capacity to disinfect and heal but also in its potential to unravel the mysteries of cellular responses to controlled plasma exposure [2–4]. Plasma devices operating under atmospheric pressure and at low temperatures offer a compelling avenue for treating fragile materials and living tissues without inducing thermal damage [2]. Recently, CAP has garnered increasing attention in healthcare, particularly within the realms of dermatology and surgery. CAP demonstrates therapeutic potential in addressing bacterial infections, impaired wound healing, and chronic wounds among patients [5]. Notably, CAP exhibits antibacterial properties capable of combating antibiotic-resistant strains, such as methicillin-resistant Staphylococcus aureus (MRSA), while preserving adjacent tissue integrity [6]. Furthermore, studies have elucidated CAP's ability to modulate wound healing processes regulating key molecular pathways [7].

Research pertaining to the medical applications of cold plasma commonly falls under two principal categories: direct plasma treatment and indirect plasma treatment [3,8]. Direct treatment methodologies involve the treated tissue serving as an electrode interface, facilitating plasma generation between its surface and the electrode of the plasma device, typically utilizing dielectric barrier discharge (DBD) mechanisms. Conversely, indirect treatment methodologies entail plasma ignition within a conduit housing a flowing process gas, such as argon, helium, or air, thereby facilitating the conveyance of active species to the target substrate. Various plasma delivery systems, including plasma torches, plasma jets, plasma needles, and plasma pencils, have been developed and deployed in clinical settings.

In general, cold plasma systems engender a complex milieu of radical species, UV radiation, and charge carriers, each capable of modulating biological responses upon direct exposure to tissues or cells. When operated with ambient air, cold plasma systems yield substantial quantities of reactive oxygen species, encompassing ozone (O_3), superoxide (O_2^-), and hydroxyl radicals ($\cdot OH$), alongside reactive nitrogen species (RNS), such as nitric oxide (NO) and nitrogen dioxide (NO_2) [9,10]. These redox-active species are postulated to underlie most CAP-mediated effects on cellular, tissue, and microbial physiology. Notably, in biological contexts, living cells and tissues typically exhibit a hydrated state or are immersed within a liquid milieu. Consequently, CAP interactions primarily transpire within this liquid environment, diverging from conventional dry plasma applications such as surface sterilization [11,12].

This Special Issue of Biomedicine, *"Plasma Applications in Biomedicine"*, brings together cutting-edge research addressing various aspects of this multidisciplinary field. This collection of articles delves into the potential therapeutic applications of CAPs and plasma-activated liquids (PALs) across diverse medical domains [12,13]. As we explore the diverse landscapes of plasma applications in biomedicine, it is essential to acknowledge the depth and breadth of ongoing research in this field. The presented works, within the confines of this Special Issue, serve as glimpses into the expansive potential of CAPs and PALs, echoing the broader literature. The intersection of physics and medicine, epitomized by plasma medicine, holds promise not only for addressing current healthcare challenges but also for unlocking new frontiers in diagnostics, treatment, and our understanding of the complex interplay between plasmas and living organisms.

The manuscripts featured in this Special Issue showcase the versatility of plasma technology, highlighting its role in addressing challenging medical scenarios. As we navigate through the abstracts, the diverse applications of CAPs and PALs become apparent ranging from endodontics and recurrent aphthous stomatitis to wound healing, tissue regeneration, cancer treatment, and even antibacterial solutions for bone infection and vaginal health.

In "Non-Thermal Atmospheric Pressure Plasma Application in Endodontics", Ana Bessa Muniz et al. conduct a comprehensive literature review, emphasizing the potential of non-thermal atmospheric pressure plasma (NTPP) in endodontics. The antimicrobial efficacy of NTPP against key endodontic microorganisms is explored, revealing promising results for clinical practice [14].

Norma Guadalupe Ibáñez-Mancera et al. present a pilot study on the "Healing of Recurrent Aphthous Stomatitis by Non-Thermal Plasma". Their research demonstrates the effectiveness of NTP in reducing pain and inflammation and promoting rapid tissue regeneration in patients with recurrent aphthous stomatitis [15].

Moving beyond oral health, Lilith Elmore et al. investigate the "Healing Potential of J-Plasma Scalpel-Created Surgical Incisions". Their study compares the outcomes of incisions made by a plasma scalpel with those created by a steel scalpel, revealing no significant difference in the appearance or physiology of wound healing. This work opens avenues for exploring the potential of cold plasma in surgical procedures [16].

Hyun-Jin Kim et al.'s "Analyses of the Chemical Composition of Plasma-Activated Water" shed light on the unique chemical compositions of plasma-activated water (PAW). The study suggests that PAW, with its antibacterial properties, could serve as a novel vaginal cleanser, offering protection against pathogens while preserving beneficial bacteria [17].

The "Synergistic antimicrobial effect of cold atmospheric plasma and redox-active nanoparticles" is explored by Artem M. Ermakov et al. In their study, the combination of cold argon plasma and metal oxide nanoparticles demonstrates a powerful inhibitory effect on bacterial growth, providing a potential avenue for antimicrobial treatments [18].

Dennis Feibel et al.'s research on the "Gas Flow-Dependent Modification of Plasma Chemistry in μAPP Jet-Generated Cold Atmospheric Plasma" highlights the importance of modulating gas flow for therapeutic use. Their findings suggest that adjusting the gas flow in micro-scaled atmospheric pressure plasma jets could influence plasma chemistry and optimize biological outcomes, paving the way for tailored clinical applications [19].

Mahsa Bagheri et al.'s preclinical evaluation, "Can Cold Atmospheric Plasma Be Used for Infection Control in Burns? A Preclinical Evaluation", investigates the use of cold atmospheric plasma for infection control in burns. The study reveals that CAP is effective against Pseudomonas aeruginosa, a common infectious agent in burn wounds. While CAP's efficacy is lower than that of some conventional treatments, its potential role in burn wound management is underscored [20].

Madline P. Gund et al.'s "Effects of Cold Atmospheric Plasma Pre-Treatment of Titanium on the Biological Activity of Primary Human Gingival Fibroblasts" explores the benefits of CAP treatment on titanium surfaces. The study suggests that CAP-treated titanium exhibits enhanced fibroblast coverage without altering biological behavior, offering insights into potential applications in implantology [21].

The second contribution by the research group, "Cold Atmospheric Plasma Improves the Colonization of Titanium with Primary Human Osteoblasts: An In Vitro Study", by Madline P. Gund et al., investigates the effects of cold atmospheric plasma (CAP) treatment on titanium surfaces and its subsequent impact on primary human osteoblast colonization. The findings contribute to the growing body of evidence supporting the utility of cold atmospheric plasmas in biomedical applications, particularly in enhancing osteoblast colonization on implant surfaces [21].

Andreas Nitsch et al.'s investigation into the "Selective Effects of Cold Atmospheric Plasma on Bone Sarcoma Cells and Human Osteoblasts" provides crucial insights into the differential impact of CAP on malignant and non-malignant bone cells. The selective effect of CAP on sarcoma cells suggests its potential clinical application in oncology [22].

Finally, in their contribution, "Enrichment of Bone Tissue with Antibacterially Effective Amounts of Nitric Oxide Derivatives by Treatment with Dielectric Barrier Discharge Plasmas Optimized for Nitrogen Oxide Chemistry", Dennis Feibel et al. enrich bone tissue with antibacterially effective amounts of nitric oxide derivatives using dielectric barrier discharge plasmas. This innovative approach presents a potential strategy for combating bacterial complications during bone healing [23].

The diverse studies presented in this Special Issue collectively underscore the transformative potential of plasma applications in biomedicine. While much progress has been made, these findings also highlight the need for continued research to fully unlock the therapeutic benefits and mechanisms of action underlying cold atmospheric plasmas and plasma-activated liquids.

We extend our gratitude to all the authors who have contributed to this Special Issue, providing valuable insights and advancing our understanding of the exciting and rapidly evolving field of plasma applications in biomedicine.

Conflicts of Interest: The authors declare no conflicts of interest.

References

1. Bernhardt, T.; Semmler, M.L.; Schafer, M.; Bekeschus, S.; Emmert, S.; Boeckmann, L. Plasma Medicine: Applications of Cold Atmospheric Pressure Plasma in Dermatology. *Oxid. Med. Cell. Longev.* **2019**, *2019*, 3873928. [CrossRef]
2. Fridman, G.; Friedman, G.; Gutsol, A.; Shekhter, A.B.; Vasilets, V.N.; Fridman, A. Applied plasma medicine. *Plasma Process. Polym.* **2008**, *5*, 503–533. [CrossRef]
3. Kong, M.G.; Kroesen, G.; Morfill, G.; Nosenko, T.; Shimizu, T.; van Dijk, J.; Zimmermann, J.L. Plasma medicine: An introductory review. *New J. Phys.* **2009**, *11*, 115012. [CrossRef]
4. Yousfi, M.; Merbahi, N.; Pathak, A.; Eichwald, O. Low-temperature plasmas at atmospheric pressure: Toward new pharmaceutical treatments in medicine. *Fundam. Clin. Pharmacol.* **2014**, *28*, 123–135. [CrossRef] [PubMed]
5. Heinlin, J.; Morfill, G.; Landthaler, M.; Stolz, W.; Isbary, G.; Zimmermann, J.L.; Shimizu, T.; Karrer, S. Plasma medicine: Possible applications in dermatology. *J. Dtsch. Dermatol. Ges.* **2010**, *8*, 968–976. [CrossRef] [PubMed]
6. Zimmermann, J.L.; Shimizu, T.; Schmidt, H.U.; Li, Y.F.; Morfill, G.E.; Isbary, G. Test for bacterial resistance build-up against plasma treatment. *New J. Phys.* **2012**, *14*, 073037. [CrossRef]
7. Arndt, S.; Unger, P.; Wacker, E.; Shimizu, T.; Heinlin, J.; Li, Y.F.; Thomas, H.M.; Morfill, G.E.; Zimmermann, J.L.; Bosserhoff, A.K.; et al. Cold Atmospheric Plasma (CAP) Changes Gene Expression of Key Molecules of the Wound Healing Machinery and Improves Wound Healing and in vivo. *PLoS ONE* **2013**, *8*, e79325. [CrossRef] [PubMed]
8. Malyavko, A.; Yan, D.Y.; Wang, Q.H.; Klein, A.L.; Patel, K.C.; Sherman, J.H.; Keidar, M. Cold atmospheric plasma cancer treatment, direct versus indirect approaches. *Mater. Adv.* **2020**, *1*, 1494–1505. [CrossRef]
9. Graves, D.B. The emerging role of reactive oxygen and nitrogen species in redox biology and some implications for plasma applications to medicine and biology. *J. Phys. D Appl. Phys.* **2012**, *45*, 263001. [CrossRef]
10. Suschek, C.V.; Opländer, C. The application of cold atmospheric plasma in medicine: The potential role of nitric oxide in plasma-induced effects. *Clin. Plasma Med.* **2016**, *4*, 1–8. [CrossRef]
11. Kaushik, N.K.; Ghimire, B.; Li, Y.; Adhikari, M.; Veerana, M.; Kaushik, N.; Jha, N.; Adhikari, B.; Lee, S.J.; Masur, K.; et al. Biological and medical applications of plasma-activated media, water and solutions. *Biol. Chem.* **2018**, *400*, 39–62. [CrossRef] [PubMed]
12. Bruggeman, P.J.; Kushner, M.J.; Locke, B.R.; Gardeniers, J.G.E.; Graham, W.G.; Graves, D.B.; Hofman-Caris, R.C.H.M.; Maric, D.; Reid, J.P.; Ceriani, E.; et al. Plasma-liquid interactions: A review and roadmap. *Plasma Sources Sci. Technol.* **2016**, *25*, 053002. [CrossRef]
13. Graves, D.B. Low temperature plasma biomedicine: A tutorial review. *Phys. Plasmas* **2014**, *21*, 080901. [CrossRef]
14. Muniz, A.B.; Vegian, M.R.D.; Leite, L.D.P.; da Silva, D.M.; Milhan, N.V.M.; Kostov, K.G.; Koga-Ito, C.Y. Non-Thermal Atmospheric Pressure Plasma Application in Endodontics. *Biomedicines* **2023**, *11*, 1401. [CrossRef]

15. Ibáñez-Mancera, N.G.; López-Callejas, R.; Toral-Rizo, V.H.; Rodríguez-Méndez, B.G.; Lara-Carrillo, E.; Peña-Eguiluz, R.; do Amaral, R.C.; Mercado-Cabrera, A.; Valencia-Alvarado, R. Healing of Recurrent Aphthous Stomatitis by Non-Thermal Plasma: Pilot Study. *Biomedicines* **2023**, *11*, 167. [CrossRef] [PubMed]
16. Elmore, L.; Minissale, N.J.; Israel, L.; Katz, Z.; Safran, J.; Barba, A.; Austin, L.; Schaer, T.P.; Freeman, T.A. Evaluating the Healing Potential of J-Plasma Scalpel-Created Surgical Incisions in Porcine and Rat Models. *Biomedicines* **2024**, *12*, 277. [CrossRef]
17. Kim, H.J.; Shin, H.A.; Chung, W.K.; Om, A.S.; Jeon, A.; Kang, E.K.; An, W.; Kang, J.S. Analyses of the Chemical Composition of Plasma-Activated Water and Its Potential Applications for Vaginal Health. *Biomedicines* **2023**, *11*, 3121. [CrossRef]
18. Ermakov, A.M.; Afanasyeva, V.A.; Lazukin, A.V.; Shlyapnikov, Y.M.; Zhdanova, E.S.; Kolotova, A.A.; Blagodatski, A.S.; Ermakova, O.N.; Chukavin, N.N.; Ivanov, V.K.; et al. Synergistic Antimicrobial Effect of Cold Atmospheric Plasma and Redox-Active Nanoparticles. *Biomedicines* **2023**, *11*, 2780. [CrossRef]
19. Feibel, D.; Golda, J.; Held, J.; Awakowicz, P.; von der Gathen, V.; Suschek, C.V.; Oplaender, C.; Jansen, F. Gas Flow-Dependent Modification of Plasma Chemistry in μAPP Jet-Generated Cold Atmospheric Plasma and Its Impact on Human Skin Fibroblasts. *Biomedicines* **2023**, *11*, 1242. [CrossRef]
20. Bagheri, M.; von Kohout, M.; Zoric, A.; Fuchs, P.C.; Schiefer, J.L.; Oplaender, C. Can Cold Atmospheric Plasma Be Used for Infection Control in Burns? A Preclinical Evaluation. *Biomedicines* **2023**, *11*, 1239. [CrossRef]
21. Gund, M.P.; Naim, J.; Lehmann, A.; Hannig, M.; Linsenmann, C.; Schindler, A.; Rupf, S. Effects of Cold Atmospheric Plasma Pre-Treatment of Titanium on the Biological Activity of Primary Human Gingival Fibroblasts. *Biomedicines* **2023**, *11*, 1185. [CrossRef] [PubMed]
22. Nitsch, A.; Sieb, K.F.; Qarqash, S.; Schoon, J.; Ekkernkamp, A.; Wassilew, G.I.; Niethard, M.; Haralambiev, L. Selective Effects of Cold Atmospheric Plasma on Bone Sarcoma Cells and Human Osteoblasts. *Biomedicines* **2023**, *11*, 601. [CrossRef] [PubMed]
23. Feibel, D.; Kwiatkowski, A.; Oplander, C.; Grieb, G.; Windolf, J.; Suschek, C.V. Enrichment of Bone Tissue with Antibacterially Effective Amounts of Nitric Oxide Derivatives by Treatment with Dielectric Barrier Discharge Plasmas Optimized for Nitrogen Oxide Chemistry. *Biomedicines* **2023**, *11*, 244. [CrossRef] [PubMed]

biomedicines

Review

Non-Thermal Atmospheric Pressure Plasma Application in Endodontics

Ana Bessa Muniz [1], Mariana Raquel da Cruz Vegian [1], Lady Daiane Pereira Leite [1], Diego Morais da Silva [1], Noala Vicensoto Moreira Milhan [1], Konstantin Georgiev Kostov [2] and Cristiane Yumi Koga-Ito [1,*]

[1] Department of Environment Engineering and Sciences Applied to Oral Health Graduate Program, São José dos Campos Institute of Science and Technology, São Paulo State University (UNESP), São José dos Campos 12247-016, SP, Brazil; bessa.muniz@unesp.br (A.B.M.)

[2] Department of Physics, Faculty of Engineering in Guaratinguetá, São Paulo State University (UNESP), Guaratinguetá 12516-410, SP, Brazil

* Correspondence: cristiane.koga-ito@unesp.br

Abstract: The failure of endodontic treatment is frequently associated with the presence of remaining microorganisms, mainly due to the difficulty of eliminating the biofilm and the limitation of conventional irrigation solutions. Non-thermal atmospheric pressure plasma (NTPP) has been suggested for many applications in the medical field and can be applied directly to biological surfaces or indirectly through activated liquids. This literature review aims to evaluate the potential of NTPP application in Endodontics. A search in the databases Lilacs, Pubmed, and Ebsco was performed. Seventeen manuscripts published between 2007 and 2022 that followed our established inclusion criteria were found. The selected manuscripts evaluated the use of NTPP regarding its antimicrobial activity, in the direct exposure and indirect method, i.e., plasma-activated liquid. Of these, 15 used direct exposure. Different parameters, such as working gas and distance from the apparatus to the substrate, were evaluated in vitro and ex vivo. NTPP showed a disinfection property against important endodontic microorganisms, mainly *Enterococcus faecalis* and *Candida albicans*. The antimicrobial potential was dependent on plasma exposure time, with the highest antimicrobial effects over eight minutes of exposure. Interestingly, the association of NTPP and conventional antimicrobial solutions, in general, was shown to be more effective than both treatments separately. This association showed antimicrobial results with a short plasma exposure time, what could be interesting in clinical practice. However, considering the lack of standardization of the direct exposure parameters and few studies about plasma-activated liquids, more studies in the area for endodontic purposes are still required.

Keywords: non-thermal plasma; endodontics; root canal; microorganism

Citation: Muniz, A.B.; Vegian, M.R.d.C.; Pereira Leite, L.D.; da Silva, D.M.; Moreira Milhan, N.V.; Kostov, K.G.; Koga-Ito, C.Y. Non-Thermal Atmospheric Pressure Plasma Application in Endodontics. *Biomedicines* **2023**, *11*, 1401. https://doi.org/10.3390/biomedicines11051401

Academic Editors: Christoph Viktor Suschek, Peter E. Murray and Gianluca Gambarini

Received: 25 February 2023
Revised: 15 March 2023
Accepted: 24 April 2023
Published: 9 May 2023

1. Introduction

Microbial presence is the primary etiological cause of pulp and periapical infections [1]. Root canal infections are mediated by microbial species that form biofilms. Multispecies composition combined with the radicular system complexity makes the canal system disinfection a complex procedure. In this context, microbial persistence is the factor that mostly contributes to failures in endodontic treatment. It allows root canal infection recidivism. This reality negatively affects the quality of life of the patients [2]. Among the several agents involved in persistent endodontic infections are *Enterococcus faecalis* and *Candida albicans*. The pathogenicity of these microorganisms is related to their ability to survive in the dentine tubules for an extended period, keeping them viable even in scarcity [3].

Endodontic treatment success primarily depends on the capacity to eliminate the intracanal infection source [4]. Therefore, root canal disinfection represents a necessary procedure before canal filling. It includes conventional mechanical–chemist methods, such as debridement and irrigation with a chemical agent.

The use of non-thermal atmospheric pressure plasmas (NTPPs) in the biomedical sciences has emerged as a promising alternative in several areas. Their properties, mostly related to the generation of reactive oxygen and nitrogen species (RONS) [5], allow microbial deactivation, making NTPPs advantageous for the sterilization of implants and medical/dental instruments [6–8]. In addition to the antimicrobial properties, NTPPs have demonstrated anti-inflammatory effects, favoring tissue repair [9,10]. Different works focused on cancer treatment have also investigated NTPPs as a promising tool to act as an adjuvant therapy against cancer [11,12]. In addition, antifungal, antibacterial, antiparasitic and antiviral properties have been described [13,14].

Cells, tissues or materials can be exposed to NTPPs in two different ways: direct, when NTPP in the form of a plasma plume is applied to a given substrate, or indirect, when a liquid is activated by an NTPP prior to its application to the substrate [15]. Both direct and indirect methods have shown good perspectives in dentistry [16]. Since root canal reinfection is recurrent and the conventional methods and techniques have failures and risks [17,18], the use of NTPPs seems to represent an alternative to promote disinfection. Therefore, this review aims to discuss the main problems related to endodontic failures and the possible role of NTPPs in this context. For this, the main advances in using NTPPs in the endodontics field will be presented and discussed.

2. Design of the Study

This review is divided into six sections. The third section will introduce the possible role of NTPPs in the endodontic field. For this, the microbiota related to endodontic infection, the conventional irrigants used in the endodontic treatment, as well as their limitations, and the main physicochemical properties of plasma will be discussed. Section 4 will explore the main text of this integrative literature review about NTPPs applied in the endodontic field, followed by a discussion in Section 5 and finally the conclusion of the work in Section 6.

In this integrative literature review, recent studies focused on the effect of direct and indirect NTPPs exposure on microorganisms of endodontic interest were selected. The following keywords, Nonthermal Plasma, Endodontics, Cold plasma and Treatment, were used in four databases (Science Direct, Lilacs, Pubmed, and Ebsco). In the general search, 279 reviews were found. From these, only research studies or literature reviews that investigated NTPPs applied to the endodontic area from 2007 to 2022 were included. Reference lists of referral studies were inspected to identify any additional relevant published data. Studies investigating the application of non-thermal atmospheric pressure plasmas in other areas of dentistry were not considered. After the evaluation of the mentioned criteria, 17 papers that included in vitro and ex vivo studies on the use of NTPPs applied to the endodontic area were selected.

3. Possible Role of NTPPs in the Endodontic Area

3.1. Infectious Endodontic Microbiota

The main cause of endodontic failure is the presence of microorganisms that lead to intra-rooted and extra-rooted infections and become resistant to disinfection protocols [19]. Although fungi, archaea and viruses contribute to the microbial diversity in endodontic infections, bacteria are the most common microorganisms present in these infections. Their diversity varies significantly according to the infection type. Moreover, different penetration levels in the endodontic invasion space can be observed in different conditions, such as caries lesions, traumatic pulp exposure, and fractures [20,21].

Root canal bacteria can be isolated as planktonic cells, suspended in the root canal liquid phase. However, aggregates or congregations can adhere to the root canal walls, forming biofilm layers. Biofilms are the bacteria growing model, where the sessile cells interact to form dynamic communities linked to a solid substrate. They are found in an extracellular polymeric matrix [22]. Endodontic infection can be classified as primary and secondary, wherein the first usually presents a great wealth of species. In contrast, in the second category, there are one or two species [2].

The bacteria frequently associated with persistent endodontic infection after treatment are usually Gram-positive. There is an equal distribution of facultative and obligate anaerobic microorganisms, with a prevalence of streptococci and enterococci [21]. Other bacterial species have been related to endodontic infections. The first endodontic infection clinic case with *Klebsiella variicola* has recently been reported [22].

The most common species involved in apical reinfection is *Enterococcus faecalis*, which is responsible for 80% of endodontic infections in humans. Its pathogenicity can be attributed to its high survival ability inside dentinal tubules [4,21,23]. Its environmental resistance is due to characteristics such as the species' capacity to form biofilms and ability to support high pH, such as the one used in intracanal medication with Ca (OH)$_2$ [24]. This pathogen tends to stand out when in a stable multi-species microbiota, becoming the dominant species [21]. It is estimated that the *E. faecalis* biofilm is about 1000 times more resistant to antimicrobial agents when compared to its planktonic counterpart [1].

For decades, endodontics microbiologists concentrated their studies on the role of bacteria in the etiopathogenesis of endodontic infection. Nevertheless, in recent years, research in this field has demonstrated an emerging interest in microorganisms from other kingdoms, such as archaea, viruses and fungi [25]. In the Fungi kingdom, *Candida* species are the most frequently related to endodontic reinfection. This yeast is present in a low proportion in endodontic infection (around 10%). However, *Candida albicans*, which is the species most frequently detected, presents a low susceptibility to the intracanal disinfectants commonly used, being able to survive inside the dentinal tubules due to the ability to form biofilms, leading to reinfection [23,25,26]. Its pathogenicity evolves dental pulp and peri-radicular tissue cells, mainly in immunocompromised patients [26].

3.2. Conventional Endodontic Irrigants

The endodontic therapy approach aims to reduce the microbial load to a subclinical level, removing the necrotic material and disinfecting the canal through the chemical–mechanical preparation using lime instrumentation, disinfection with irrigants, and subsequent intracanal medication [2].

Disinfection solutions, such as sodium hypochlorite (NaOCl) or chlorhexidine, and chelating (i.e., with EDTA) show outstanding performance during disinfection [17]. Although highly effective, NaOCl solutions have limited penetration capacity in the apical part and the dentinal tubules, mainly in bi- or multi-rooted teeth with more complex intra-rooted anatomy [4].

For better efficacy, techniques such as applying ultrasound to reduce the surface tension or the pressure increment in the irrigation moment have been proposed. However, during this procedure, fluid extravasation can injure the healthy periapical [4,18]. This is related to the fact that sodium hypochlorite is highly toxic and can lead to an inflammatory process in the vital tissues, leading to tissue necrosis. It occurs due to the high sodium hypochlorite oxidative potential. Furthermore, NaOCl can modify dentin properties and characteristics, such as microhardness and resistance. These changes can affect the stability of the dental element [17,18].

In this sense, an innovative non-thermal atmospheric plasma jet technique can represent an alternative to the conventional treatment [18]. Studies have reported the efficacy of a plasma jet against biofilms inside the root canal, which suggest that this technology can be helpful as an adjuvant in conventional disinfection therapies, seeking to decrease the discussed limitations. One of the advantages of plasma is its capacity to reach microorganisms lodged in narrow niches, in addition to low toxicity to eukaryotic cells [23].

3.3. Main Physicochemical Properties Related to NTPPs

Plasma is a complex mixture of electrons, ions and neutral species and is generally considered to be the fourth state of matter. Plasma is abundant in nature (stars, auroras, lightning, etc.) and can also be produced in the laboratory at different conditions using electrical discharges. Many plasma devices, such as plasma torches and microwave discharges, are operated at high temperatures. It is also possible to generate non-thermal plasma in which the gas temperature is quite low. Especially attractive for applications is the non-thermal plasma produced at atmospheric pressure, because it does not require the use of vacuum systems [27]. NTPPs have two crucial features that make them applicable in medical and dentistry fields: plasma has high plasma has biological effects due to the reactive species, and due to the reactive species, and it does not damage the tissue structure [4].

Usually, non-thermal plasmas are generated by applying short electrical pulses to a discharge gap, where through gas ionization unstable reactive species are generated while the temperature is kept stable [27]. NTPPs are obtained from a low-temperature gas (<40 °C), which does not induce thermal injuries to the tissues [17]. These gases are partially ionized, containing highly reactive particles such as excited atoms, oxygen and nitrogen reactive species, U.V. irradiation, photons and electrons [1,2,28]. The main reactive oxygen species (ROS) include ozone, atomic oxygen, superoxide, hydrogen peroxide and hydroxyl radicals [29]. ROS present key roles in cell signaling pathways and redox reactions, with potential for biomedical applications [30].

Gases such as Helium (He), Argon (Ar), Nitrogen (N), Oxygen (O), atmospheric air and a mixture of the previously cited ones are frequently used for NTPP generation. NTPPs compounded by Ar/O_2 and Ar/air have been considered the most efficient [23].

4. NTPPs Applied to Endodontics

The aim of root canal treatment is the reduction in viable bacteria in the local. Bacterial persistence during canal filling is one of the risk factors that can lead to apical periodontitis post-treatment [20]. The application of NTPPs has demonstrated promising results in association with conventional endodontic treatment regarding their antimicrobial effects [23]. There are two modalities of treatment that have been explored in vivo, in vitro and ex vivo: direct and indirect. While in the direct modality the area of interest on the substrate is directly exposed to NTPPs, in the indirect one different liquids which have been previously activated by plasma are then applied to the substrates. The indirect plasma application is based on the fact that when plasma is in contact with liquid its reactive species interact with the liquid molecules, producing other long-lasting reactive species in the liquid. The plasma-activated liquids' PALs can be frozen and stored for long periods of time before being used for biomedical treatments. The available literature on the use of NTPPs in Endodontics (Table 1) will be discussed in different subtopics.

Table 1. Studies on the use of NTPP in Endodontics.

Author/ Year	Type of Study	Plasma Exposure Method	Microorganism	Working Gas	Exposure Time	Distance from Apparatus to the Substrate	Other Antimicrobial Agent	Main Findings
Jiyang et al. [31]	In vitro and ex vivo	Direct	*Bacillus atrophaeus*	He and He $(1\%)O_2$ He $(1\%)O_2$	1 min 3 min	5 mm	No	$He/(1\%)O_2$ NTPP is more antimicrobial than He-NTPP alone against *B. atrophaeus*
Pan et al. [3]	In vitro	Direct	*E. faecalis*	$Ar/O_2[2\%]$	2–10 min	5 mm	No	NTPP was effective against *E. faecalis*. The exposure time of 8 or 10 min had significantly higher antimicrobial efficacy
Bansode et al. [32]	In vitro	Direct	*E. faecalis*	He He/O_2	2 min	2–3 cm	Chlorhexidine (CLX)	A significant reduction in the biofilm viability was observed after chlorohexidine or NTPP treatment

Table 1. *Cont.*

Author/ Year	Type of Study	Plasma Exposure Method	Microorganism	Working Gas	Exposure Time	Distance from Apparatus to the Substrate	Other Antimicrobial Agent	Main Findings
Du et al. [33]	In vitro	Direct	*E. faecalis and multispecies biofilms*	He and He/O$_2$	2–5 min	5 mm	CLX	Modified nonequilibrium plasma was more effective in killing *E. faecalis* and multispecies biofilms at both 2 and 5 min than conventional plasma. No significant difference was detected between nonequilibrium plasma and CHX groups
Habib, Hottel and Hong [4]	Ex vivo	Direct	*E. faecalis*	Argon	2 min	Not indicated	No	NTPP presented significant antimicrobial effects. It was as effective as 6% sodium hypochlorite
Jablonowski et al. [34]	Ex vivo	Direct	*E. faecalis*	Argon	3 min	2 mm	NaOCl and CLX	All treatments led to significant reduction of *E. faecalis* compared to NaOCl. NTPP was the most effective treatment
Schaudinn et al. [27]	Ex vivo	Direct	*Biofilm*	He/O$_2$	30 min	Not indicated	NaOCl	NTPP showed lower biofilm removal than 6% NaOCl
Üreyen Kaya et al. [35]	Ex vivo	Direct	*E. faecalis*	He/O$_2$	2 min	1 mm	NaOCL, Ozone	The highest antibacterial activity was observed in the NaOCl, NTPP and ozone groups. In the middle third of the root canal wall NTPP presented superior efficacy than NaOCl
Doria et al. [36]	In vitro	Direct	*Candida albicans*	Argon and Argon+ compressed air	10 min	30 mm	No	Cell viability tests indicated that only about 8% of the yeast cells treated with Argon+ compressed air plasma could survive, proliferate and/or generate other cells. This treatment was the most effective in *Candida albicans* biofilm inactivation
Herbst et al. [37]	Ex vivo	Direct	*E. faecalis*	Argon	30 s (CLX) 60 s (NTPP)	1 mm	CLX	The highest antimicrobial action was observed in the association of NTPP and CLX, followed by NTPP and CLX alone
Simoncelli et al. [17]	In vitro and model	Direct and indirect	*E. faecalis*	He	3 min (direct) 1 min (indirect)	5 mm	NaOCl and CLX	The highest level of bacterial inactivation was observed in the dry environment by direct exposure, but a relevant bacterial load reduction was also obtained when the root canal system was irrigated with PAW
Armand et al. [1]	Ex vivo	Direct	*E. faecalis*	He and He/(0.5%)O$_2$	2–8 min	2 mm	PDT	All the modalities showed a significant reduction in bacteria after treatment, and He/O$_2$ NTPP was the most effective against *E. faecalis*, followed by photodynamic therapy and He plasma, respectively

Table 1. *Cont.*

Author/ Year	Type of Study	Plasma Exposure Method	Microorganism	Working Gas	Exposure Time	Distance from Apparatus to the Substrate	Other Antimicrobial Agent	Main Findings
Ledernez et al. [18]	In vitro	Direct	*E. faecalis, Streptococcus mutans Staphylococcus aureus Pseudomonas aeruginosa Escherichia coli*	He/(1%)O_2	3 min	1–3 mm	No	After treatment, the median size of the bacterium-free area in Petri dishes of E. faecalis was 0.25 mm^2, which corresponded to five times the area of the plasma nozzle and matched the target surface area in root canal treatment. This value was even larger for other investigated bacteria. *E. coli* presented the largest median bacterium-free surface area (2.5 mm^2)
Li et al. [31]	In vitro	Direct	*E. faecalis*	Ar/O_2	3, 6, 9 and 12 min	10 mm	Ca(OH)$_2$, 2% CLX gel and Ca(OH)$_2$/CLX	There were no detectable live bacteria after 12 min of NTPP treatment
Sallewong et al. [24]	Ex vivo	Direct	*E. faecalis*	He/O_2	1 min	Not indicated	NaOCl	NTPP showed significant bacterial reduction as well as NaOCl and NaOCl + NTPP. The NaOCl + NTPP group significantly reduced *E. faecalis* in the deeper dentin level compared to the other groups
Li et al. [31]	In vitro	Indirect	*E. faecalis*	Compressed air	10–90 s	20 mm	No	PAW treatment inhibited *E. faecalis* biofilm and decreased quorum-sensing-related virulence genes expression
Kerli-kowski et al. [23]	Ex vivo	Direct	*Candida albicans*	Ar/O_2	6–12 min	1–2 mm	CLX NaOCL OCT	NTPP presented the highest disinfection efficiency among all the treatments, after 6 and 12 min of exposure

4.1. Direct Method of NTPP Application

One of the first ex vivo studies using NTPP inside a root canal with a microbial biofilm from saliva showed that the application of plasma for 5 min reduced the total amount of microorganisms until 1 mm deep into the dentin root canal when compared to the non-treated canal. The NTPP was generated from a mixture of He and O_2 (1%). Additionally, the authors performed an in vitro test with *Bacillus atrophaeus* that showed 100% disinfection. The authors mentioned, in addition, that the biofilm removal level and the dispersion level inside the root canal differed depending on the treatment parameters [31].

To evaluate *E. faecalis'* inactivation by NTPP, Pan et al. [3] submitted 85 single-rooted teeth to treatment with argon and oxygen plasma (98% + 2%) with a gas flux of 5 L/min for periods varying between 2 and 10 min. After, they evaluated the viability of bacteria cells with Scanning Electron Microscopy (SEM) and Laser microscopy. The authors found that the inactivation level increased over the plasma exposure time. In this scenario, the periods between 8 and 10 min were the most efficient in achieving reliable disinfection. The SEM images showed the biofilm rupture in the tubules and on the dentin surface after 10 min of NTPP exposure.

Some investigations using NTPP on the most prevalent species in endodontic infection have demonstrated promising results in the disinfection process. A reduction in *E. faecalis* planktonic cells was observed after He-generated NTPP exposure for 2 min [32].

The Plasma Creator device model RC-2 has been used to study plasma application with chlorhexidine. For this, a gas tube containing He and O_2 (1%) and chlorhexidine (2% solution) was connected to the device. To test this model, 120 bovine dentin disks were experimentally infected with *E. faecalis* or microcosm biofilm collected from a clinical infection. The results indicated that the plasma associated with chlorhexidine was more effective than the isolated treatment. It showed an 80% reduction in bacterial viability when applied for 5 min [33].

Ledernez et al. [18] evaluated the in vitro NTPP effect against microorganisms of medical interest, such as *E. faecalis*, *S. mutants*, *S. aureus*, *P. aeruginosa* and *E. coli*. The working gas was a mixture of He and O_2 with a flux of 1 L/min. The exposure was performed for 3 min at 1–3 mm from the Petri dish surface. It was observed that the disinfection efficacy varied according to the species. *E. faecalis* was the most resistant, while *E. coli* was the most susceptible to the plasma treatment. It was also found that the halo area (0.25 mm^2) corresponding to the decontamination was higher than the target area in the root treatment.

A recent study evaluated the effect of a gas mixture of He and O_2 plasma applied on ex vivo *E. faecalis* biofilms. These biofilms were grown inside single-root teeth. Thus, Saleewong et al. [24] evaluated the plasma effect with a gas flux of 0.5 L/min in 54 teeth. The dental elements were allocated to five experimental groups (NaOCl, NTPP, NaOCl + NTPP, only the gas, and the negative control). The results demonstrated a significant reduction when using the NTPP for 5 min, or for 1 min in association with NaOCl (5%). Moreover, the association of both decontamination techniques led to a greater depth level of decontamination in the dentin.

To evaluate the treatment with NTPP in depth dentinal tubules, an NTPP generated with dielectric barrier discharge (DBD) was applied at 1 mm from the prepared root. The human teeth had a length of 14 mm, and the treatment was performed for 5 min [35]. The authors verified that NTPP had a disinfection capacity almost similar to NaOCl (2.5%) and seemed to be more effective in the middle part of the dentin. Minor effects were observed in the coronal and apical thirds.

Other investigations about NTPP antimicrobial effects in different depths of root dentine were performed. Herbst et al. [37] used argon plasma for 60 s isolated or in association with CHX (2%). The exposition time was 30 s inside roots from human premolar teeth. They observed that the combination of CHX and NTPP was more effective than isolated treatments in the general disinfection, inside the coronary part (0–300 μm) and also in the deeper layers (500–800 μm). These results suggest that an antimicrobial reduction to a maximum depth of 800 μm can be reached by using NTPP as an adjuvant.

The comparison of NTPP and other antimicrobial agents was also evaluated in a study with infection root canal simulation. For this, the teeth were infected with *E. faecalis* cells. Ex vivo treatments with Ar NTPP, NaOCl (0.6%) or CHX (0.1%) were performed for 3 min [34]. The treatment with NTPP had visible destructive effects and was better at the elimination of the residual bacteria compared to the control and the conventional irrigants. However, the visible destruction of the cellular walls and the most effective reduction in the microbial load were observed with the association of the treatments. In this study, the disinfection action of NTPP could be considered superior to the CHX and equal to NaOCl 0.6%.

Another study compared ex vivo teeth treated with the He/O_2 NTPP for 30 min with teeth treated with conventional irrigants. Microorganism reduction was observed in both cases. According to the authors, the concentration of the irrigants and the application time of the irrigants and plasma are key factors. For example, treatments with NaOCl at 6% were shown to be 4x more potent when compared to the He and O_2 (1%) NTPP, both applied for 30 min in each extracted human teeth root canal [27].

Armand et al. [1] submitted 60 single-rooted human teeth to He and O_2 combined NTPP under 4 L/min with an injector nozzle of 2 mm. The injector nozzle was inserted in the root canal for a time between 4 and 8 min. The results were compared to the control group, and the group treated only with He. All NTPP groups significantly reduced the microbial load, but the mixed gas worked better than the pure He NTPP.

Interestingly, a study showed that the treatment with Ar/O_2 (2%) NTPP on the root canal reduced 98.8% of *E. faecalis* colonies after 8 min of exposure. When the microbial population was evaluated after seven days of the treatment, a reinfection was detected. The authors related that it would be necessary to have 30 min of NTPP exposure to prevent the reinfection [38]. Nevertheless, NTPP was shown to be a fast and efficient alternative when compared to the traditional treatment composed of intracanal medicine, which could stand for 1 or 2 weeks. Another study where NTAPP was operated with the same gases verified that treatment for 12 min eliminated three-week *E. coli* biofilms from the root canal [28].

Resistant microorganisms commonly seen in oral infections are not restricted to bacterial species, but also appear in fungi. Between them, some strains of *C. albicans* can lead to a persistent infection process. It is estimated that fungal infection can be present in approximately one in every ten endodontics infections. From these, 9% are observed in primary disorders and 9.3% in secondary ones. However, the role of fungi in the pathogenesis of endodontic infection is still unclear [39].

In this sense, Doria et al. [36] investigated the effect of NTPP using a mixture of Ar and air on *C. albicans* biofilm. The authors worked with three groups: control, NTPP and the plasma air mixture. The microorganism reduction was 85% and 88.1% in the plasma and air plasma groups, respectively. The fungal viability was 33% in the plasma group and only 8% in the association group (air plasma), demonstrating that this technology could represent an exciting alternative for medical-dentistry applications.

Kerlikowski et al. [23] evaluated Ar/air NTPP in the proportion 99:1 with a flux of 5 L/min. The exposure was performed in incisive maxillary and mandibular premolars infected with *C. albicans*. The study compared the effect of pure gas, saline solution (0.9%), CHX at 2%, NaOCl at 5.25%, octenidine (0.1%) and Ar/O_2 NTPP. The most efficient treatments were single NTPP, followed by the treatments where plasma was associated with a common irrigant solution. The treatment consisted of irrigant contact for 6 min and NTPP exposure for 6 more min. When considering monotherapies, the application of plasma for 12 min stands out compared to the conventional irrigants.

4.2. Indirect Method of NTPP Application

To investigate He NTPP for endodontic disinfection, some authors used a handheld Plasma Gun (PG) [17]. The authors evaluated the direct and indirect modalities of treatment. For the indirect procedure, distilled water was activated, generating plasma-activated water (PAW). PAW was prepared through the exposure of 100 µL of sterile water to NTPP for 1, 3 and 5 min in 96-well-plates and immediately used in the root canals. The root canals were contaminated by *E. faecalis* in a wet and dry environment. They verified a higher microbial load reduction in a humid climate when the PAW was activated for 5 min. In the same study, it was observed that NTPP direct treatment in a dry environment showed the best decontamination results among all the procedures, including 3 min of NaOCl (0.6%) and CHX (0.2%).

Another recent study using PAW activated with compressed air on *E. faecalis* verified that there was a planktonic cell reduction with treatment for at least 45 s. A decreased capacity of the microorganisms to grow biofilm and to express virulence was also observed after the treatment [28].

5. Discussion

NTPP is attractive because it combines conventional treatment efficiency and outstanding safety. Different generation sources may be related to NTPP [17]. The direct application was demonstrated to be heterogeneous, mainly due the differences between the used parameters. These parameters include time of exposure, work gas, and power supply characteristics [1].

The works analyzed in this study showed a large work gas variety. The most used work gas was pure helium [17,32] or its association with oxygen [1,17,18,24,27,31,33]. The use of argon was described in seven studies [3,23,28,34,36,37]. Only one study evaluated the exclusive usage of ozone to generate NTPP [35].

The exact mechanism that promotes the microorganism inactivation and destruction is still unclear. It is assumed that U.V. irradiation and charged particles promote microorganism disruption through atomic oxygen or irradiation, damaging the extracellular biofilm matrix, which is compounded by a polymer [3]. Reactive species and atoms obtained from molecular oxygen are essential due to their capacity to penetrate the cells [1]. However, a recent study refutes the theory that plasma can act in cell wall destruction after spectrophotometric analysis did not find any intracellular liquid extravasation clues, such as nucleic acids [40].

Preliminary studies about the effect of PAW on *E. faecalis* planktonic cells [28] and *C. albicans* [41] showed an inhibitory activity of these microorganisms with a decreased expression of virulence genes [28] In this way, PAW is also a promising adjunct treatment for the endodontic area, as well as what it has shown in other dentistry areas [42].

To our knowledge, plasma and plasma-activated liquids present different mechanisms of action. In both cases, the bactericidal effects can be attributed to a combination of physical-chemical products. Direct exposure is mainly related to the formation of ROS and U.V., while RNS has been observed in higher amounts in PAWs. RNS and ROS can damage cellular components such as the cellular walls, proteins, and genetic material. However, the solubility effect of reactive species found in PAW differs from species generated in the gas phase. Moreover, environmental interactions can be observed in PAW because changes may happen to composition and efficacy during the treatment time [16,42,43]. It is worth mentioning that RNS are typical in aerobic environments, so the bacteria may preset the natural protection system against such aggression, represented by enzymes, molecules, and neutralized proteins [43].

Saleewong et al. [24] pointed to a high penetrability ability of NTPP in the gas phase to reach the microorganism after the irrigant usage. This study also showed that without previous electric or electromagnetic activation to ionize the gas, He exposure did not show a bactericidal effect. The reactive species generated in NTPP, such as atomic oxygen, ozone, and hydroxyl radicals, seemed to confer disinfectant capacity.

In the reviewed studies, the minimal exposure time was 2 min [3,32,38] while the most frequent application time was equal to or superior to 5 min [23,24,31,35]. The maximum application time was 30 min, as Schaudin et al. [27] described. The researchers assigned a direct relationship between the exposure time and the disinfection capacity. According to Bansode et al. [32], an exposure time of 2 min was enough to reduce the microbial load significantly. On the other hand, Du et al. [33] estimated that after 2 min the microbial load could be reduced by 54%, while a reduction of 80% could be reached after 5 min, when compared to CHX.

On the other hand, several studies have demonstrated that the ideal time of exposure is greater than 8 min [3,28,38]. Pan et al. [3] noticed that the microbial inactivation increased with exposure time, with the period between 8 and 10 min being the most efficient for disinfection, with biofilm rupture in the dentine surface tubules and a reduction in biofilm viability after 10 min.

Li et al. [28] evidenced the destruction of biofilm architecture and the decrease in its thickness after NTPP exposure, where plasma achieved the ideal intra-root disinfection after 12 min. Wang et al. [38] obtained great disinfection results at 6–8 min of Ar/O$_2$ NTPP. The authors observed that the disinfection through the root was heterogeneous, so in the apical region it was still possible to find considerable viable amounts of microorganisms. Because of this, they concluded that the best result to avoid reinfection would be NTPP exposure for 30 min, which could lead the reinfection risk to zero.

According to Ledernez et al. [18], there are differences in the microorganism susceptibility to NTPP action, i.e., the antimicrobial effect of NTPP may change according to the microbial species. *E. faecalis* has been presented as the most resistant, while *E. coli* has been considered the most susceptible to NTPP treatment. *Streptococcus* and *Staphylococcus* spp. have shown intermediate susceptibility. NTPP may be efficient on young and mature *E. faecalis* forms [1]. The vulnerability can be attributed to different compositions of the cell wall. Usually, Gram-positive bacteria have a thicker polyglycan layer when compared to Gram-negative bacteria, which can act as a shield against NTPP action [43].

The tooth canal anatomy can be complex. Moreover, most of the research focused on NTPP to the endodontic area until the present has been based on ex vivo models using single-rooted teeth. Armand et al. [1] warned that this dental type is less complex and has a straighter canal compared to multichannel teeth. Therefore, the NTPP effect on multichannel teeth may present a different behavior.

The surface complexity on which the treatment is performed must be considered once the plane surface does not represent the canal cavity structure and can show outstanding results in a short treatment time [23].

According to Schaudinn et al. [27], NTPP is, in theory, able to reach the whole canal extension because the reachable area is superior to the intra-root surface. Using devices such as applicator nozzles, it is estimated that the decontamination area corresponds to five times the nozzle area and reaches a size equivalent to 0.25 mm^2, which could promote enough space to disinfect the walls of the whole tooth canal [18].

However, NTPP efficacy has been shown to be heterogeneous through the canal. Üreyen Kaya et al. [35] highlighted the high NTPP ability to eliminate microorganisms on dentinal tubules and canal walls. Despite this, the same could not be observed in the coronal and apical areas. According to the authors, the highest disinfection capacity could be attributed to the highest NTPP formation under the nozzle at 5–6 mm. This area corresponds to the medial region.

NTPP's disinfectant effect could be potentialized by associating two or more techniques. As presented in the other section, Herbst et al. [37] showed that in all treatments a significant reduction in the microbial load compared to the control and the highest decrease occurred in the combination of CHX and NTPP at 0–300 μm. There was a low disinfection capacity in depth between 500 and 800 μm when combining treatments compared to a single treatment. The authors attributed this result to the root canal anatomic features. On the other hand, Saleewong et al. [24] observed that NTPP association with a common disinfectant (NaOCl) could increase the efficacy, mainly in deeper dentin levels.

It is essential to highlight that NaOCl in high concentrations, as Schaudinn et al. [27] used in their study, could not be clinically viable. A high irrigant concentration and high time of application can provide the best antimicrobial capacity, but also increases the oxidative potential. In this way, a high NaOCl concentration can compromise dentine chemical properties and promote high aggression to periapical tissues [44]. In this scenario, NTPP could be helpful for not significantly damaging the microhardness and roughness of dentine [28].

Aiming to demonstrate the efficacy of NTPP in realistic conditions, Simoncelli et al. [17] tested different protocols of root canal irrigation with PAW and the direct NTPP exposure using a plasma pistol. The results indicated that the exposure to PAW is enough to promote a significant reduction in the microbial load after 1 min of irrigation. Otherwise, the best results in bacteria inactivation were obtained in a dry environment and with direct NTPP exposure. These findings corroborate the results of Fridman et al. [15], which also verified the most effective bacterial inactivation on bacterial colonies from the skin after direct treatment. The authors attributed this effectiveness in sterilization to charged particles, microtherm effects, and vacuum UV radiation that were generated at the bacterial surface.

In view of the mentioned studies with microorganisms of endodontic interest, the use of NTPP has been demonstrated to be an effective adjunct to conventional treatments, although the best exposure parameters, considering the particularities of root canals, still need to be studied. It is important to emphasize that besides additional in vitro studies to set up the best parameters, clinical studies are extremely important to verify this effectiveness in vivo. The lack of clinical studies in the literature, and consequently in this review, represents a limitation of the present work.

Few in vitro studies have been conducted with PAW in the endodontic field, but they also showed promising findings with good perspectives to be used in the area. A possible advantage of using PAW would be that it may keep the antimicrobial effect for over one month after its generation, when stored at at least $-80\ ^\circ$C [45,46]. In this way, the dentists would not necessarily need to have the plasma device in the clinic, which can be a limitation of using NTPP in dental offices. Moreover, the liquid form from PAW could be a solution to root canal anatomy complexity, which can also be pointed out as a limitation of NTPP. Additionally, treatment with PAW would probably be more comfortable for the patient than NTPP, depending on the time of exposure of the direct treatment. Therefore, new in vitro and in vivo studies with PAW are welcome and required to establish its real usefulness in the endodontic area.

6. Conclusions

Keeping in mind that conventional methods and techniques have limitations which often lead to root canal reinfection, NTPP could be used for endodontic disinfection as an adjunct treatment to root canal disinfection. Its efficacy in eliminating endodontic pathogens in vitro and ex vivo has been successfully demonstrated, mainly against *E. faecalis* and *C. albicans*.

The combination of NTPP and conventional antimicrobial methods has shown, in general, higher antimicrobial activity than the isolated treatments. This association may be an important method for decreasing the plasma exposure time and enhancing antimicrobial effectiveness, since long exposure and root canal anatomy complexity can build limiting factors to direct plasma exposure as a single treatment in clinical practice.

Future studies, including clinical investigations, are needed to confirm the effectiveness of NTPP and to set up the best parameters in which this technology can contribute to endodontic disinfection.

Author Contributions: A.B.M., M.R.d.C.V., L.D.P.L., D.M.d.S., N.V.M.M., C.Y.K.-I. and K.G.K. contributed to the writing, reviewing, and editing of the manuscript. All authors have read and agreed to the published version of the manuscript.

Funding: The São Paulo Research Foundation (FAPESP) (Grant 19/05856-7), The National Council for Scientific and Technological Development (CNPq, 309762/2021-9), and Coordenação de Aperfeiçoamento de Pessoal de Nível Superior (CAPES), Finance Code 001.

Institutional Review Board Statement: Not applicable.

Informed Consent Statement: Not applicable.

Data Availability Statement: The data presented in this study are available on request from the corresponding author. The data are not publicly available due to privacy and ethical restrictions.

Conflicts of Interest: The authors declare no conflict of interest.

References

1. Armand, A.; Khani, M.; Asnaashari, M.; Ahmadi, A.; Shokri, B. Comparison study of root canal disinfection by cold plasma jet and photodynamic therapy. *Photodiagnosis Photodyn. Ther.* **2019**, *26*, 327–333. [CrossRef] [PubMed]
2. Neelakantan, P.; Romero, M.; Vera, J.; Daood, U.; Khan, A.U.; Yan, A.; Cheung, G.S.P. Biofilms in endodontics—Current status and future directions. *Int. J. Mol. Sci.* **2017**, *18*, 1748. [CrossRef]
3. Pan, J.; Sun, K.; Liang, Y.; Sun, P.; Yang, X.; Wang, J.; Becker, K.H. Cold plasma therapy of a tooth root canal infected with Enterococcus faecalis biofilms in vitro. *J. Endod.* **2013**, *39*, 105–110. [CrossRef] [PubMed]
4. Habib, M.; Hottel, T.L.; Hong, L. Antimicrobial effects of non-thermal atmospheric plasma as a novel root canal disinfectant. *Clin. Plasma Med.* **2014**, *2*, 17–21. [CrossRef]
5. Kim, S.J.; Chung, T.H. Cold atmospheric plasma jet-generated RONS and their selective effects on normal and carcinoma cells. *Sci. Rep.* **2016**, *6*, 20332. [CrossRef] [PubMed]
6. Whittaker, A.G.; Graham, E.M.; Baxter, R.L.; Jones, A.C.; Richardson, P.R.; Meek, G.; Baxter, H.C. Plasma cleaning of dental instruments. *J. Hosp. Infect.* **2004**, *56*, 37–41. [CrossRef]

7. Sung, S.J.; Huh, J.B.; Yun, M.J.; Chang, B.M.W.; Jeong, C.M.; Jeon, Y.C. Sterilization effect of atmospheric pressure non-thermal air plasma on dental instruments. *J. Adv. Prosthodont.* **2013**, *5*, 2–8. [CrossRef]
8. Monetta, T.; Scala, A.; Malmo, C.; Bellucci, F. Antibacterial Activity of Cold Plasma−Treated Titanium Alloy. *Plasma Med.* **2011**, *1*, 205–214. [CrossRef]
9. Rezaeinezhad, A.; Eslami, P.; Mirmiranpour, H.; Ghomi, H. The effect of cold atmospheric plasma on diabetes-induced enzyme glycation, oxidative stress, and inflammation; in vitro and in vivo. *Sci. Rep.* **2019**, *9*, 19958. [CrossRef]
10. Mirpour, S.; Fathollah, S.; Mansouri, P.; Larijani, B.; Ghoranneviss, M.; Mohajeri Tehrani, M.; Amini, M.R. Cold atmospheric plasma as an effective method to treat diabetic foot ulcers: A randomized clinical trial. *Sci. Rep.* **2020**, *10*, 10440. [CrossRef]
11. Metelmann, H.R.; Seebauer, C.; Miller, V.; Fridman, A.; Bauer, G.; Graves, D.B.; Von Woedtke, T. Clinical experience with cold plasma in the treatment of locally advanced head and neck cancer. *Clin. Plasma Med.* **2018**, *9*, 6–13. [CrossRef]
12. Keidar, M.; Walk, R.; Shashurin, A.; Srinivasan, P.; Sandler, A.; Dasgupta, S.; Trink, B. Cold plasma selectivity and the possibility of a paradigm shift in cancer therapy. *Br. J. Cancer* **2011**, *105*, 1295–1301. [CrossRef] [PubMed]
13. Guo, D.; Liu, H.; Zhou, L.; Xie, J.; He, C. Plasma-activated water production and its application in agriculture. *J. Sci. Food Agric.* **2021**, *101*, 4891–4899. [CrossRef] [PubMed]
14. Nakamura-Silva, R.; Macedo, L.M.D.; Cerdeira, L.; Oliveira-Silva, M.; Silva-Sousa, Y.T.C.; Pitondo-Silva, A. First report of hypermucoviscous Klebsiella variicola subsp. variicola causing primary endodontic infection. *Clin. Microbiol. Infect.* **2021**, *27*, 303–304. [CrossRef]
15. Fridman, G.; Brooks, A.D.; Balasubramanian, M.; Fridman, A.; Gutsol, A.; Vasilets, V.N.; Friedman, G. Comparison of direct and indirect effects of non-thermal atmospheric-pressure plasma on bacteria. *Plasma Process. Polym.* **2007**, *4*, 370–375. [CrossRef]
16. Milhan, N.V.M.; Chiappim, W.; Sampaio, A.D.G.; Vegian, M.R.D.C.; Pessoa, R.S.; Koga-Ito, C.Y. Applications of plasma-activated water in dentistry: A review. *Int. J. Mol. Sci.* **2022**, *23*, 4131. [CrossRef]
17. Simoncelli, E.; Barbieri, D.; Laurita, R.; Liguori, A.; Stancampiano, A.; Viola, L.; Colombo, V. Preliminary investigation of the antibacterial efficacy of a handheld Plasma Gun source for endodontic procedures. *Clin. Plasma Med.* **2015**, *3*, 77–86. [CrossRef]
18. Ledernez, L.; Engesser, F.; Altenburger, M.J.; Urban, G.A.; Bergmann, M.E. Effect of Transient Spark Disinfection on Various Endodontics-Relevant Bacteria. *Plasma Med.* **2019**, *9*, 121–128. [CrossRef]
19. Prada, I.; Micó-Muñoz, P.; Giner-Lluesma, T.; Micó-Martínez, P.; Collado-Castellano, N.; Manzano-Saiz, A. Influence of microbiology on endodontic failure. Literature review. *Med. Oral Patol. Oral Cir. Bucal* **2019**, *24*, e364. [CrossRef]
20. Siqueira Jr, J.F.; Rôças, I.N. Diversity of endodontic microbiota revisited. *J. Dent. Res.* **2009**, *88*, 969–981. [CrossRef]
21. Dioguardi, M.; Di Gioia, G.; Illuzzi, G.; Arena, C.; Caponio, V.C.A.; Caloro, G.A.; Lo Muzio, L. Inspection of the microbiota in endodontic lesions. *Dent. J.* **2019**, *7*, 47. [CrossRef] [PubMed]
22. Narayanan, L.L.; Vaishnavi, C. Endodontic microbiology. *J. Conserv. Dent.* **2010**, *13*, 233–239. [CrossRef] [PubMed]
23. Kerlikowski, A.; Matthes, R.; Pink, C.; Steffen, H.; Schlüter, R.; Holtfreter, B.; Jablonowski, L. Effects of cold atmospheric pressure plasma and disinfecting agents on Candida albicans in root canals of extracted human teeth. *J. Biophotonics* **2020**, *13*, e202000221. [CrossRef]
24. Saleewong, K.; Wanachantararak, P.; Louwakul, P. Efficacy of cold atmospheric pressure plasma jet against Enterococcus faecalis in apical canal of human single-rooted teeth: A preliminary study. In Proceedings of the International Conference on Materials Research and Innovation, Bangkok, Thailand, 17–21 December 2018. [CrossRef]
25. Mergoni, G.; Percudani, D.; Lodi, G.; Bertani, P.; Manfredi, M. Prevalence of Candida species in endodontic infections: Systematic review and meta-analysis. *J. Endod.* **2018**, *44*, 1616–1625. [CrossRef] [PubMed]
26. Yoo, Y.J.; Kim, A.R.; Perinpanayagam, H.; Han, S.H.; Kum, K.Y. Candida albicans virulence factors and pathogenicity for endodontic infections. *Microorganisms* **2020**, *8*, 1300. [CrossRef]
27. Schaudinn, C.; Jaramillo, D.; Freire, M.O.; Sedghizadeh, P.P.; Nguyen, A.; Webster, P.; Jiang, C. Evaluation of a nonthermal plasma needle to eliminate ex vivo biofilms in root canals of extracted human teeth. *Int. Endod. J.* **2013**, *46*, 930–937. [CrossRef]
28. Li, Y.; Sun, K.; Ye, G.; Liang, Y.; Pan, H.; Wang, G.; Fang, J. Evaluation of cold plasma treatment and safety in disinfecting 3-week root canal Enterococcus faecalis biofilm in vitro. *J. Endod.* **2015**, *41*, 1325–1330. [CrossRef]
29. Kong, M.G.; Kroesen, G.; Morfill, G.; Nosenko, T.; Shimizu, T.; Van Dijk, J.; Zimmermann, J.L. Plasma medicine: An introductory review. *New J. Phys.* **2009**, *11*, 115012. [CrossRef]
30. Jha, N.; Ryu, J.J.; Choi, E.H.; Kaushik, N.K. Generation and role of reactive oxygen and nitrogen species induced by plasma, lasers, chemical agents, and other systems in dentistry. *Oxid. Med. Cell. Longev.* **2017**, *2017*, 7542540. [CrossRef]
31. Jiang, C.; Vernier, P.T.; Chen, M.T.; Wu, Y.H.; Gundersen, M.A.; Wang, L.L. Pulsed atmospheric-pressure cold plasma for endodontic disinfection. In Proceedings of the IEEE 35th International Conference on Plasma Science, Karlsruhe, Germany, 15–19 June 2008. [CrossRef]
32. Bansode, A.S.; Beg, A.; Pote, S.; Khan, B.; Bhadekar, R.; Patel, A.; Mathe, V.L. (Non-Thermal Atmospheric Plasma for Endodontic Treatment. In Proceedings of the International Conference on Biomedical Electronics and Devices, Barcelona, Spain, 11–14 February 2013. [CrossRef]
33. Du, T.; Shi, Q.; Shen, Y.; Cao, Y.; Ma, J.; Lu, X.; Haapasalo, M. Effect of modified nonequilibrium plasma with chlorhexidine digluconate against endodontic biofilms in vitro. *J. Endod.* **2013**, *39*, 1438–1443. [CrossRef]

34. Jablonowski, L.; Koban, I.; Berg, M.H.; Kindel, E.; Duske, K.; Schroeder, K.; Kocher, T. Elimination of E. faecalis by a new non-thermal atmospheric pressure plasma handheld device for endodontic treatment. A preliminary investigation. *Plasma Process. Polym.* **2013**, *10*, 499–505. [CrossRef]
35. Üreyen Kaya, B.; Kececi, A.D.; Güldaş, H.E.; Çetin, E.S.; Öztürk, T.; Öksuz, L.; Bozduman, F. Efficacy of endodontic applications of ozone and low-temperature atmospheric pressure plasma on root canals infected with Enterococcus faecalis. *Lett. Appl. Microbiol.* **2014**, *58*, 8–15. [CrossRef] [PubMed]
36. Doria, A.C.O.C.; Sorge, C.D.P.C.; Santos, T.B.; Brandão, J.; Gonçalves, P.A.R.; Maciel, H.S.; Pessoa, R.S. Application of post-discharge region of atmospheric pressure argon and air plasma jet in the contamination control of Candida albicans biofilms. *Res. Biomed. Eng.* **2015**, *31*, 358–362. [CrossRef]
37. Herbst, S.R.; Hertel, M.; Ballout, H.; Pierdzioch, P.; Weltmann, K.D.; Wirtz, H.C.; Preissner, S. Bactericidal efficacy of cold plasma at different depths of infected root canals in vitro. *Open Dent. J.* **2015**, *9*, 486–491. [CrossRef] [PubMed]
38. Wang, R.; Zhou, H.; Sun, P.; Wu, H.; Pan, J.; Zhu, W.; Fang, J. The effect of an atmospheric pressure, DC nonthermal plasma microjet on tooth root canal, dentinal tubules infection and reinfection prevention. *Plasma Med.* **2011**, *1*, 143–155. [CrossRef]
39. Alberti, A.; Corbella, S.; Taschieri, S.; Francetti, L.; Fakhruddin, K.S.; Samaranayake, L.P. Fungal species in endodontic infections: A systematic review and meta-analysis. *PLoS ONE* **2021**, *16*, e0255003. [CrossRef]
40. Theinkom, F.; Singer, L.; Cieplik, F.; Cantzler, S.; Weilemann, H.; Cantzler, M.; Zimmermann, J.L. Antibacterial efficacy of cold atmospheric plasma against Enterococcus faecalis planktonic cultures and biofilms in vitro. *PLoS ONE* **2019**, *14*, e0223925. [CrossRef]
41. Laurita, R.; Barbieri, D.; Gherardi, M.; Colombo, V.; Lukes, P. Chemical analysis of reactive species and antimicrobial activity of water treated by nanosecond pulsed DBD air plasma. *Clin. Plasma Med.* **2015**, *3*, 53–61. [CrossRef]
42. Zhang, Q.; Liang, Y.; Feng, H.; Ma, R.; Tian, Y.; Zhang, J.; Fang, J. A study of oxidative stress induced by non-thermal plasma-activated water for bacterial damage. *Appl. Phys. Lett.* **2013**, *102*, 203701. [CrossRef]
43. Mai-Prochnow, A.; Zhou, R.; Zhang, T.; Ostrikov, K.; Mugunthan, S.; Rice, S.A.; Cullen, P.J. Interactions of plasma-activated water with biofilms: Inactivation, dispersal effects and mechanisms of action. *NPJ Biofilms Microbiomes* **2021**, *7*, 11. [CrossRef]
44. Marion, J.J.; Manhaes, F.C.; Bajo, H.; Duque, T.M. Efficiency of different concentrations of sodium hypochlorite during endodontic treatment. Literature review. *Dental Press Endod.* **2012**, *2*, 32–37.
45. Shen, J.; Tian, Y.; Li, Y.; Ma, R.; Zhang, Q.; Zhang, J.; Fang, J. Bactericidal effects against S. aureus and physicochemical properties of plasma activated water stored at different temperatures. *Sci. Rep.* **2016**, *6*, 28505. [CrossRef]
46. Tsoukou, E.; Bourke, P.; Boehm, D. Temperature stability and effectiveness of plasma-activated liquids over an 18 months period. *Water* **2020**, *12*, 3021. [CrossRef]

Article

Healing of Recurrent Aphthous Stomatitis by Non-Thermal Plasma: Pilot Study

Norma Guadalupe Ibáñez-Mancera [1], Régulo López-Callejas [2], Víctor Hugo Toral-Rizo [3], Benjamín Gonzalo Rodríguez-Méndez [2,*], Edith Lara-Carrillo [3], Rosendo Peña-Eguiluz [2], Regiane Cristina do Amaral [4], Antonio Mercado-Cabrera [2] and Raúl Valencia-Alvarado [2]

[1] Interdisciplinary Center for Health Sciences CICS-UST, Instituto Politécnico Nacional, Av. Luis Enrique Erro S/N, Unidad Profesional Adolfo López Mateos, Zacatenco 07738, Mexico
[2] Plasma Physics Laboratory, Instituto Nacional de Investigaciones Nucleares, Carretera México Toluca S/N, Ocoyoacac 52750, Mexico
[3] Orocenter Clinic, Facultad de Odontología, Universidad Autónoma del Estado de México, Av. Paseo Tollocan esq. Jesús Carranza, Colonia Universidad, Toluca de Lerdo 50130, Mexico
[4] Departamento de Odontología, Universidade Federal de Sergipe, Aracajú 49060-108, SE, Brazil
* Correspondence: benjamin.rodriguez@inin.gob.mx

Abstract: Recurrent aphthous stomatitis (RAS) is a common disease in the oral cavity characterized by recurrent ulcers (RU). Usually, these cause acute pain without definitive treatment. The present study determines the efficacy of non-thermal plasma (NTP) for treating RU. NTP is applied to the patient's RU using a radiofrequency generator connected to a point reactor. The power density applied to the ulcer is 0.50 W/cm^2, less than 4 W/cm^2, which is the maximum value without biological risk. Each patient received two treatments of three minutes each and spaced 60 min apart at a distance of 5 mm from the RU. From a sample of 30 ulcers in patients treated for RU with an average age of 37 years, they stated that the pain decreased considerably and without the need for ingestion of analgesics and antibiotics. Regeneration took place in an average of three days. The NTP proved to be an excellent therapeutic alternative for the treatment of RU since it has a rapid effect of reducing pain and inflammation, as well as adequate tissue regeneration.

Keywords: healing therapy; recurrent aphthous stomatitis; non-thermal plasma; recurrent ulcers

Citation: Ibáñez-Mancera, N.G.; López-Callejas, R.; Toral-Rizo, V.H.; Rodríguez-Méndez, B.G.; Lara-Carrillo, E.; Peña-Eguiluz, R.; do Amaral, R.C.; Mercado-Cabrera, A.; Valencia-Alvarado, R. Healing of Recurrent Aphthous Stomatitis by Non-Thermal Plasma: Pilot Study. *Biomedicines* **2023**, *11*, 167. https://doi.org/10.3390/biomedicines11010167

Academic Editor: Christoph Viktor Suschek

Received: 15 November 2022
Revised: 29 December 2022
Accepted: 31 December 2022
Published: 9 January 2023

1. Introduction

Recurrent aphthous stomatitis (RAS) is a disease that affects approximately 20% of the population, although the impact varies according to ethnic group and socioeconomic status [1,2]. It is characterized by lesions with a loss of continuity of the epithelium that manifest repeatedly, recurrent ulcers (RU), which are representative due to their frequency and symptomatology, constituting 20% of stomatological emergencies [3]. RAS is not a deadly disease but affects the patient's quality of life. The lesions observed clinically are round or oval ulcers on the buccal mucosa and are covered by a pale or yellowish membrane with an erythematous outline. According to the size of the ulcers, they are classified into three types (Figure 1): minor RAS is the most common type where there are 1 to 5 ulcers less than one centimeter in diameter and they take seven to fourteen days to heal, major RAS, corresponds to 1 to 10 ulcers of more than one centimeter in diameter, which heal in up to thirty days; and herpetiform RAS corresponds to multiple ulcers (from 10 to 100) of 1 to 3 mm in diameter, which take between fifteen and thirty days to heal [1,4–7] The RU compromises the mucosa of the lips, cheeks, background of the sack, oropharynx tongue, the floor of the mouth, palate and gums [4,5].

The etiology of RAS remains unclear, so various factors have been proposed: immuno logical, hereditary, genetic, allergic, nutritional and microbial, as causative agents and there are no efficient curative treatments. Various treatments for RUs have been reported to heal these types of injuries or reduce pain [8–11]. Regarding pharmacological therapies prescribed, many medications include antiseptics, anti-inflammatories, antibiotics, corticosteroids, or thalidomide [2,12–15] and plant-based medicine [16]. In patients with constant and aggressive outbreaks (major canker sores), the pain is severe and typical treatment cannot provide symptom relief; therapy is indicated via corticosteroids (prednisone), among other drugs [15]. There are currently other alternatives, such as laser therapy [17–21], ozone therapy [22–24]; biological treatments such as Anti-TNF-alpha [6] and probiotics [25], which provide good results but are still need not an ideal alternative.

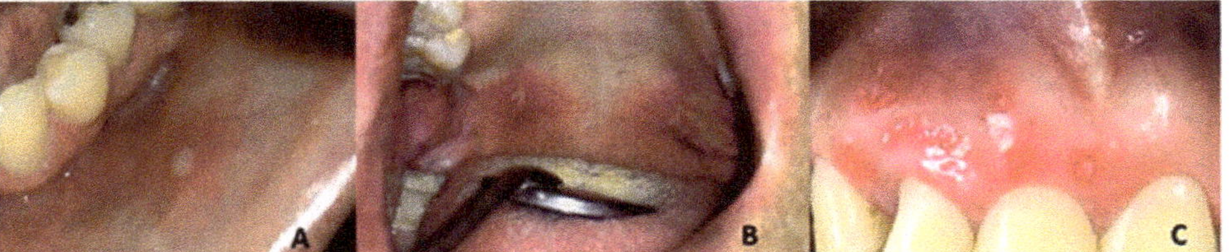

Figure 1. Clinical characteristics of: (**A**) minor RAS, (**B**) major RAS, (**C**) herpetiform RAS.

From the physical point of view, plasma is defined as the fourth state of matter. Depending on the thermal balance between electrons and atoms or gas molecules, it can be classified as a thermal plasma or non-thermal plasma [26]. Non-thermal plasma (NTP) can be generated by electrical discharges or gas bombardment with a high-energy electron beam at atmospheric pressure and, therefore, at room temperature [27]. In non-thermal plasmas, the temperature of ions, atoms and/or molecules, both excited and neutral, is close to room temperature. In contrast, the temperature of electrons is in the order of thousands of Kelvin. Thus, the electrons strengthen kinetic energy and the existence of the plasma through the formation of different energy states of heavy particles under the direct influence of the ionization process of some atoms and/or molecules present in the gas. In addition, high-temperature electrons can separate molecular gases such as oxygen and nitrogen, generating various chemical reactions. Chemical species in NTP include: the reactive oxygen species (ROS), such as atomic oxygen (O_2), superoxide (O_2^-), ozone (O_3), hydroxyl radical (OH) and hydrogen peroxide (H_2O_2); and the reactive nitrogen species (RNS), such as nitric oxide (NO), nitrogen dioxide (NO_2), nitrogen trioxide (NO_3), nitrous oxide (N_2O), nitrogen oxide (N_2O_4) and positive ions such as N_2^+, of considerable importance in biomedicine [28–32].

Nowadays, application of NTP is used in biomedical procedures [33], such as disinfection [34], coagulation [35] and wound healing [36], among others. In particular, various research groups have reported the effects of reducing the healing time of wounds and ulcers. The present study applied NTP to RU to identify tissue repair time and symptomatology in patients with RAS.

2. Materials and Methods

Fourteen patients were selected who attended the Orocentro Clinic of the Dentistry Faculty from the Mexico State Autonomous University due to ulcers diagnosed as RU. The patients were over 18 years old, of both genders, without systemic compromise and agreed to participate in the study by signing an informed consent. Patients with oral soft tissue diseases, maxillary orthopedic, orthodontic treatment and those with maladjusted dental prostheses were excluded.

A longitudinal study was carried out to evaluate the tissue repair time of recurrent ulcers and the presence of pain. Once it was completed and the patient medical history had established the diagnosis of RU, data of size and location of ulcer were recorded, as well as the evaluation time and pain intensity. Pain intensity was measured using the 10-point visual analog scale (0 = "no pain at all", 10 = "worst possible pain") [37]. Once the patient had agreed to participate in the study and the informed consent was signed, the application of NTP with helium gas continued for three minutes at a distance of 5 mm, the ulcer was measured again and the intensity of the pain was assessed. After the first application, the NTP application was repeated under the same conditions (three minutes at a distance of 5 mm) and the ulcer was subsequently measured. Pain intensity was assessed similarly. The patients were evaluated every 24 h after the NTP application, with ulcer measurement and pain assessment, until tissue regeneration with surface continuity was identified.

The proposed atmospheric pressure NTP system is constituted by a radiofrequency (RF) power generator directly coupled to an LC resonant load, avoiding use of a matching network, where L is a non–linear inductor and C is the resultant capacitance of a coaxial cable connected to an ergonomic plasma reactor. The RF power generator is based on the 13.56 MHz class E RF module, which is supplied by three dc voltage sources, one of them providing an adjustable voltage magnitude represented by V_{DC}. The available electric power at the RF module output terminals P_{RF} is controlled by the magnitude of the applied dc voltage (V_{DC}) [38].

For the NTP application generated with helium gas, the operating parameters reported in previous works were considered [34,36]. These are the radiofrequency (RF) and an electrical power generator of 20 W at a frequency of 13.56 MHz was applied to the plasma reactor, with which a power density applied to the ulcer is established based on the geometry of the reactor. The irradiance applied to the patient was 0.50 W/cm^2, less than 4 W/cm^2. With this irradiance, it is guaranteed that there is no biological risk, according to the data of the International Commission on Non-Ionizing Radiation Protection [39]. In the applications, 0.75 LPM of helium gas flows through the reactor and the exposure time in each of the two sessions per patient was 180 s. Once the plasma was generated, the tip of the reactor was kept at a distance of 5 mm from the ulcer (see Figure 2).

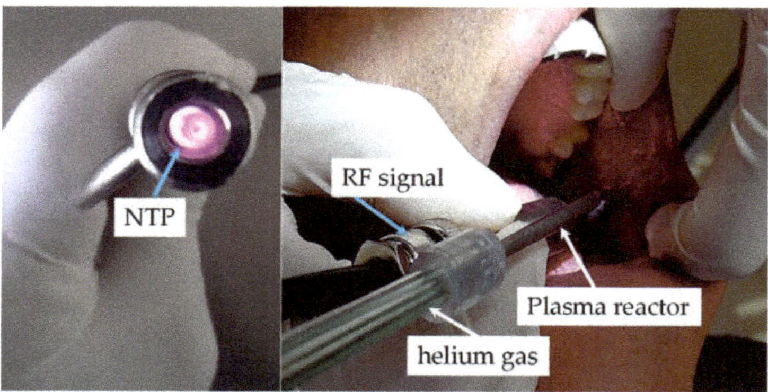

Figure 2. Plasma reactor and non-thermal plasma application in the patient.

For the statistical analysis, descriptive and comparative studies were performed on ulcer type (Fisher's exact test). Comparative analyzes (ANOVA one-way test, followed by Tukey) for pain and lesion size were also achieved.

3. Results

We measured the size of the ulcer and the erythematous contour in each case and took a photographic record. The sample consisted of 14 patients, of whom nine were women (64.3%) and five were men (35.7%). The age range was from 18 to 67 years old, with a mean of 37 years old. The patients presented from one to five ulcers, resulting in 30 ulcers. Of these, 19 (64%) were minor RU, 10 (33%) corresponded to major RU and 1 (3%) was herpetiform RU. The distribution of oral RU cases is shown in Table 1. The size of the ulcers ranged between 2 and 19 mm, with a mean of 7 mm. Before the application of the NTP, all patients reported pain on a numerical verbal scale between 0 and 10 (where 0 is no pain and 10 is the most intense pain); 75% of the patients reported pain, with a mean of 7.

Table 1. Sociodemographic and risk factors for injuries according to the type of ulcer in this study.

	Recurrent Herpetiform Ulcers	Major Recurrent Ulcers	Minor Recurrent Ulcers	Fisher's Exact
Gender				
Male			10	0.01
Female	1	10	9	
Age				
18–30		2	9	
31–50	1		7	0.005
51–67		8	3	
Marital status				
Married	1	8	6	0.02
Single		2	13	
Scholarship				
High school		2	3	
Basic	1	8	6	0.04
Bachelor's degree			10	
Occupation				
Unemployed		3	1	
Student		2	3	0.003
Homework	1	5	1	
Employee			14	
Medication consumption				
Not	1	2	11	0.08
Yes		8	8	
Smoking				
Cigarette		3	4	0.73
No	1	7	15	
Alcohol consumption				
2 to 4 times a month		1		
Never	1	8	4	0.02
Once a year		1	8	
Once a month			7	
Ulcers evolution time				
Two to five days	1	4	10	
Six to ten days		6	8	0.69
One day			1	
Ulcers recurrence period				
Zero			1	
One for year	1	5	2	
One for month		5	5	0.07
One for week			3	
More than a year			8	

When socioeconomic and risk factors are associated with the type of ulcer, a statistically significant association is verified for recurrence, alcohol use, medication use, occupation, education, marital status, age and gender ($p < 0.05$—Fisher's exact) (see Table 1).

After performing the previous review in all cases after the second NTP application, six cases (20%) showed tissue regeneration in the ulcer, considering them repaired at zero days. In this exact measurement, 18 (60%) of the cases reported asymptomatic and the other 12 (40%) cases reported decreased pain with a mean of 4. At 24 h after the two applications with NTP, 11 cases (37%) showed total repair and the remaining 19 cases (63%) showed a reduction in the ulcer size of between 1 and 6 mm with a mean of 3 mm. Regarding the sensation of pain, when assessing the 30 cases, 27 (90%) reported asymptomatic, two cases (7%) reported pain with a value of 1 and the other case (3%) presented pain decreasing from a value from 7 to 4 on the visual analog scale. Ulcer repair time ranged from one hour to 7 days, with a mean of 3 days. In a single case, the ulcer persisted after applying NTP; in addition to the patient reporting pain, the patient reported positive for smoking. The time it took for the ulcers to regenerate showed a correlation only with the ulcer's initial size of 0.703 (p = 0.000). Table 2 shows the evolution of the cases regarding ulcer size, pain and erythema size.

Table 2. Size, pain and erythema in the initial period and after plasma application, one hour, one day and one week.

Mean and Standard Deviation of Ulcer Size	
Initial	6.9 ± 4.51 A
One hour	3.96 ± 3.1 B
One day	2 ± 2 BC
One week	0.03 ± 0.18 C
p	<0.0001

Mean and Standard Deviation of Pain	
Initial	7.1 ± 2.5 A
One hour	1.5 ± 2.3 B
One day	0.2 ± 0.7 C
One week	0.06 ± 0.3 C
p	<0.0001

Mean and Standard Deviation of Erythema Size	
Initial	2.1 ± 1.1 A
One hour	1.4 ± 1 B
One day	0.5 ± 0.6 C
One week	0.03 ± 0.18 C
p	<0.0001

Different letters = statistically significant differences. It was verified that, after one hour of plasma application, there was a substantial reduction in size and pain (ANOVA one-way test followed by Tukey).

Figure 3 shows, in the usual manner, the evolution of three cases with RU treated with NTP, which resolved within a maximum of 24 h.

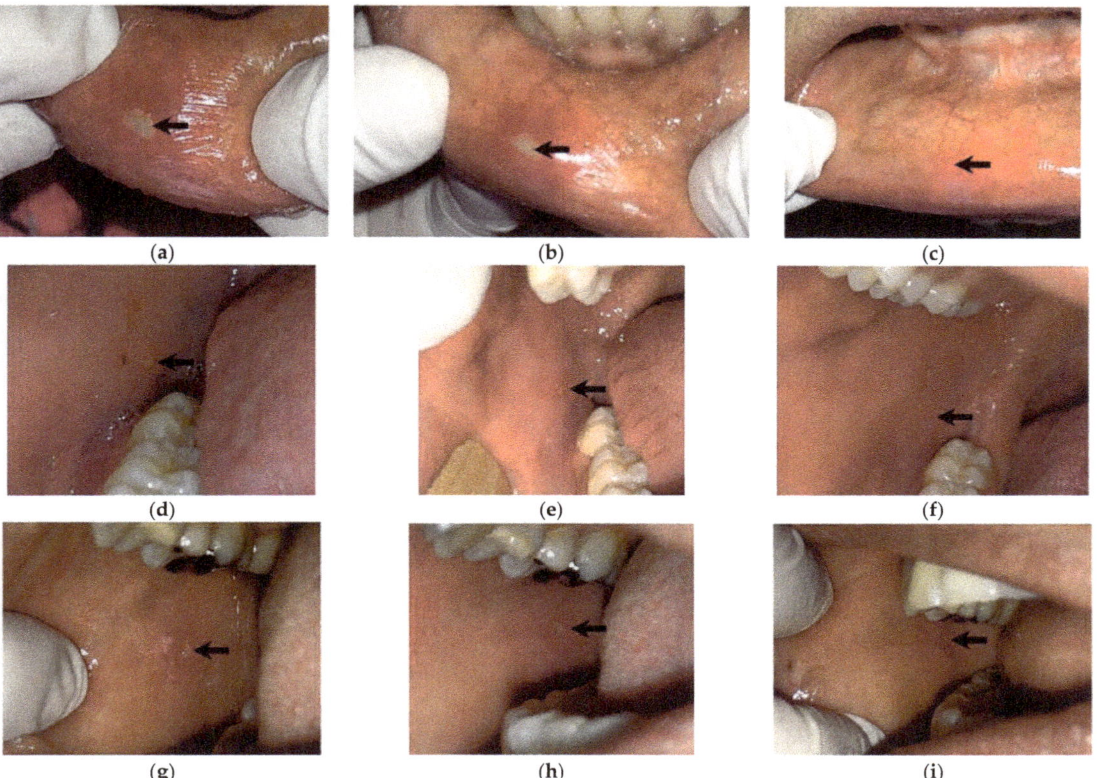

Figure 3. RU in the buccal mucosa, horizontally showing the evolution of the ulcers in three patients; (**a**,**d**,**g**) correspond to the initial state of the RU; (**b**,**e**,**h**) show regeneration after the first application of NTP; finally, (**c**,**f**,**i**) show the healthy tissue 24 h after starting the treatment.

Figure 4 shows three cases of patients with RU who healed on average after three days of NTP application.

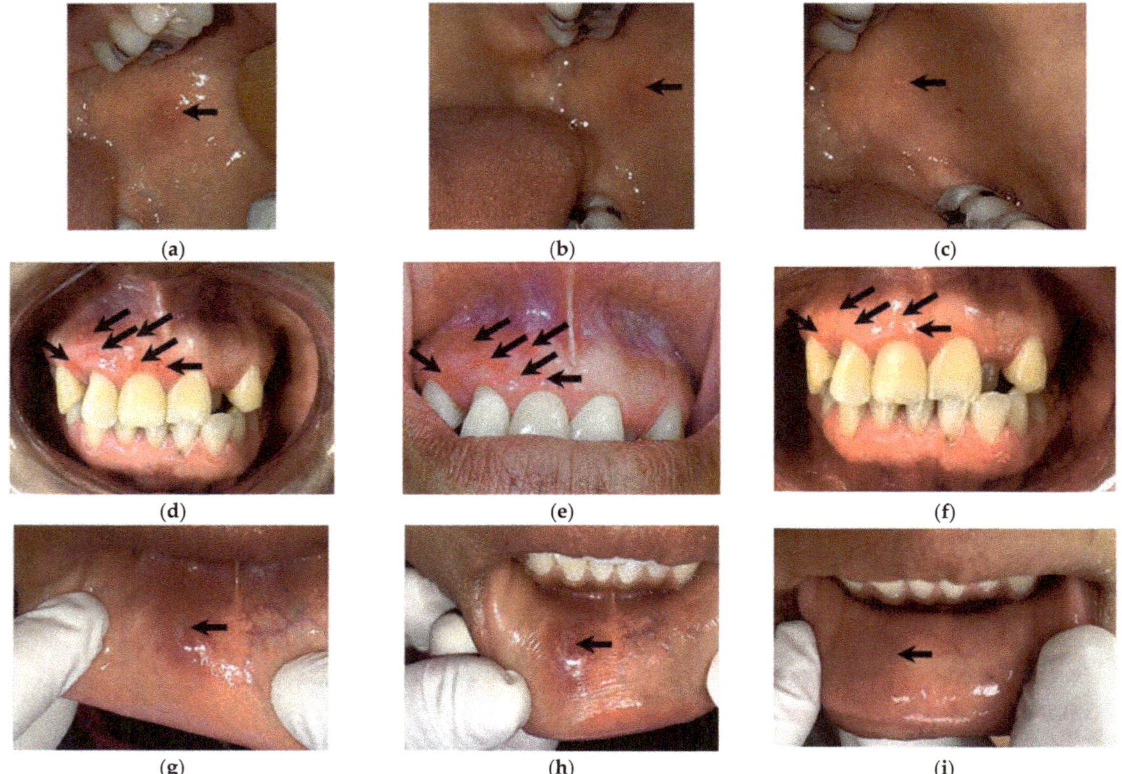

Figure 4. RU is located in different parts of the oral cavity (indicated with arrows): (**a,d,g**) at patient admission. (**b,e,h**) after the first application of the NTP. Finally, (**c,f,i**) show follow-up when the wound has healed (indicated with arrows).

4. Discussion

Today, no specific drug or treatment is available for recurrent aphthous stomatitis (RAS). RAS healing is a highly specialized process for repairing damaged/injured tissues through various therapies. Failures in the normal healing process of these ulcers lead to abnormal scar formation and a chronic state that is more susceptible to infection. Wounds caused by RU affect patients' quality of life, which is why there is an urgent demand for the development of competitive therapies. There are new developments in advanced technologies for the care of RU, for example, laser technology [40–42], nano-therapy [43,44], etc. Those studies can improve therapeutic results by focusing on tissue regeneration with minimal side effects. Reports show that evolution time and recurrence periods have been reduced [45,46]. The pain caused due to the ulcers experienced by patients with RAS turns out to be the main problem and usually is treated with antihistaminic agents, analgesics, or even anesthetics to reduce pain and promote patients' oral function [18]. Recent research suggests that ROS/RNS are central players in the actions of antimicrobial and antiparasitic drugs, cancer therapies, wound healing therapies and treatments involving the cardiovascular system. Understanding how ROS/RNS act in established therapies may help guide future efforts in exploiting novel non-thermal plasma-based medical therapies [47,48].

NTP at atmospheric pressure was applied to the ulcer in the cheek, labial mucosa, tongue, oropharynx, mouth floor, soft palate and gingiva. In, patients with RAS and treated with NTP, its anti-inflammatory and healing effect is evident since ulcer size and pain are significantly reduced. A notable improvement was observed since 86% of the patients did not report pain after one hour of application, being statistically significant ($p = 0.0001$). Additionally, at 24 h, 90% of the cases reported being asymptomatic, 7% reported pain with a value of 1 and 3% (one case) reported pain with a value of 4. In this regard, it is essential to point out that the two cases that reported pain with a 1 value correspond to a patient with a significant RAS whose ulcers measured 15 and 19 mm. The case that reported pain with a value of 4 is a smoker and the only case that took seven days to heal; therefore, it is crucial to advise smoking patients with RAS to stop smoking during treatment.

With an average time to repair of three days, considering that more than 50% of the cases healed within two days, it was observed that the application of NTP is efficient for RU management. In contrast, studies using laser treatments reported ten days as a time of recovery [20,49] and others 15 days [50]. Some authors report a decrease in pain and erythema after three days of RU laser treatment; in 75% of cases, they have identified total re-epithelialization five days after beginning treatment [51,52]. However, with these results using lasers, there are still three to five days of epithelial exposure, pain and decreased function remaining for RAS patients. Another of the most widely used treatments for RU is topical corticosteroids, which take five to six days to heal a mouth ulcer [45,53].

Regarding the use of these medications, it is essential to consider the adverse effects that can develop as candidiasis. In addition to treatments based on biological products, they have recently been used for wound healing, reducing pain and recurrence frequency in patients with RAS [54,55]. The NTP application highlights that no drug was applied as an adjuvant.

The initial size of the RU showed a correlation with the time elapsed for tissue repair. That is, the larger the initial size of the ulcer, the longer the repair took, which is explained by the amount of epithelium that needs to be formed. Regarding the type of RAS, it is suggested to repeat the treatment with two applications of NTP at 24 h because the significant RU and herpetiform took four to six days to heal completely.

The findings of this study demonstrated the absence of adverse reactions in the tissues. That is, managing recurrent ulcers with NTP was safe for all treated patients since no case showed adverse effects at six months of follow-up and there was significantly reduced pain levels and repair time after the application of NTP. This could be explained due to the mechanism of action of NTP. Most authors agree that treatment with NTP accelerates wound healing by decreasing the level of inflammatory factors and bacterial load and has anti-neoplastic effects [56]. Reactive oxygen and nitrogen species (RONS) generated by NTP can stimulate blood circulation, causing more significant cellular activity. Thus, restoration potentials are improved by accelerating tissue repair and promoting accelerated tissue regeneration. Several studies demonstrate that the role of RONS, particularly H_2O_2, superoxide ($O_2{}^-$) and nitric oxide (NO), is essential in wound healing beyond the simple oxidative elimination of pathogens. The importance of these elements lies in infection prevention, the induction of angiogenesis, the keratinocyte's increase in differentiation and migration, and the increase in the proliferation of fibroblasts and collagen synthesis. Furthermore, NO has been demonstrated to participate in an intercellular communication network during the healing process, regulating the behavior of macrophages, keratinocytes and fibroblasts [57]. That is, tissue repair and regeneration steps are highly influenced by these intracellular RONS. Therefore, the application of NTP can be considered to cure RU.

5. Conclusions

The study showed that the application of non-thermal plasma in recurrent aphthous stomatitis reduces the levels of pain sensation in the patient, accelerates the healing of the RU and does not produce adverse effects. The technology allowed a quick recovery of the patient and, consequently, a reactivation of daily life. The NTP application in the RUs enables its repair. NTP is an excellent therapeutic alternative for RUs that afflict patients since it provides rapid control of pain and inflammation, accelerating the natural healing process. Due to its application characteristic, it does not bother patients.

Author Contributions: N.G.I.-M., contributed to the conception, experimental design, NTP application, data acquisition, interpretation, writing and critical review of the manuscript; R.L.-C., contributed to the conception, experimental design, NTP application, data acquisition, interpretation, writing and critical review of the manuscript; V.H.T.-R. contributed to the conception, data acquisition and interpretation, performed statistical analyses and drafted and critically reviewed the manuscript; E.L.-C., contributed to the conception, design and data acquisition, performed statistical analysis; B.G.R.-M., experimental design and writing of the manuscript; R.C.d.A. contributed to the statistical analysis, R.P.-E., experimental design and writing of the manuscript; A.M.-C., experimental design and writing of the manuscript; R.V.-A., data acquisition and interpretation and drafted the manuscript; all authors gave final approval and agree to be accountable for all aspects of the work. All authors have read and agreed to the published version of the manuscript.

Funding: This research received no external funding.

Institutional Review Board Statement: The study was conducted in accordance with the Declaration of Helsinki and approved by the Research Ethics Committee of the Center for Research and Advanced Studies in Dentistry of the Faculty of Dentistry of the Autonomous University of the State of Mexico (CEICIEAO-2018-001).

Informed Consent Statement: Informed consent was obtained from all subjects involved in the study.

Data Availability Statement: Not applicable.

Acknowledgments: We appreciate the work of the technicians M.T. Torres-Martínez, P. Angeles-Espinosa, I. Contreras-Villa and M. Lugo-Hernández in the construction of the different equipment systems used in this project. As well as the OROCENTRO Clinic patients who gave their consent for the application of atmospheric pressure non-thermal plasma in the treatment. This study was financially supported by research project CB-001 from National Institute of Nuclear Research and OROCENTRO Clinic of Odontology Faculty from UAEMex. The authors declare no potential conflicts of interest with respect to the authorship and/or publication of this article.

Conflicts of Interest: The authors declare no conflict of interest.

References

1. Akintoye, S.; Greenberg, M. Recurrent aphthous stomatitis. *Dent. Clin. N. Am.* **2014**, *58*, 281–297. [CrossRef] [PubMed]
2. Lau, C.B.; Smith, G.P. Recurrent aphthous stomatitis: A comprehensive review and recommendations on therapeutic options. *Dermatol. Ther.* **2022**, *35*, e15500. [CrossRef] [PubMed]
3. Borilova Linhartova, P.; Janos, J.; Slezakova, S.; Bartova, J.; Petanova, J.; Kuklinek, P.; Fassmann, A.; Dusek, L.; Izakovicova Holla, L. Recurrent aphthous stomatitis and gene variability in selected interleukins: A case–control study. *Eur. J. Oral Sci.* **2018**, *126*, 485–492. [CrossRef] [PubMed]
4. Han, M.; Fang, H.; Li, Q.L.; Cao, Y.; Xia, R.; Zhang, Z.H. Effectiveness of laser therapy in the management of recurrent aphthous stomatitis: A systematic review. *Scientifica* **2016**, *2016*, e9062430. [CrossRef] [PubMed]
5. Monteiro, S.I.; Costa, A.M.; de Vasconcelos, B.C.; Amarante, M.V.; Teixeira, P.T.; de Vasconcelos Gurgel, B.C.; Dantas da Silveira, É.J. Recurrent aphthous ulceration: An epidemiological study of etiological factors, treatment and differential diagnosis. *An. Bras. Dermatol.* **2018**, *93*, 341–346. [CrossRef]
6. Sánchez-Bernal, J.; Conejero, C.; Conejero, R. Recurrent aphthous stomatiti. *Actas Dermosifiliogr.* **2020**, *111*, 471–480. [CrossRef]
7. Gasmi, B.A.; Noor, S.; Menzel, A.; Gasmi, A. Oral aphthous: Pathophysiology, clinical aspects and medical treatment. *Arch. Razi Inst.* **2021**, *76*, 1155–1163. [CrossRef]
8. Bergmeier, L.A.; Fortune, F. Clinical management of oral mucosal disease: A literature review. In *Oral Mucosa in Health and Disease*; Bergmeier, L., Ed.; Springer International Publishing: Cham, Switzerland, 2018; pp. 161–171. [CrossRef]
9. Saikaly, S.K.; Saikaly, T.S.; Saikaly, L.E. Recurrent aphthous ulceration: A review of potential causes and novel treatments. *J. Dermatol. Treat.* **2018**, *29*, 542–552. [CrossRef]
10. Zhang, Y.; Ng, K.; Kuo, C.; Wu, D. Chinese herbal medicine for recurrent aphthous stomatitis: A protocol for systematic review and meta-analysis. *Medicine* **2018**, *97*, e13681. [CrossRef]
11. Fitzpatrick, S.G.; Cohen, D.M.; Clark, A.N. Ulcerated Lesions of the Oral Mucosa: Clinical and Histologic Review. *Head Neck Pathol.* **2019**, *13*, 91–102. [CrossRef]
12. Szyszkowska, B.; Łepecka-Klusek, C.; Kozłowicz, K.; Jazienicka, I.; Krasowska, D. The influence of selected ingredients of dietary supplements on skin condition. *Postepy. Dermatol. Alergol.* **2014**, *31*, 174–181. [CrossRef]
13. Belenguer-Guallar, I.; Jiménez-Soriano, Y.; Claramunt-Lozano, A. Treatment of recurrent aphthous stomatitis. A literature review. *J. Clin. Exp. Dent.* **2014**, *6*, e168–e174. [CrossRef]

14. Arafa, M.G.; Ghalwash, D.; El-Kersh, D.M.; Elmazar, M.M. Propolis-based niosomes as oromuco-adhesive films: A randomized clinical trial of a therapeutic drug delivery platform for the treatment of oral recurrent aphthous ulcers. *Sci. Rep.* **2018**, *8*, 18056. [CrossRef]
15. Hará, Y.; Shiratuchi, H.; Kaneko, T.; Sakagami, H. Search for drugs used in hospitals to treat stomatitis. *Medicines* **2019**, *6*, 19. [CrossRef]
16. Rezvaninejad, R.; Navabi, N.; Khoshroo, M.R.; Torabi, N.; Atai, Z. Herbal medicine in treatment of recurrent aphthous stomatitis: A literature review. *J. Iran. Dent. Assoc.* **2017**, *29*, 127–134. [CrossRef]
17. Anand, V.; Gulati, M.; Govila, V.; Anand, B. Low level laser therapy in the treatment of aphthous ulcer. *J. Dent. Res.* **2013**, *24*, 267–270. [CrossRef]
18. Nasry, S.A.; El Shenawy, H.M.; Mostafa, D.; Ammar, N.M. Different modalities for treatment of recurrent aphthous stomatitis. A randomized clinical trial. *J. Clin. Exp. Dent.* **2016**, *8*, e517–e522. [CrossRef]
19. Suter, V.G.A.; Sjölund, S.; Bornstein, M.M. Effect of laser on pain relief and wound healing of recurrent aphthous stomatitis: A systematic review. *Lasers Med. Sci.* **2017**, *32*, 953–963. [CrossRef]
20. Jahromi, N.Z.; Ghapanchi, J.; Pourshahidi, S.; Zahed, M.; Ebrahimi, H. Clinical evaluation of high and low-level laser treatment (CO_2 vs InGaAlP diode laser) for Recurrent Aphthous Stomatitis. *J. Dent.* **2017**, *18*, 17–23.
21. Melis, M.; Di Giosia, M.; Colloca, L. Ancillary factors in the treatment of orofacial pain: A topical narrative review. *J. Oral Rehabil.* **2019**, *46*, 200–207. [CrossRef]
22. Tiwari, S.; Avinash, A.; Katiyar, S.; Iyer, A.A.; Jain, S. Dental applications of ozone therapy: A review of literature. *Saudi J. Dent. Res.* **2017**, *8*, 105–111. [CrossRef]
23. Sabbah, F.; Nogales, C.G.; Zaremski, E.; Martinez-Sanchez, G. Ozone therapy in dentistry—Where we are and where we are going to? *Rev. Esp. Ozonot.* **2018**, *8*, 37–63.
24. Bader, K.A.Z. Management of denture-related traumatic ulcers using ozone. *J. Prosth. Dent.* **2019**, *121*, 76–82. [CrossRef]
25. Nirmala, M.; Smitha, S.G.; Kamath, G. A study to assess the efficacy of local application of oral probiotic in treating recurrent aphthous ulcer and oral Candidiasis. *Indian J. Otolaryngol. Head Neck Surg.* **2019**, *71*, S113–S117. [CrossRef] [PubMed]
26. Dubin, D.H.E. Effect of correlations on the thermal equilibrium and normal modes of a non-neutral plasma. *Phys. Rev. E* **1996**, *53*, 5268–5290. [CrossRef]
27. Hama-Aziz, K.H.; Miessner, H.; Mueller, S.; Kalass, D.; Moeller, D.; Khorshid, I.; Rashid, M.A.M. Degradation of pharmaceutical diclofenac and ibuprofen in aqueous solution, a direct comparison of ozonation, photocatalysis, and non-thermal plasma. *Chem. Engen. J.* **2017**, *313*, 1033–1041. [CrossRef]
28. Kajiyama, H.; Utsumi, F.; Nakamura, K.; Tanaka, H.; Toyokuni, S.; Hori, M.; Kikkawa, F. Future perspective of strategic non-thermal plasma therapy for cancer treatment. *J. Clin. Biochem. Nutr.* **2017**, *60*, 33–38. [CrossRef]
29. Laroussi, M. Plasma medicine: A brief introduction. *Plasma* **2018**, *1*, 47–60. [CrossRef]
30. Dubuc, A.; Monsarrat, P.; Virard, F.; Merbahi, N.; Sarrette, J.P.; Laurencin-Dalicieux, S.; Cousty, S. Use of cold-atmospheric plasma in oncology: A concise systematic review. *Ther. Adv. Med. Oncol.* **2018**, *10*, 1–12. [CrossRef]
31. Jang, J.Y.; Hong, Y.J.; Lim, J.; Choi, J.S.; Choi, E.H.; Kang, S.; Rhim, H. Cold atmospheric plasma (CAP), a novel physicochemical source, induces neural differentiation through cross-talk between the specific RONS cascade and Trk/Ras/ERK signaling pathway. *Biomaterials* **2018**, *156*, 258–273. [CrossRef]
32. Park, J.; Lee, H.; Lee, H.J.; Kim, G.C.; Kim, S.S.; Han, S.; Song, K. Non-thermal atmospheric pressure plasma is an excellent tool to activate proliferation in various mesoderm-derived human adult stem cells. *Free Radic. Biol. Med.* **2019**, *134*, 374–384. [CrossRef]
33. Wu, S.; Cao, Y.; Lu, X. The state of the art of applications of atmospheric-pressure nonequilibrium plasma jets in dentistry. *IEEE Trans. Plasma Sci.* **2016**, *44*, 134–151. [CrossRef]
34. Betancourt-Ángeles, M.; Peña-Eguiluz, R.; López-Callejas, R.; Domínguez-Cadena, N.A.; Mercado-Cabrera, A.; Muñoz-Infante, J.; Rodríguez-Méndez, B.G.; Valencia-Alvarado, R.; Moreno-Tapia, J.A. Treatment in the healing of burns with a cold plasma source. *Int. J. Burns Trauma* **2017**, *7*, 142–146.
35. de Masi, G.; Gareri, C.; Cordaro, L.; Fassina, A.; Brun, P.; Zaniol, B.; Cavazzana, R.; Martines, E.; Zuin, M.; Marinaro, G.; et al. Plasma coagulation controller: A low-power atmospheric plasma source for accelerated blood coagulation. *Plasma Med.* **2018**, *8*, 245–254. [CrossRef]
36. López-Callejas, R.; Peña-Eguiluz, R.; Valencia-Alvarado, R.; Mercado-Cabrera, A.; Rodríguez-Méndez, B.G.; Serment-Guerrero, J.H.; Cabral-Prieto, A.; González-Garduño, A.C.; Domínguez-Cadena, N.A.; Muñoz-Infante, J.; et al. Alternative method for healing the diabetic foot by means of a plasma needle. *Clin. Plasma Med.* **2018**, *9*, 19–23. [CrossRef]
37. Martin, W.J.J.M.; Skorpil, N.E.; Ashton-James, C.E.; Tuinzing, D.B.; Forouzanfar, T. Effect of vasoconstriction on pain after mandibular third molar surgery: A single-blind, randomized controlled trial. *Quintessence Int.* **2016**, *47*, 589–596. [CrossRef]
38. Peña-Eguiluz, R.; Serment-Guerrero, J.H.; Azorín-Vega, E.P.; Mercado-Cabrera, A.; Flores-Fuentes, A.A.; Jaramillo-Sierra, B.; Hernández-Arias, A.N.; Girón-Romero, K.; López-Callejas, R.; Rodríguez-Méndez, B.G.; et al. Development and characterization of a non-thermal plasma source for therapeutic treatments. *IEEE. Trans. Biomed. Eng.* **2021**, *68*, 1467–1476. [CrossRef]
39. International Commission on Non-Ionizing Radiation Protection. Guidelines on limits of exposure to ultraviolet radiation of wavelengths between 180 nm and 400 nm (incoherent optical radiation). *Health Phys.* **2004**, *87*, 171–186. [CrossRef]
40. dos Santos, J.A.; Normando, A.G.C.; de Toledo, I.P.; Melo, G.; de Luca Canto, G.; Santos-Silva, A.R.; Silva-Guerra, E.N. Laser therapy for recurrent aphthous stomatitis: An overview. *Clin. Oral Investig.* **2020**, *24*, 37–45. [CrossRef]

41. Rao, Z.; Tomar, D.; Kaushik, S.; Mittal, S.; Kumar, P. Application of soft tissue laser in the management of recurrent apthous stomatitis: A placebo controlled study. *Int. J. Health Sci.* **2022**, *6*, 2590–2598. [CrossRef]
42. Nagieb, C.S.; Harhash, T.A.E.; Fayed, H.L.; Ali, S. Evaluation of diode laser versus topical corticosteroid in management of Behcet's disease-associated oral ulcers: A randomized clinical trial. *Clin. Oral Investig.* **2022**, *26*, 697–704. [CrossRef] [PubMed]
43. Makvandi, P.; Josic, U.; Delfi, M.; Pinelli, F.; Jahed, V.; Kaya, E.; Ashrafizadeh, M.; Zarepour, A.; Rossi, F.; Zarrabi, A.; et al. Drug delivery (nano) platforms for oral and dental applications: Tissue regeneration, infection control, and cancer management. *Adv. Sci.* **2021**, *8*, e2004014. [CrossRef] [PubMed]
44. Gao, H.; Wu, N.; Nini Wang, N.; Li, J.; Sun, J.; Peng, Q. Chitosan-based therapeutic systems and their potentials in treatment of oral diseases. *Int. J. Biol. Macromol.* **2022**, *222*, 3178–3194. [CrossRef] [PubMed]
45. Scully, C.; Porter, S. Oral mucosal disease: Recurrent aphthous stomatitis. *Br. J. Oral Maxillofac. Surg.* **2008**, *46*, 198–206. [CrossRef]
46. Yarom, N.; Zelig, K.; Epstein, J.B.; Gorsky, M. The efficacy of minocycline mouth rinses on the symptoms associated with recurrent aphthous stomatitis: A randomized, double-blind, crossover study assessing different doses of oral rinse. *Oral Surg. Oral Med. Oral Pathol. Oral Radiol.* **2017**, *123*, 675–679. [CrossRef]
47. Pekbağrıyanık, T.; Dadas, F.K.; Enhoş, Ş. Effects of non-thermal atmospheric pressure plasma on palatal wound healing of free gingival grafts: A randomized controlled clinical trial. *Clin. Oral Investig.* **2021**, *25*, 6269–6278. [CrossRef]
48. Suresh, M.; Hemalatha, V.T.; Sundar, N.M.; Nisha, A. Applications of cold atmospheric pressure plasma in dentistry—A review. *J. Pharm. Res. Int.* **2022**, *34*, 45–55. [CrossRef]
49. Yilmaz, H.G.; Albaba, M.R.; Caygur, A.; Cengiz, E.; Boke-Karacaoglu, F.; Tumer, H. Treatment of recurrent aphthous stomatitis with Er,Cr:YSGG laser irradiation: A randomized controlled split mouth clinical study. *J. Photochem. Photobiol. B Biol.* **2017**, *170*, 1–5. [CrossRef]
50. Aggarwal, H.; Singh, M.P.; Nahar, P.; Mathur, H.; Sowmya, G.V. Efficacy of low-level laser therapy in treatment of recurrent aphthous ulcers—A sham controlled, split mouth follow up study. *J. Clin. Diagn. Res.* **2014**, *8*, 218–221. [CrossRef]
51. Lalabonova, H.; Daskalov, H. Clinical assessment of the therapeutic effect of low-level laser therapy on chronic recurrent aphthous stomatitis. *Biotechnol. Biotechnol. Equip.* **2014**, *28*, 929–933. [CrossRef]
52. de Sousa, A.C.T.; da Rocha, Í.B.P.; de Carvalho, A.F.M.; de Freitas Coelho, N.P.M.; Feitosa, M.C.P.; Barros, E.M.L.; Arisawa, E.A.L.; de Amorim, M.R.L. Comparative study between low level laser and therapeutic ultrasound in second intertion ulcers repair in mice. *J. Laser Med. Sci.* **2018**, *9*, 14–38. [CrossRef]
53. Hamishehkar, H.; Nokhodchi, A.; Ghambarzadeh, S.; Kouhsoltani, M. Triamcinolone acetonide oromucoadhesive paste for treatment of aphthous stomatitis. *Adv. Pharm. Bull.* **2015**, *5*, 277–282. [CrossRef]
54. Liu, D.; Zhang, T.; Zhou, H.; Meng, Y.; Wu, C.; Sun, Y.; Xu, Y.; Deng, X.; Wang, H.; Jiang, L. Role of biologics in refractory recurrent aphthous stomatitis. *J. Oral Path. Med.* **2022**, *51*, 694–701. [CrossRef]
55. Shavakhi, M.; Sahebkar, A.; Shirban, F.; Bagherniya, M. The efficacy of herbal medicine in the treatment of recurrent aphthous stomatitis: A systematic review of randomized clinical trials. *Phytother. Res.* **2022**, *36*, 672–685. [CrossRef]
56. Amini, M.R.; Hosseini, M.S.; Fatollah, S.; Mirpour, S.; Ghoranneviss, M.; Larijani, B.; Mohajeri-Tehrani, M.R.; Khorramizadeh, M.R. Beneficial effects of cold atmospheric plasma on inflammatory phase of diabetic foot ulcers; a randomized clinical trial. *J. Diabetes Metab. Disord.* **2020**, *19*, 895–905. [CrossRef]
57. Jha, N.; Ryu, J.J.; Choi, E.H.; Kaushik, N.K. Generation and role of reactive oxygen and nitrogen species induced by plasma, lasers, chemical agents, and other systems in dentistry. *Oxid. Med. Cell. Longev.* **2017**, *2017*, 7542540. [CrossRef]

Article

Evaluating the Healing Potential of J-Plasma Scalpel-Created Surgical Incisions in Porcine and Rat Models

Lilith Elmore [1], Nicholas J. Minissale [2], Lauren Israel [1], Zoe Katz [1], Jordan Safran [1], Adriana Barba [3], Luke Austin [4], Thomas P. Schaer [3] and Theresa A. Freeman [1,*]

[1] Department of Orthopaedic Research, Thomas Jefferson University, Philadelphia, PA 19107, USA; jordan.safran@students.jefferson.edu (J.S.)

[2] School of Osteopathic Medicine, Rowan University, Stratford, NJ 08084, USA

[3] Department of Clinical Studies, New Bolton Center, School of Veterinary Medicine, University of Pennsylvania, Kennett Square, PA 19348, USA; tpschaer@vet.upenn.edu (T.P.S.)

[4] Rothman Orthopaedic Institute, Philadelphia, PA 19107, USA

* Correspondence: theresa.freeman@jefferson.edu

Citation: Elmore, L.; Minissale, N.J.; Israel, L.; Katz, Z.; Safran, J.; Barba, A.; Austin, L.; Schaer, T.P.; Freeman, T.A. Evaluating the Healing Potential of J-Plasma Scalpel-Created Surgical Incisions in Porcine and Rat Models. *Biomedicines* **2024**, *12*, 277. https://doi.org/10.3390/biomedicines12020277

Academic Editors: Willibald Wonisch and Christoph Viktor Suschek

Received: 1 December 2023
Revised: 16 January 2024
Accepted: 20 January 2024
Published: 25 January 2024

Abstract: Cold atmospheric plasma devices generate reactive oxygen and nitrogen species that can be anti-microbial but also promote cell migration, differentiation, and tissue wound healing. This report investigates the healing of surgical incisions created using cold plasma generated by the J-Plasma scalpel (Precise Open handpiece, Apyx Medical, Inc.) compared to a steel scalpel in in vivo porcine and rat models. The J-Plasma scalpel is currently FDA approved for the delivery of helium plasma to cut, coagulate, and ablate soft tissue during surgical procedures. To our knowledge, this device has not been studied in creating surgical incisions but only during deeper dissection and hemostasis. External macroscopic and histologic grading by blinded reviewers revealed no significant difference in wound healing appearance or physiology in incisions created using the plasma scalpel as compared with a steel blade scalpel. Incisions created with the plasma scalpel also had superior hemostasis and a reduction in tissue and blood carryover. Scanning electron microscopy (SEM) and histology showed collagen fibril fusion occurred as the plasma scalpel incised through the tissue, contributing to a sealing effect. In addition, when bacteria were injected into the dermis before incision, the plasma scalpel disrupted the bacterial membrane as visualized in SEM images. External macroscopic and histologic grading by blinded reviewers revealed no significant difference in wound healing appearance or physiology. Based on these results, we propose additional studies to clinically evaluate the use of cold plasma in applications requiring hemostasis or when an increased likelihood of subdermal pathogen leakage could cause surgical site infection (i.e., sites with increased hair follicles).

Keywords: cold atmospheric plasma; surgical site infection; periprosthetic joint infection; plasma scalpel

1. Introduction

Many surgical procedures require a means to control or stop bleeding (hemostasis) into the surgical site. Hemostasis during surgery is commonly accomplished using mono- and bipolar electrocautery devices that generate intense heat to cauterize blood vessels. Unfortunately, the use of these devices is necessary but can impair healing, cause skin damage, produce charring, and result in tissue scarring [1]. In contrast, cold plasma technology has been explored as an alternative way to control bleeding through platelet activation and blood coagulation [2,3] without intense heating or extensive tissue damage. Cold plasma is a complex mixture of biologically active components, including charged particles, an electric current, UV radiation, and reactive oxygen and nitrogen species (ROS/RNS). The reactive species produced by cold plasma have also been shown to have bactericidal properties and enhance wound healing [4,5]. A cold plasma jet creates plasma using a high-powered electric current generator and helium gas to deliver the plasma to the tissue. By comparison, current traditional electrocautery devices, such as a Bovie device, use the same type of generator to deliver a high-energy radiofrequency (RF) spark directly to the tissue, causing localized heating and tissue damage [6].

Thus, this study was initiated to determine if the use of a cold plasma scalpel to create surgical incisions could achieve hemostasis without the side effects of electrocautery and undergo wound healing comparable to a steel scalpel. To our knowledge, cold plasma has not been studied in creating surgical incisions but only during deeper dissections. Cold plasma has been used to dissect osteoseptocutaneous flaps, fasciocutaneous flaps, and musculocutaneous flaps without any observed complications [7]. However, the optimal settings to create full-thickness skin incisions have not previously been defined or reported. Other biomedical applications of cold plasma have been investigated, including infection control, chronic wound healing, coagulation, deep dissection in plastic and vascular surgery, and eradication of malignant disease [8–12].

Cold plasma, with a range of plasma devices, has been employed in various medical fields, such as oncology, gastroenterology, dermatology, and plastic surgery. Of note, the term "cold plasma" is generically applied to a wide variety of devices that generate cold plasma using differing gas sources, including helium, oxygen, and air, or multiple settings for power or intensity and rates of gas flow. This results in the ability to fine tune cold plasma's composition of ions and reactive species and intensity of energy for specific applications. Specifically, when cold plasma is generated at low power and is applied for short durations, wound healing and immune cell function can be stimulated, while at higher powers, bacterial infections or cancer cells can be eradicated [13–15]. It is generally accepted that the differential production of ROS/RNS accounts for the biological effects described above [16]. Other actions attributed to the cold plasma-generated reactive species when interacting with tissues include the modification of proteins, activation of molecular signaling pathways [17–19], cell proliferation, and tissue regeneration [20], which can also be stimulated by cold plasma. At higher power intensities or when treatment times are increased, DNA damage, lipid peroxidation [21], and cell death have also been reported [3,22,23].

Current methods of preoperative surface skin sterilization are inadequate to eradicate bacteria resident in dermal appendages (hair follicles and sweat and sebaceous glands) that lie beneath the skin surface. As the incision is created, pathogens from these structures can be released into the surgical site via scalpel transfer or through the flow of lymph wound fluid from the tissue [24,25]. The use of a cold plasma scalpel to create an incision has the potential to limit this, as the reactive species generated will make contact with the bacteria as it passes through the tissue. Thus, by creating surgical incisions with cold plasma, the antimicrobial nature of the reactive species generated can be exploited to enhance surgical site sterility [13,15].

Taken together, we propose that surgical incisions that limit the blood, wound fluid, and subdermal pathogens that flow into the surgical site could be beneficial to infection prevention, especially in areas with large numbers of hair follicles and sweat glands that harbor greater numbers of subdermal pathogens. However, compromising the wound healing process would abrogate these benefits. Therefore, as a first step, this study first determines the settings required to operate a cold plasma scalpel to create an incision smoothly with the least tissue damage. Next, as wound healing is of paramount importance, we compare wound healing to the gold standard of incisions created with a steel scalpel in porcine and rat models.

2. Materials and Methods

Ex vivo trials for surgical plasma setting optimization—We chose to employ the J-Plasma Precise Open (Apyx Medical, Clearwater, FL, USA) cold plasma scalpel, an FDA approved helium plasma device used to coagulate and ablate soft tissue during open surgical procedures. To determine the optimal J-Plasma power (100% = 40 W) and gas flow (L/min) settings to create a full depth incision, an ex vivo model of skin was employed. Freshly harvested porcine or rat skin tissue samples from an unrelated study were acquired and kept on ice until the experimental procedure. Samples were placed on a silicone block over a ground plate connected to the J-Plasma device and moistened with 10 mM phosphate-buffered saline (PBS), pH 7. Incisions were created at variable gas flow and powers using the J-Plasma scalpel. At the end of the experiments, tissues were fixed in 4% paraformaldehyde > 24 h, dehydrated in graded ethanol (70 to 100%), and processed for either SEM or embedded in paraffin. Subsequently, 5 μm sections were acquired for histological analysis using Masson's trichrome and hematoxylin and eosin (H&E) stains. Rat cadavers were also employed for the ex vivo testing of plasma settings. Rats were placed on the ground plate and 1.5-inch incisions were created at 30 or 40% (15 or 20 W) power at gas flows of 3 or 4 L/min.

Analysis of chemical species in plasma-treated liquid—To characterize the concentration of the reactive species present in J-Plasma, liquid saline was treated with J-Plasma, and its nitrite and hydrogen peroxide concentrations were determined. A solution of 3 mL of 0.9% sodium chloride was added to a flat-bottomed glass dish and surface treated with J-Plasma 10 mm away at either 30 or 40% power with gas flow settings of 3, 4, or 5 L/min for 3 min. The treated liquid was collected and prepared in a 96-well plate for the quantification of species. For the detection of nitrite species, 50 μL of Griess reagent (Sigma Aldrich, St. Louis, MO, USA) was added to 50 μL of J-Plasma-treated saline, incubated in the dark for 30 min, and measured at 548 nm on a Tecan Infinite M1000 spectrometer. For the detection of hydrogen peroxide, 100 μL of potassium iodide and 50 μL of 10 mM phosphate buffer at a pH of 7 was added to 50 μL of J-Plasma treated saline, incubated for 30 min, and measured at 390 nm. Values were converted to μM using a standard concentration curve.

Ex vivo tissue preparation for scanning electron microscopy (SEM) analysis—J-Plasma versus steel scalpel incisions were also assessed by SEM to visualize their effects on collagen fibrils and bacteria during incision creation. Incisions were created in excised skin from porcine or rat models, then fixed in 4% paraformaldehyde for at least 24 h before serial dehydration with a graded series of ethanol from 70 to 100%, with each phase lasting at least 3 h, then dried 48 h or longer. The samples were then sputter coated with platinum and imaged with a Tabletop Hitachi scanning electron microscope. The tissue surrounding the incision was then excised and prepared for SEM imaging as described above. SEM images were used to compare incisions.

Ex vivo rat bacteria injection incision trial—Fresh rat cadavers were injected with 300 μL saline containing 108 colony-forming units (CFUs)/mL of Staphylococcus aureus (ATCC 25923) into the dermal region of the skin. Incisions were created minutes after using either a steel or plasma scalpel, after which the surrounding tissue was immediately excised and fixed in 4% paraformaldehyde overnight. Tissues were then prepared for SEM. SEM images were used to visualize bacterial structural integrity in incisions created with J-Plasma versus a steel scalpel.

In vivo porcine trial for surgical plasma optimization—Following International Animal Care and Use Committee (IACUC) approval consistent with ARRIVE guidelines by the University of Pennsylvania, studies were performed in a porcine model at Penn Vet New Bolton Center Hospital for Large Animals. The porcine skin has remarkable similarities to human skin and is used widely in dermatological and wound healing studies [26]. For this comparative study, we used one castrated male 35 kg Yorkshire pig. The animal was kept in a 6 × 6 ft pen with straw bedding and visual, olfactory, and auditory contact to herd mates in adjacent pens. The pig was premedicated intramuscularly with 0.2 mg/kg butorphanol (Turbogesic, Zoetis, Parsippany, NJ, USA), 0.3 mg/kg midazolam (Midazolam, West-Ward, Eatontown, NJ, USA) and 0.015 mg/kg dexmedetomidine (Dexdomitor, Zoetis, Parsippany, NJ, USA) mixed in one syringe. After onset of appropriate sedation approximately 15 min after drug administration, an intravenous catheter was placed into the auricular vein and 2 mg/kg ketamine (Ketathesia, Henry Schein, Melville, NY, USA) was administered intravenously. Anesthesia was maintained with isoflurane (Isothesia, Henry Schein) in oxygen. Intravenous crystalloid solution (Normosol, Abbott Laboratories, Abbott Park, IL, USA) was administered at a rate of 3 mL/kg/h. The surgical site was shaved and prepped for aseptic surgery. Plasma full-thickness incisions were made along the animal's dorsal flank with the J-Plasma device at 40% power, gas flow 3 L/min. Plasma, bipolar electrocautery, and steel scalpel (#11) full-thickness incisions were made 10 cm apart. Incisions were closed using #3 polyglactin 910 (Vicryl, Johnson and Johnson, Bridgewater, NJ, USA) in a simple continuous pattern for the subcutaneous tissue layer followed by #2 poliglecaprone 25 (Monocryl, Johnson and Johnson, Bridgewater, NJ, USA) in an interrupted pattern and dressed with a sterile dressing. The animal received buprenorphine (0.01 mg/kg IV or IM) pre- and post-operatively (SID-QID); fentanyl (2.5 mcg/kg/h transdermal), which was removed after 72 h, and flunixin meglumine (1.1 mg/kg IV or IM) SID for 3 days beginning the day of surgery for pain management. Wound healing was documented using photography, and aseptic full-thickness skin biopsy samples from each incision were taken at post-operative day 19 or 33 under general anesthesia. Biopsies from both time points were fixed >24 h in 4% paraformaldehyde (Sigma, St. Louis, MO, USA).

In vivo rat wound healing trial—Sixteen male Wistar rats weighing about 400 g were housed under standard conditions following IACUC guidelines with the approval of the Thomas Jefferson University IACUC committee. Each rat acted as its own control, comparing the standard-of-care steel scalpel incision on one side of the dorsum with the J-Plasma Precise Open instrument (Apyx Medical, Clearwater, FL, USA) incision on the opposite side. After anesthesia with isoflurane, animals were administered a 1 mg/kg dose of slow-release buprenorphine. Rat dorsa were shaved with electric clippers, and the skin was prepped with iodine and allowed to dry prior to marking 2 cm lines on the skin with a surgical marker on the lateral dorsum. On the left dorsum, an 11-blade steel scalpel was used to make a full-thickness incision, and on the right dorsum, the J-Plasma with a power level of 40% and gas flow of 3 L/min was used to make a second incision. Full-thickness incisions were made in dermal layers only, with their depth not penetrating beyond the myofascial tissue. Both wounds were sutured with non-absorbable 5-0 ethicon (Med-Vet International, Mettawa, IL, USA). Animals were monitored daily post-operatively, and photographs of the external appearance of the wounds were taken daily to monitor the progression of wound healing for each incision. If the rat chewed on the wound or otherwise bothered with it, that animal was excluded. Animals were euthanized with CO_2 at days 3, 8, or 20 post-operatively (n $\geq$ 4 rats for each time point), and the tissue around the incision was excised and fixed >24 h in 4% paraformaldehyde (Sigma, St. Louis, MO, USA)

Grading of external healing of incisions in rat model—The images acquired daily of incisions to monitor external wound healing were provided to 4 blinded graders. The Southampton wound grading system [27] was modified to adjust their alphabetical descriptions, which were assigned a numerical value for quantitation purposes (Table 1) This system was used to grade images of both scalpel and plasma wounds from days 3 and 8 post-incision. Images from post-incision day 20 scalpel and plasma incisions were also given to the reviewers, and the Stony Brook Scar Evaluation Scale was used to grade these images [28].

Histology—After fixation with 4% paraformaldehyde for 24 h, samples underwent paraffin infiltration using the Tissue-Tek VIP processor. The orientation of the sample when embedding in paraffin was bisected perpendicular to, and in the center of, each incision, then embedded with the bisected side down into a paraffin wax block for sectioning (Thermo Fisher Scientific Inc., Waltman, MA, USA). Serial 5 µm transverse sections originating from the center of the incision were stained with toluidine blue, H&E, or trichrome stains (Thermo Fisher Scientific Inc., Waltman, MA, USA). Slides were then mounted with Permount (Thermo Fisher Scientific Inc., Waltman, MA, USA).

Table 1. Modified Southampton Scoring System.

Southampton Scoring System	
Grade	**Appearance**
0 Normal Healing	
(1) Normal Healing with mild bruising or erythema	0.25 some bruising
	0.50 considerable bruising
	0.75 mild erythema
(2) Erythema plus other signs of inflammation	0.25 at one point
	0.50 around sutures
	0.75 along wound
	0.95 around wound
(3) Clear or hemoserous discharge	0.25 at one point only (<2 cm)
	0.50 along wound (>2 cm)
	0.75 large volume
	0.95 prolonged (>3 days)
(4) Pus/purulent discharge	0.25 at one point only (<2 cm)
	0.50 along wound (>2 cm)
(5) Deep or severe wound infection with or without tissue breakdown	

Image Analysis—Microscopic images were acquired using a Motic BA300Pol light microscope and the Jenoptik M50 ProgRes camera (Jupitor, FL, USA) or Nikon Eclipse (Nikon, Inc., Melville, NY, USA) with the Q-imaging Eclipse (Q-imaging, BC, Canada) at 4× and 10× magnification as 24-bit TIFF files. To objectively assess and quantitate wound healing, automated image analysis was performed at days 8 (correlating to the end of proliferative phase of wound healing) and 20 (end-stage wound healing) post-incision. Images of trichrome- and toluidine blue-stained sections were imported to Image-Pro Plus Ver. 7 (Media Cybernetics, Silver Spring, MD, USA) for analysis. As outlined by Sulthana et al. and others [29,30], to assess healing status, analyses of histological parameters should include quantifications of the amount of granulation tissue, immature vs. mature collagen, cellularity, and immune infiltrate. Instead of giving arbitrary qualitative values for minimal, moderate, and profound amounts, we used a quantitative assessment by the measurement of 3–4 images at multiple random locations along each incision for each type of measurement from at least 3 different animals/timepoint/treatment. Most analyses were paired such that incisions from both devices were performed on each rat dorsum.

As healing was not uniform along the 2 cm incisions, several sections (2–3 slides) were taken at multiple locations (4 or 5) along the incision length at 200–300-micron intervals from the center in both directions. The width of granulation tissue was measured on sections from areas that had complete re-epithelialization. Using the manual draw tool to define the edges of the granulation tissue by their color of staining (light blue compared to the darker blue normal dermal tissue), the average, minimum, and maximum distance between two lines was then calculated automatically by the Image Pro Plus software Ver. 7 (Media Cybernetics, Silver Spring, MD, USA) to generate the average width (as illustrated by red arrows) of the granulation tissue. For granulation tissue width, collagen maturity was assessed by extracting the green color channel from RGB trichrome images of day 8 wounds, yielding a grayscale image. A standardized rectangular region of interest (ROI) was created for all images around the wound site. Set thresholds were applied to all images to assess the areas comprising either the darker mature collagen or the lighter regions of immature collagen. Cellular infiltrate within granulation tissue was calculated using the toluidine blue-stained sections and a custom macro in ImagePro to count blue cells based on intensity and size filters. Further analysis of toluidine blue histology was conducted to count mast cells within the granulation tissue, with parameters set to account for the larger cell size and darker purple/pink intensity of mast cells. Conditions for each macro were determined by testing for accuracy on multiple images before applying it to all images in the dataset.

Statistical Analysis—All data were entered into the GraphPad Prism 8 (GraphPad Software, San Diego, CA, USA) statistical software. Images from the blinded grading of external wounds or histology sections were assessed by two-tailed t-tests; when images were from the same animal, the test was paired. Data were tested for normality using the Shapiro–Wilk normality test (alpha = 0.05, and a p-value of <0.05 was determined as statistical significance. For the blinded review of images, a Chi-square test was used.

3. Results

3.1. Determining the Optimal Settings for the J-Plasma Scalpel

As the interaction of cold plasma depends on tissue type, we determined it was paramount to establish cold plasma power settings that created a smooth incision through the skin. Secondly, we hypothesized that producing the greatest amount of ROS/RNS for the incision would enhance the likelihood that it would have microbicidal capabilities. Thus, various combinations of power and gas flow settings on the J-Plasma device were tested. It was determined the scalpel could not smoothly incise the tissue at powers less than 30%, and powers higher than 40% caused tissue charring. Similarly, gas flows of less than 2 L/min did not create a satisfactory incision. Therefore, we proceeded with only 30 or 40% power to determine ROS/RNS generation (nitrite and peroxide). Concentrations were measured in a 0.9% saline solution after treatment with the J-Plasma device at either 30 or 40% power at gas flows of 3, 4, or 5 L/min. We found that 30% power at a gas flow of 3 L/min generated the highest average concentration of H_2O_2 (24 ± 0.09 µM) and nitrite (37.49 ± 0.32 µM) (Figure 1a,c). No detectable peroxide species and low nitrite species were observed at gas flows of 4 and 5 L/min. A gas flow of 3 L/min also generated the highest peroxide (1033 ± 4.38 µM) and nitrite (136.3 ± 0.81 µM) concentrations at 40% power. Gas flows of 4 and 5 L/min yielded significantly lower concentrations of these species (Figure 1b,d). Overall, the highest amounts of peroxide and nitrite were generated at 40% power with a gas flow of 3 L/min. An image series of the plasma scalpel creating an incision is shown in Figure 1e.

To determine the effect on tissues after incision, ex vivo tissue samples from porcine or rat models were treated with a steel scalpel or a J-Plasma scalpel at 30 or 40% power with a gas flow of 3 L/min (Figure 1f). Trichrome staining of the incision made by the steel scalpel shows blue-stained, wavy collagen fibrils characteristic of the loose connective tissue of the dermis (top image). In contrast, tissues cut with the J-Plasma scalpel have areas of red staining along the incision (left side). We previously described this effect of plasma-generated reactive species modifying the affinity of the acid dyes in the trichrome stain [16]. Briefly, the basic nature of collagen is altered, causing a change in the affinity of the dyes such that the blue staining is replaced by red. Thus, the red color represents a rough estimate of reactive species penetration into the tissue. Histology showed that the depth of red staining (penetration of the species) was greater at 30% power compared to 40% power. This could have occurred because the scalpel moves slower through the tissue at 30% power, and therefore, the tissue is exposed to the plasma for a longer time, as it had been noted that making incisions at 40% power was much smoother and quicker than at 30%. Another observation was that the red-stained collagen fibrils also seemed fused together when compared to the blue-stained individual thread-like fibrils of collagen observed in the image of the tissue created with the steel blade scalpel.

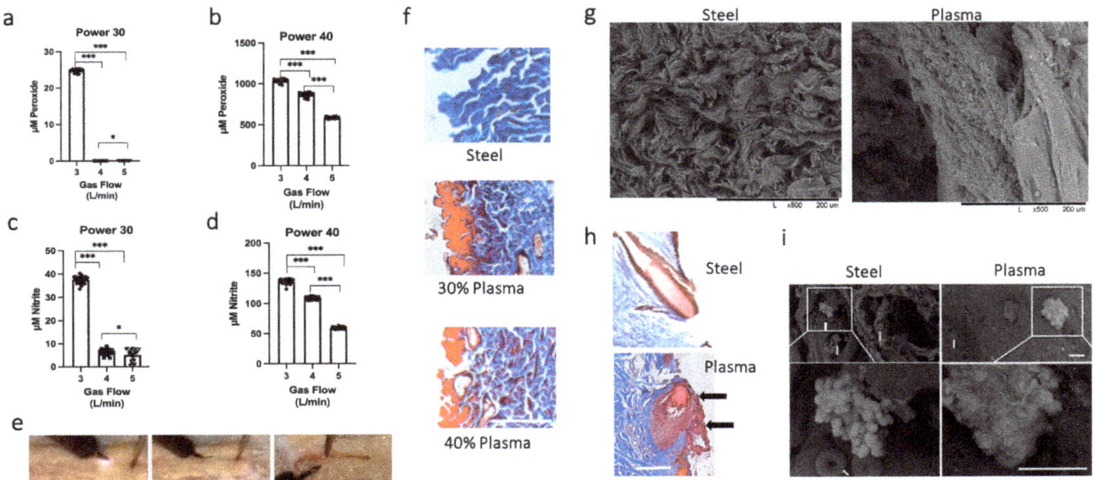

Figure 1. Comparison of reactive species and incision characteristics of steel versus J-Plasma scalpels. J-Plasma at gas flows of 3, 4, or 5 L/min was tested for the presence of peroxide and nitrite at 30% (**a,c**) or 40% power (**b,d**), respectively. (* = $p < 0.05$, *** = $p < 0.001$) (**e**) Series of images taken as plasma scalpel creates incision through porcine skin. The first picture taken as plasma ignites at initiation, followed by gliding through tissue, with the last image being the resultant incision with a smooth cut. (**f**) Comparison of reactive species penetration by trichrome staining of tissue cut with steel or plasma scalpels at 30% or 40% power. (**g**) Scanning electron microscopy (SEM) comparing porcine tissue cut with steel or plasma scalpels. (**h**) Trichrome stain of incision cut with steel or plasma scalpels at 40% power exhibits "sealing" of hair follicle by collagen fibril fusion (black arrows). (**i**) SEM of rat dermal tissue incision after injection of S. aureus at low magnification shows individual collagen fibril strands in steel scalpel incision vs. fused smoothed surface after plasma incision. Individual red blood cells (white arrows) are also more numerous in steel scalpel incision. White boxes indicate small colonies of bacteria shown at higher magnification in the lower panel, whereas a deterioration of the bacterial colony surface is evident after plasma treatment (mag. bar on histology sections = 1 mm).

3.2. Scanning Electron Microscope Results Support Plasma Fusion of Collagen Fibrils and Disruption of Bacterial Structure

To further investigate the fusion effect, scanning electron microscopy (SEM) of the cut collagen fibrils along the incisions was performed. Incisions created with steel or the plasma scalpel at the optimal setting of 40% power at 3 L/min were analyzed (Figure 1g). The SEM images clearly show the loose cut ends of the collagen fibrils after being incised with the steel scalpel. Conversely, after being incised by the plasma scalpel, the collagen fibrils have a smooth sheet-like appearance, confirming fibril fusion. Figure 1h shows the additional histology of a hair follicle in the plane of the incision with each scalpel type. Once again, the incision created with the plasma scalpel has the appearance of creating a fusion or sealing of the follicle (arrows) when compared to the incision created with the steel scalpel.

To assess what would happen to bacteria that come into contact with the plasma scalpel during the process of making the incision, we injected live S. aureus bacteria into the rat dermis before creating an incision. Within minutes after injection, an incision was made through the injected tissue with either the steel or plasma scalpel, after which the whole area was incised and prepared for SEM. SEM images of the incision made with the steel scalpel again clearly show the wavy individual fibrous "threads" of collagen. In contrast, as observed by histology, the collagen fibrils from incisions made with the plasma scalpel exhibited a smooth, sheet-like wall without any loose fibrils and with fusion of the collagen fibril ends. This smooth, fused topography, in addition to the decreased number of red blood cells (white arrows), indicates both tissue sealing, which may prevent lymph seepage into the surgical site, and the hemostasis of small vessels. Closer inspection at higher magnification to identify the effect on the injected bacteria showed bacterial clumps along the sealed collagen fibril interfaces (white boxes). These bacterial colonies were unaffected by the steel scalpel, whereas bacteria in the incision made by the J-Plasma scalpel showed a marked deterioration of surface topology (Figure 1i).

3.3. In Vivo Incisions in a Porcine Model Comparing Electrocautery and J-Plasma to a Steel Scalpel

To test the ability of the J-Plasma scalpel to create a hemostatic incision, an in vivo porcine model was used. One animal was used as a proof of concept to assess hemostasis and the gross healing of incisions created with the J-Plasma scalpel compared to a steel scalpel and electrocautery device. An image of each incision type is shown immediately after making the incision (Figure 2a) and after being closed with sutures, with each device being shown next to their respective incisions (Figure 2b). As expected, the scalpel incision resulted in the most amount of blood upon incision. In comparison, the incision with electrocautery resulted in almost no blood around the incision, and the incision made by the J-Plasma scalpel produced only minimal traces of blood. Additionally, blood remained on the steel scalpel post-incision creation (Figure 2c), in contrast to the J-Plasma scalpel (Figure 2c1), where blood was not visible.

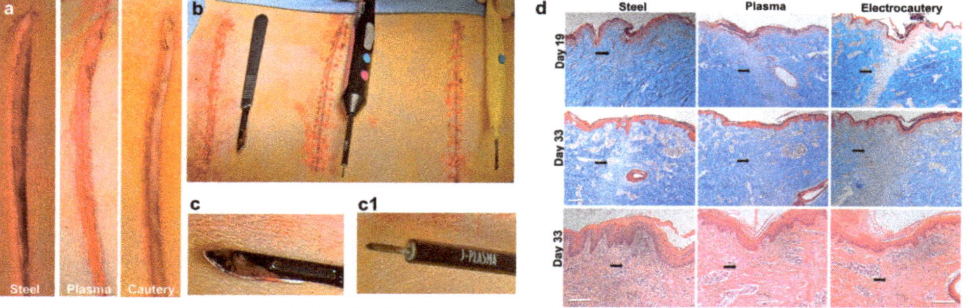

Figure 2. Comparison of incisions by steel, plasma, or electrocautery scalpels on porcine dorsa. External visualization of incisions immediately after incising with standard steel scalpel, J-Plasma, or electrocautery before (**a**) and after suturing (**b**) showing blood and fluid carry over on steel (**c**) versus plasma (**c1**) scalpels. (**d**) H&E and trichrome-stained histology images highlight differences in wound healing via granulation tissue appearance (black arrows) between wound healing at day 19 and day 33 post-incision with each device. 10× magnification (mag. bar = 1 mm).

3.4. J-Plasma Scalpel Showed Improved Tissue Healing Compared to Electrocautery in the Porcine Model

A histological analysis of the wound healing of the incisions created in the porcine model was performed on tissue sections from the incisions obtained on day 19 (biopsies) and day 33 after sacrifice using trichrome and H&E staining (Figure 2d; black arrows indicate granulation tissue, which stains light blue or white) in tissue incised by the electrocautery device. Conversely, the granulation tissue formation on sections treated by the steel scalpel or plasma scalpel was similar. After post-operative day 33, the incisional wounds created by the plasma or steel scalpels presented with well-arranged collagen. Overall, wound healing from the steel scalpel and J-Plasma incisions were similar in appearance and clearly more advanced than the tissue incised using electrocautery. At higher magnifications, H&E-stained sections from day 33 clearly show the newly remodeled collagen in the granulation area of the steel and plasma scalpel-incised tissue and the absence of this collagen in the tissue subjected to electrocautery (black arrows).

3.5. Plasma Scalpel Exhibits No Differences in External Wound Appearance When Compared to Steel Scalpel in the Rat Model

A more detailed analysis of wound healing was employed to test the effectiveness of the plasma scalpel after incision-making in a larger rat study. If future studies are to be planned and carried out in human subjects, cosmesis will likely be an important factor for both the surgeons and study participants. For this reason, a modified Southampton wound scoring system was employed for both day 3 and day 8 wounds. Representative images of external wound appearance are shown in Figure 3a at days 3, 8, and 20 post-incisions. Wound scores from four blinded graders resulted in no significant difference for day 3 ($p = 0.44$), with an average score for the steel scalpel of 1.5 ± 0.32 and 1.8 ± 0.34 for the plasma scalpel. On day 8, the steel scalpel scores averaged 0.50 ± 0.21, and plasma scores averaged 0.62 ± 0.23, and again scores were not significantly different ($p = 0.70$) (Figure 3b). On day 20, the average wound grades were 3.63 ± 0.50 for the steel scalpel and 3.06 ± 0.29 for the plasma scalpel. The results from day 20 also had no significant differences between the scars' appearances ($p = 0.21$) as graded by the Stony Brook Scar Evaluation Scale. Overall, there was no significant difference between the external appearance of wounds and scars generated after incision with either the J-Plasma scalpel or the steel scalpel.

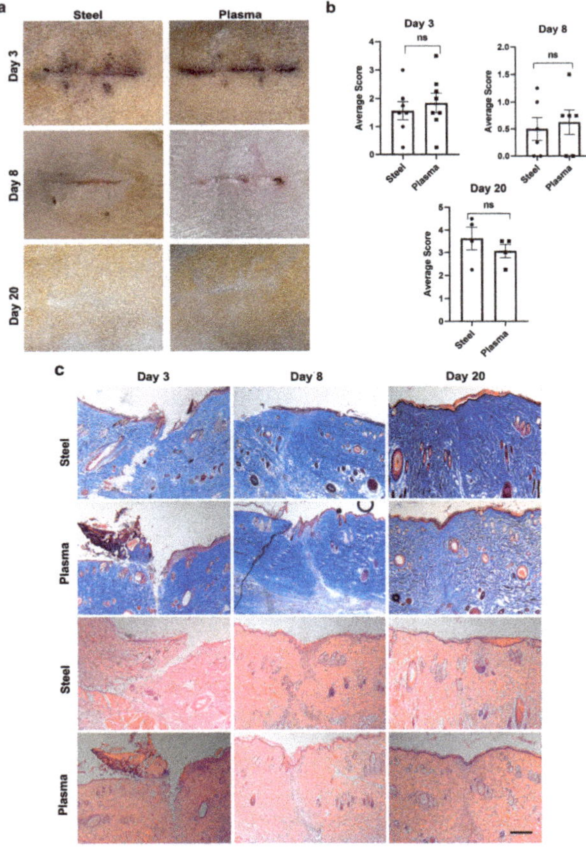

Figure 3. External wound grading and histological evidence of wound healing between incisions in an in vivo rat model made with steel or J-Plasma scalpels. (**a**) Representative images of days 3, 8, and

20 depicting external wounds caused by steel or plasma scalpels cutting 2 cm incisions. (**b**) Graphs showing the results from blinded wound grading using the Southampton Scoring System for day 3 ($p = 0.44$), day 8 ($p = 0.70$), and the Stony Brook Scar Evaluation Scale for day 20 ($p = 0.21$). (**c**) Trichrome- (upper 2 panels) and H&E- (lower 2 panels) stained sections give representative comparisons of wound healing at 3-, 8-, and 20-days post-incision with steel or J-Plasma scalpels (mag. bar = 1 mm; ns = not significant).

Histology on sections from wounds 3, 8, or 20 days after incision was used to visualize the stages of wound healing (Figure 3c) using trichrome (top 2 panels) and H&E (bottom 2 panels) staining. This was used to quantify the external appearance and wound healing in another re-producible manner. The progression of wound healing comparing the steel scalpel to the plasma scalpel showed little or no variation between the two modalities at any stage, including early healing (day 3), granulation (day 8), and the fully healed stage (day 20).

3.6. Quantitation of Histological Wound Healing Showed No Significant Differences between Steel and J-Plasma Scalpels in the Rat Model

We quantified and compared the progression of the wound healing of wounds made by either a steel scalpel or J-Plasma using trichrome-stained sections on days 8 and 20 and analyzed the width of granulation tissue using ImagePro Plus (Figure 4a). Day 3 was excluded due to the lack of re-epithelialization in both steel and plasma scalpel wounds. As is evident from the external wounds' appearance (Figure 4a), on day 8, incisions made with either scalpel type had areas with greater (white ^) or less (*) healing. Histology using trichrome staining shows a section from corresponding areas of worse (Figure 4b; white *, lack of epithelialization) and better (Figure 4c; white ^, thickened area of new epithelialization) areas of re-epithelialization along incisions from both steel and plasma scalpels made on the same rat. The steel scalpel wounds had an average width of 3.57 ± 0.35 mm, and a slightly smaller width was measured for wounds made with J-Plasma (2.78 ± 0.3 mm, $p = 0.054$; Figure 4d). Each dot represents a tissue section analyzed (2–3/rat, paired). Additional quantification of day 8 wounds to compare the ratio of mature to immature collagen was conducted on ImagePro Plus (Figure 4c). As a measure of tissue remodeling, our image analysis of mature collagen and immature fibrils (Figure 4e) showed that day 8 scalpel incisions contained $38.9 \pm 4.82\%$ mature collagen compared to plasma incisions with $44.6 \pm 5.48\%$ mature collagen, which also resulted in no significant difference between treatment groups ($p = 0.22$). At day 20 post-incision, granulation tissue from the wound bed in both incision types looked very similar (Figure 4f), and measurements confirmed this, with average granulation tissue width measuring 1.77 ± 0.62 mm for the steel scalpel and 1.89 ± 0.47 mm for J-Plasma (Figure 4g; $p = 0.82$).

3.7. Total Cell Infiltrate in Granulation Tissue Is Similar in Both Steel and Plasma Wound Beds on Days 8 and 20 Post-Incision, but Steel Incisions Have a Significantly Greater Number of Mast Cells

Toluidine blue is a stain for mast cells, which turn a dark or pink/purple color; their granular morphology is identifiable with light microscopy, while all nucleated cells stain light blue. Thus, this stain was used to quantitate total cellular infiltrate into the granulation tissue (demarcated by yellow drawn lines) on day 8 (Figure 5a,b) and day 20 (Figure 5c,d) post-incision. As previously mentioned, on day 8, there is a range of healing along each incision, and examples of more and less healing are shown for each scalpel type. Counted cells are pseudo-colored red to illustrate examples of automated counting with the image analysis macro. When the average number of cells in the granulation area were counted in images taken from at least 4 sections/rat (n = 3 rats) and a paired analysis was performed, no significant difference was observed (Figure 5b). A similar analysis performed on day 20 post-incision (Figure 5c; without the pseudo-coloring) also showed no significant difference (Figure 5d). Mast cells in toluidine blue-stained sections were larger and darker; when the same sections were analyzed using filters that isolated these cells, significantly more were observed in the granulation tissues from day 20 steel scalpel incisions (Figure 5e). A mean of 1.1 ± 0.31 mast cells/total cells were present when incisions were created with a steel scalpel (Figure 5f), while incisions made with the plasma scalpel had significantly less (0.25 ± 0.15 mast cells/total cells; $p = 0.005$).

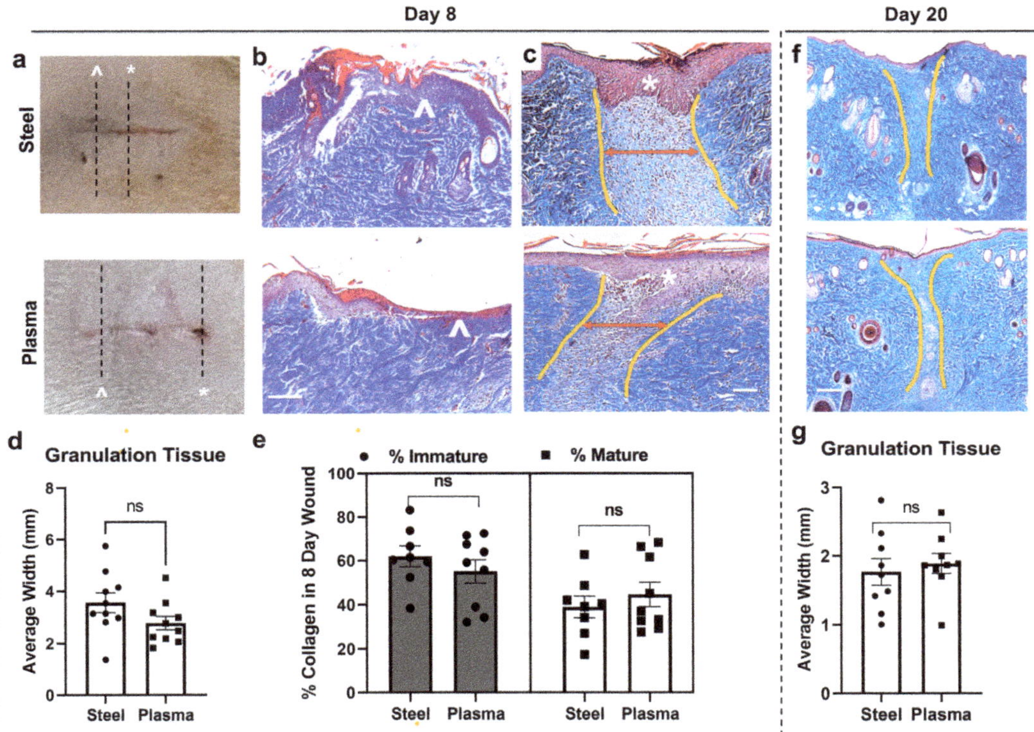

Figure 4. Comparison of granulation tissue amounts and composition in wound healing after steel or plasma scalpel incision. (**a**) Images of external wounds show areas with more (^) or less (*) healing along incision sites (2 cm). (**b**) Trichrome stain of wounds from steel or plasma scalpels on day 8 show differences in re-epithelialization corresponding to * and ^. (**c**) Lighter blue granulation tissue outlined by yellow lines were measured for average width (red two-sided arrows). (**d**) No significant (ns) difference was observed in a pairwise comparison of images (3–4 images/rat; $p = 0.86$; n = 3). (**e**) Analysis of trichrome-stained images for darker blue, mature collagen (peri-wound) or immature, light blue collagen in granulation area confirms no significant difference ($p = 0.22$). The same paired analysis of granulation tissue images on day 20 post-incision (**f**) also resulted in no significant difference (**g**) ($p = 0.82$, all mag. bars = 1 mm).

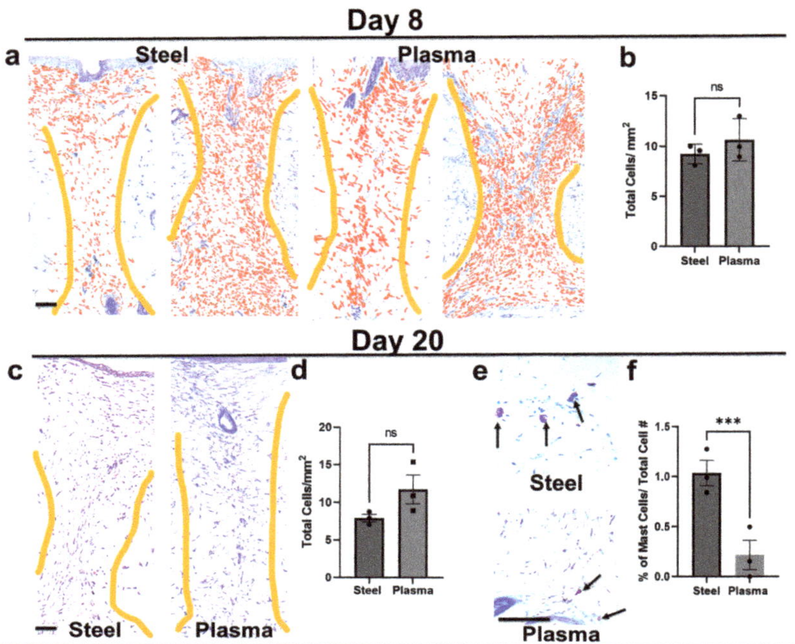

Figure 5. Comparison of cellular infiltration and mast cells in granulation tissue of steel or plasma scalpel wounds on days 8 and 20 post-incision. (**a**) Examples of more- and less-healed toluidine blue-stained granulation tissue from day 8 wound beds used to analyze cellular infiltrate by automated image analysis (red color indicates counted cells and yellow lines define granulation tissue area; mag. bar = 1 mm). (**b**) Graphed results indicate no significant (ns) differences ($p = 0.054$). (**c**) Comparisons of day 20 post-incision images used to analyze total cell count (mag. bar = 1 mm). (**d**) This analysis also resulted in no significant difference. (**e**) Toluidine blue images of larger pink/purple-stained mast cells observed on day 20 (arrows) (mag. bar = 1 mm). (**f**) Graphed results and paired *t*-tests indicate significantly more mast cells in the granulation tissue of steel scalpel wounds day 20 post-incision (***, $p = 0.005$). Average quantitation of 3-5 images for 3 separate rats (shown by black dot or squares) on each graph.

4. Discussion

The purpose of this study was to determine the impact of cold plasma (J-Plasma) during surgical incision on the skin while evaluating wound healing and bactericidal activity when compared to the current standard of care of the surgical steel scalpel. We hypothesized that cold-activated plasma could be used to make full-thickness skin incisions and decrease bacterial burden within deep surgical sites without negatively impacting wound healing. This study found that full-thickness skin incisions by J-Plasma heal both externally and internally in a manner similar to wounds created by a steel blade scalpel with the added benefit of increased hemostasis.

A recent study by Pinelli et al. demonstrated optimal settings for deep dissection using the J-Plasma device for fibula free flap (J-Plasma at 4 L/min flow and 30% power), radial free forearm flap (J-Plasma at 3 L/min flow and 20% power), and latissimus dorsi flap (J-Plasma at 4 L/min and 50% power) harvests [7]. Our results, similar to their settings for human tissue dissection, suggest that 3 L/min flow and 40% power is the optimal choice for the maximum generation of nitrite and peroxide while also allowing for precise control with no visible tissue eschar.

Prior studies have demonstrated cold plasma to be effective in hemostasis [2,3,31,32]. While the standard steel scalpel must rely on the subject's natural coagulation cascade and adequate surgical technique for hemostasis, cold plasma exhibits hemostatic qualities. Additionally, if required, bipolar electrocautery is routinely used in surgical settings to control hemostasis; however, as we have shown in our porcine model, this is much less conducive to efficient wound healing when creating the surgical incision through the dermis. The 2017 study by Bekeschus et al. hypothesized that plasma-derived ROS/RNS oxidize platelets, thereby increasing their activation, finally mediating hemostasis. The study found profound hemostatic activity utilizing a cold plasma device during liver resection in a murine model with a result comparable to electrocauterization. Our study demonstrates similar findings, with increased hemostasis utilizing J-Plasma to create full-thickness skin incisions compared to the standard steel scalpel in both an in vivo porcine model and in vivo rat model. Our study also demonstrates comparable hemostasis when comparing electrocautery during skin incision in an in vivo porcine model without the formation of visible necrotic tissue and better tissue healing (Figure 2). Utilizing a cold plasma scalpel could improve workflow efficiency during incision and lead to less blood loss while reducing the steps and the number of instruments used (i.e., superficial and deep scalpel blades, as well as a separate electrocautery device). The dissemination of pathogens such as C. acnes into deeper tissues layers during skin incision is a recognized clinical problem associated with increased surgical site infections and has been previously reported [33].

Numerous studies have shown that cold plasma (supplied by various plasma devices) may accelerate wound healing [34–36] and decrease bacterial inoculate of both natural wound flora and infected wounds [37,38]. In our study, J-Plasma incisions were found to have no difference in external wound appearance using previously published scoring systems [28,39] when compared to steel scalpel incisions at all time points throughout the study. Histologic analyses also found no difference at each time point with regards to granulation tissue, mature to immature collagen ratio, or collagen tissue width on day 8 across both treatment groups in the rat study.

A recent study by Wang et al. [40] observed the efficacy of non-thermal plasma, also known as cold plasma, on healing in an acute rat wound model. Two full-thickness dorsal cutaneous wounds of rats were treated with either a non-thermal helium plasma jet or helium. Their study found that compared to the control group, the wound area in the treatment group was significantly smaller on days 5, 7, and 14. The time for wound healing in the treatment group was also 2 days shorter compared to the control group. In addition, H&E staining of the scar sample on day 21 revealed that the scar width was not only smaller but also exhibited superior re-epithelialization compared to the control group. Our study varies from Wang et al. as the wounds were not treated multiple times with varying lengths of cold plasma application times but rather created by it. Our study does demonstrate, however, that compared to the standard of care, J-Plasma does not cause increased collagen formation or a change in the mature to immature collagen ratio and does not cause increased scar formation or delay wound healing compared to full-thickness incisions created by the steel scalpel.

Another study by Lee et al. [41] identified decreased numbers of mast cells and eosinophils along with lower epidermal thickness after the treatment of mice skin with non-thermal plasma (NTP). They determined that NTP suppresses mast cell activation, which is important for the allergic response, and ameliorates an atopic dermatitis-like skin inflammatory disease in mice. Our study agrees with these previous findings as scalpel incisions were found to have significantly higher counts of mast cells per millimeter than plasma incisions. High numbers of mast cells have also been observed in chronic wounds, hypertrophic scars, and keloids and have been implicated in fibrosis [42] and abnormal healing that could indicate scar tissue formation [43]. However, mast cells are also important for protecting against infection [44] and have been implicated in late-stage collagen remodeling, which fits with the time we observed the increased numbers.

Notable strengths of this study include the novel use of an FDA-approved device for the creation of full-thickness skin incisions, expanding its potential clinical utility. Our investigations employed a clinically relevant porcine model for select experiments of this study while resorting to a lower vertebrate model (rat) for the experiments that required an increased number of animals using standardized protocols regarding the collection and analysis of wounds, including their external appearance, collagen content, granulation tissue width, and H&E staining protocols.

5. Conclusions

The results of this study indicate surgical incisions created with cold plasma scalpel heal in a manner comparable to incisions created with a steel scalpel. In some instances, the creation of a surgical incision that exhibits enhanced hemostasis compared to the steel scalpel could be beneficial, specifically in areas where high levels of contamination from subdermal pathogens, such as the shoulder, groin, and spine, as the observed sealing of collagen fibrils and hemostasis may translate to a decreased rate of surgical site infection. Further studies are necessary to determine the clinical efficacy of cold plasma in human subjects.

Author Contributions: Conceptualization, L.E., N.J.M., L.A. and T.A.F.; methodology, L.E., L.A., N.J.M., A.B., T.P.S. and T.A.F.; software, T.A.F.; validation, L.E., Z.K., L.I., N.J.M. and T.A.F.; formal analysis, L.E., J.S. and T.A.F.; investigation, L.E., Z.K., L.I., N.J.M., A.B., T.P.S. and T.A.F.; resources, T.A.F.; data curation, L.E., Z.K., L.I., J.S. and T.A.F.; writing—original draft preparation, L.E., N.J.M. and T.A.F.; writing—review and editing, L.E., N.J.M., J.S. and T.A.F.; supervision, T.A.F.; project administration, T.A.F.; funding acquisition, T.A.F. All authors have read and agreed to the published version of the manuscript.

Funding: We would like to acknowledge NIH funding from National Institute of Arthritis and Musculoskeletal and Skin Diseases (NIAMS), RO1 AR069119 (Freeman). Funding from the Jefferson Research Foundation through a donation from Apyx Medical Inc.

Institutional Review Board Statement: Following International Animal Care and Use Committee (IACUC) approval consistent with ARRIVE guidelines by the University of Pennsylvania, studies were performed in a porcine model at the Penn Vet New Bolton Center Hospital for Large Animals (protocol code 806714 and 6-9-2022). Rat studies were performed with the approval of the Thomas Jefferson University IACUC committee (protocol code 02362 and 9-13-21).

Data Availability Statement: Data supporting this study is available through Jefferson University or the corresponding author.

Acknowledgments: Special thanks go to Noreen Hickok and Irving Shapiro for conversations involving infection and microbiology. Thanks also go to Shawn Roman at Apyx Medical Inc. for the use of the J-Plasma device.

Conflicts of Interest: Apyx Medical, Inc. generously provided the use of the J-Plasma device. Theresa Freeman has been the recipient of a donation was made by Apyx Medical to the Jefferson Research Foundation to support research in her laboratory. After this study was completed, Theresa Freeman became a paid consultant of Apyx Medical, Inc. Apyx Medical, Inc. had no role in the design of the study; in the collection, analyses, or interpretation of data; in the writing of the manuscript; or in the decision to publish the results. The remaining authors declare that the research was conducted in the absence of any commercial or financial relationships that could be construed as a potential conflict of interest.

References

1. Dandu, N.; Nelson, B.B.; Easley, J.T.; Huddleston, H.P.; DeFroda, S.F.; Bisazza, K.T.; Garrigues, G.E.; Yanke, A.B. Quantifying the Magnitude of Local Tendon Injury from Electrosurgical Transection. *J. Shoulder Elbow Surg.* **2022**, *31*, 832–838. [CrossRef]
2. Bekeschus, S.; Brüggemeier, J.; Hackbarth, C.; Von Woedtke, T.; Partecke, L.-I.; Van Der Linde, J. Platelets Are Key in Cold Physical Plasma-Facilitated Blood Coagulation in Mice. *Clin. Plasma Med.* **2017**, *7–8*, 58–65. [CrossRef]
3. Bekeschus, S.; Clemen, R.; Metelmann, H.-R. Potentiating Anti-Tumor Immunity with Physical Plasma. *Clin. Plasma Med.* **2018**, *12*, 17–22. [CrossRef]

4.   Brany, D.; Dvorská, D.; Halašová, E.; Škovierová, H. Cold Atmospheric Plasma: A Powerful Tool for Modern Medicine. *Int. J. Mol. Sci.* **2020**, *21*, 2932. [CrossRef]

5.   Hertwig, C.; Meneses, N.; Mathys, A. Cold Atmospheric Pressure Plasma and Low Energy Electron Beam as Alternative Nonthermal Decontamination Technologies for Dry Food Surfaces: A Review. *Trends Food Sci. Technol.* **2018**, *77*, 131–142. [CrossRef]

6.   Taheri, A.; Mansoori, P.; Sandoval, L.F.; Feldman, S.R.; Pearce, D.; Williford, P.M. Electrosurgery. *J. Am. Acad. Dermatol.* **2014**, *70*, 591.e1–591.e14. [CrossRef] [PubMed]

7.   Pinelli, M.; Starnoni, M.; De Santis, G. The Use of Cold Atmospheric Plasma Device in Flap Elevation. *Plast. Reconstr. Surg.-Glob. Open* **2020**, *8*, e2815. [CrossRef]

8.   Filis, K.; Galyfos, G.; Sigala, F.; Zografos, G. Utilization of Low-Temperature Helium Plasma (J-Plasma) for Dissection and Hemostasis during Carotid Endarterectomy. *J. Vasc. Surg. Cases Innov. Tech.* **2020**, *6*, 152–155. [CrossRef]

9.   Gan, L.; Zhang, S.; Poorun, D.; Liu, D.; Lu, X.; He, M.; Duan, X.; Chen, H. Medical Applications of Nonthermal Atmospheric Pressure Plasma in Dermatology. *JDDG J. Dtsch. Dermatol. Ges.* **2018**, *16*, 7–13. [CrossRef]

10.   Musavi, E.S.; Khorashadizadeh, S.M.; Fallah, R.; Rahmanian Sharifabad, A. Effect of Nonthermal Atmospheric Pressure Plasma on Plasma Coagulation in Healthy Persons and Patients under Treatment with Warfarin. *Contrib. Plasma Phys.* **2019**, *59*, 354–357. [CrossRef]

11.   Nguyen, L.; Lu, P.; Boehm, D.; Bourke, P.; Gilmore, B.F.; Hickok, N.J.; Freeman, T.A. Cold Atmospheric Plasma Is a Viable Solution for Treating Orthopedic Infection: A Review. *Biol. Chem.* **2018**, *400*, 77–86. [CrossRef]

12.   Weltmann, K.-D.; Von Woedtke, T. Plasma Medicine—Current State of Research and Medical Application. *Plasma Phys. Control. Fusion* **2017**, *59*, 014031. [CrossRef]

13.   Bourke, P.; Ziuzina, D.; Han, L.; Cullen, P.J.; Gilmore, B.F. Microbiological Interactions with Cold Plasma. *J. Appl. Microbiol.* **2017**, *123*, 308–324. [CrossRef]

14.   Chernets, N.; Kurpad, D.S.; Alexeev, V.; Rodrigues, D.B.; Freeman, T.A. Reaction Chemistry Generated by Nanosecond Pulsed Dielectric Barrier Discharge Treatment Is Responsible for the Tumor Eradication in the B16 Melanoma Mouse Model: DBD Generated ROS Eliminates Melanoma. *Plasma Process. Polym.* **2015**, *12*, 1400–1409. [CrossRef]

15.   Gilmore, B.F.; Flynn, P.B.; O'Brien, S.; Hickok, N.; Freeman, T.; Bourke, P. Cold Plasmas for Biofilm Control: Opportunities and Challenges. *Trends Biotechnol.* **2018**, *36*, 627–638. [CrossRef]

16.   Eisenhauer, P.; Chernets, N.; Song, Y.; Dobrynin, D.; Pleshko, N.; Steinbeck, M.J.; Freeman, T.A. Chemical Modification of Extracellular Matrix by Cold Atmospheric Plasma-Generated Reactive Species Affects Chondrogenesis and Bone Formation. *J. Tissue Eng. Regen. Med.* **2016**, *10*, 772–782. [CrossRef]

17.   Graves, D.B. Mechanisms of Plasma Medicine: Coupling Plasma Physics, Biochemistry, and Biology. *IEEE Trans. Radiat. Plasma Med. Sci.* **2017**, *1*, 281–292. [CrossRef]

18.   Steinbeck, M.J.; Chernets, N.; Zhang, J.; Kurpad, D.S.; Fridman, G.; Fridman, A.; Freeman, T.A. Skeletal Cell Differentiation Is Enhanced by Atmospheric Dielectric Barrier Discharge Plasma Treatment. *PLoS ONE* **2013**, *8*, e82143. [CrossRef] [PubMed]

19.   Von Woedtke, T.; Weltmann, K.-D. Grundlagen der Plasmamedizin. *MKG-Chirurg* **2016**, *9*, 246–254. [CrossRef]

20.   Chernets, N.; Zhang, J.; Steinbeck, M.J.; Kurpad, D.S.; Koyama, E.; Friedman, G.; Freeman, T.A. Nonthermal Atmospheric Pressure Plasma Enhances Mouse Limb Bud Survival, Growth, and Elongation. *Tissue Eng. Part A* **2015**, *21*, 300–309. [CrossRef] [PubMed]

21.   Leutner, S.; Eckert, A.; Müller, W.E. ROS Generation, Lipid Peroxidation and Antioxidant Enzyme Activities in the Aging Brain. *J. Neural Transm.* **2001**, *108*, 955–967. [CrossRef]

22.   Bekeschus, S.; Brüggemeier, J.; Hackbarth, C.; Weltmann, K.-D.; Von Woedtke, T.; Partecke, L.-I.; Van Der Linde, J. The Feed Gas Composition Determines the Degree of Physical Plasma-Induced Platelet Activation for Blood Coagulation. *Plasma Sources Sci. Technol.* **2018**, *27*, 034001. [CrossRef]

23.   Wende, K.; Bekeschus, S.; Schmidt, A.; Jatsch, L.; Hasse, S.; Weltmann, K.D.; Masur, K.; von Woedtke, T. Risk Assessment of a Cold Argon Plasma Jet in Respect to Its Mutagenicity. *Mutat. Res. Toxicol. Environ. Mutagen.* **2016**, *798–799*, 48–54. [CrossRef]

24.   Elston, M.J.; Dupaix, J.P.; Opanova, M.I.; Atkinson, R.E. Cutibacterium Acnes (Formerly Proprionibacterium Acnes) and Shoulder Surgery. *Hawaii J. Health Soc. Welf.* **2019**, *78*, 3–5. [PubMed]

25.   Saltzman, M.D.; Nuber, G.W.; Gryzlo, S.M.; Marecek, G.S.; Koh, J.L. Efficacy of Surgical Preparation Solutions in Shoulder Surgery. *J. Bone Jt. Surg.* **2009**, *91*, 1949–1953. [CrossRef]

26.   Seaton, M.; Hocking, A.; Gibran, N.S. Porcine Models of Cutaneous Wound Healing. *ILAR J.* **2015**, *56*, 127–138. [CrossRef] [PubMed]

27.   Campwala, I.; Unsell, K.; Gupta, S. A Comparative Analysis of Surgical Wound Infection Methods: Predictive Values of the CDC, ASEPSIS, and Southampton Scoring Systems in Evaluating Breast Reconstruction Surgical Site Infections. *Plast. Surg.* **2019**, *27*, 93–99. [CrossRef]

28.   Fearmonti, R.; Bond, J.; Erdmann, D.; Levinson, H. A Review of Scar Scales and Scar Measuring Devices. *Eplasty* **2010**, *10*, e43.

29.   Sultana, J.; Molla, M.R.; Kamal, M.; Shahidullah, M.; Begum, F.; Bashar, M.A. Histological Differences in Wound Healing in Maxillofacial Region in Patients with or without Risk Factors. *Bangladesh J. Pathol.* **1970**, *24*, 3–8. [CrossRef]

30.   van de Vyver, M.; Boodhoo, K.; Frazier, T.; Hamel, K.; Kopcewicz, M.; Levi, B.; Maartens, M.; Machcinska, S.; Nunez, J.; Pagani, C.; et al. Histology Scoring System for Murine Cutaneous Wounds. *Stem Cells Dev.* **2021**, *30*, 1141–1152. [CrossRef]

31. Heslin, C.; Boehm, D.; Milosavljevic, V.; Laycock, M.; Cullen, P.J.; Bourke, P. Quantitative Assessment of Blood Coagulation by Cold Atmospheric Plasma. *Plasma Med.* **2014**, *4*, 153–163. [CrossRef]
32. Kramer, A.; Lindequist, U.; Weltmann, K.-D.; Wilke, C.; von Woedtke, T. Plasma Medicine—Its Perspective for Wound Therapy. *GMS Krankenhaushygiene Interdiszip.* **2008**, *3*, Dec16.
33. Falconer, T.M.; Baba, M.; Kruse, L.M.; Dorrestijn, O.; Donaldson, M.J.; Smith, M.M.; Figtree, M.C.; Hudson, B.J.; Cass, B.; Young, A.A. Contamination of the Surgical Field with Propionibacterium Acnes in Primary Shoulder Arthroplasty. *J. Bone Jt. Surg.* **2016**, *98*, 1722–1728. [CrossRef]
34. Bekeschus, S.; Seebauer, C.; Wende, K.; Schmidt, A. Physical Plasma and Leukocytes—Immune or Reactive? *Biol. Chem.* **2018**, *400*, 63–75. [CrossRef] [PubMed]
35. Bekeschus, S.; Von Woedtke, T.; Emmert, S.; Schmidt, A. Medical Gas Plasma-Stimulated Wound Healing: Evidence and Mechanisms. *Redox Biol.* **2021**, *46*, 102116. [CrossRef] [PubMed]
36. Shahbazi Rad, Z.; Abbasi Davani, F.; Etaati, G. Determination of Proper Treatment Time for in Vivo Blood Coagulation and Wound Healing Application by Non-Thermal Helium Plasma Jet. *Australas. Phys. Eng. Sci. Med.* **2018**, *41*, 905–917. [CrossRef]
37. Darmawati, S.; Nasruddin, N.; Putri, G.S.A.; Iswara, A.; Kurniasiwi, P.; Wahyuningtyas, E.S.; Nurani, L.H.; Hayati, D.N.; Ishijima, T.; Nakatani, T.; et al. Accelerated Healing of Chronic Wounds under a Combinatorial Therapeutic Regimen Based on Cold Atmospheric Plasma Jet Using Contact and Noncontact Styles. *Plasma Med.* **2021**, *11*, 1–18. [CrossRef]
38. Klebes, M.; Ulrich, C.; Kluschke, F.; Patzelt, A.; Vandersee, S.; Richter, H.; Bob, A.; Von Hutten, J.; Krediet, J.T.; Kramer, A.; et al. Combined Antibacterial Effects of Tissue-tolerable Plasma and a Modern Conventional Liquid Antiseptic on Chronic Wound Treatment. *J. Biophotonics* **2015**, *8*, 382–391. [CrossRef]
39. Bailey, I.S.; Karran, S.E.; Toyn, K.; Brough, P.; Ranaboldo, C.; Karran, S.J. Community Surveillance of Complications after Hernia Surgery. *BMJ* **1992**, *304*, 469–471. [CrossRef]
40. Wang, X.-F.; Fang, Q.-Q.; Jia, B.; Hu, Y.-Y.; Wang, Z.-C.; Yan, K.; Yin, S.-Y.; Liu, Z.; Tan, W.-Q. Potential Effect of Non-Thermal Plasma for the Inhibition of Scar Formation: A Preliminary Report. *Sci. Rep.* **2020**, *10*, 1064. [CrossRef]
41. Lee, M.-H.; Lee, Y.S.; Kim, H.J.; Han, C.H.; Kang, S.U.; Kim, C.-H. Non-Thermal Plasma Inhibits Mast Cell Activation and Ameliorates Allergic Skin Inflammatory Diseases in NC/Nga Mice. *Sci. Rep.* **2019**, *9*, 13510. [CrossRef] [PubMed]
42. Noli, C.; Miolo, A. The Mast Cell in Wound Healing. *Vet. Dermatol.* **2001**, *12*, 303–313. [CrossRef]
43. Urb, M.; Sheppard, D.C. The Role of Mast Cells in the Defence against Pathogens. *PLoS Pathog.* **2012**, *8*, e1002619. [CrossRef] [PubMed]
44. Iba, Y.; Shibata, A.; Kato, M.; Masukawa, T. Possible Involvement of Mast Cells in Collagen Remodeling in the Late Phase of Cutaneous Wound Healing in Mice. *Int. Immunopharmacol.* **2004**, *4*, 1873–1880. [CrossRef] [PubMed]

Article

Analyses of the Chemical Composition of Plasma-Activated Water and Its Potential Applications for Vaginal Health

Hyun-Jin Kim [1,†], Hyun-A Shin [1,†], Woo-Kyung Chung [2], Ae-Son Om [2], Areum Jeon [1], Eun-Kyung Kang [1], Wen An [1] and Ju-Seop Kang [1,*]

[1] Department of Pharmacology, College of Medicine, Hanyang University, Seoul 04736, Republic of Korea; hope0211@hanyang.ac.kr (H.-J.K.); yellowhyun74@hanyang.ac.kr (H.-A.S.); areumii0904@hanyang.ac.kr (A.J.); silverk1239@hanyang.ac.kr (E.-K.K.); anwen@hanyang.ac.kr (W.A.)
[2] Department of Food and Nutrition, Hanyang University, Seoul 04736, Republic of Korea; entksv10@naver.com (W.-K.C.); aesonom@hanyang.ac.kr (A.-S.O.)
* Correspondence: jskang@hanyang.ac.kr; Tel.: +82-2-2220-0652
† These authors contributed equally to this work.

Abstract: This study aimed to elucidate the unique chemical compositions of plasma-activated water (PAW) and the potential antibacterial efficacy of PAW as a novel vaginal cleanser. We analyzed the ion compositions (four anions: F^-, Cl^-, NO_3^-, SO_4^{2-}; five cations: Na^+, NH_4^+, K^+, Mg^{2+}, Ca^{2+}) of several formulations of PAW generated at different electrical powers (12 and 24 V) at various treatment time points (1, 10, and 20 min), and stay durations (immediate, 30, and 60 min). As treatment duration increased, hypochlorous acid (HOCl), Ca^{2+}, and Mg^{2+} concentrations increased and Cl^- concentration decreased. Higher electrical power and longer treatment duration resulted in increased HOCl levels, which acts to prevent the growth of general microorganisms. Notably, PAW had no antibacterial effects against the probiotic, *Lactobacillus reuteri*, which produces lactic acid and is important for vaginal health. These findings indicate that PAW contains HOCl and some cations (Ca^{2+} and Mg^{2+}), which should help protect against pathogens of the vaginal mucosa and have a cleansing effect within the vaginal environment while not harming beneficial bacteria.

Keywords: plasma-activated water (PAW); hypochlorous acid (HOCl); probiotics (*Lactobacillus reuteri*); mucosa protection; vaginal cleansing effect

Citation: Kim, H.-J.; Shin, H.-A.; Chung, W.-K.; Om, A.-S.; Jeon, A.; Kang, E.-K.; An, W.; Kang, J.-S. Analyses of the Chemical Composition of Plasma-Activated Water and Its Potential Applications for Vaginal Health. *Biomedicines* **2023**, *11*, 3121. https://doi.org/10.3390/biomedicines11123121

Academic Editor: Willibald Wonisch

Received: 25 October 2023
Revised: 17 November 2023
Accepted: 18 November 2023
Published: 23 November 2023

1. Introduction

Plasma technology has been proven to be effective in various medical applications, including regenerative medicine for skin and dental treatments, as well as surface sanitization and sterilization of medical tools [1]. Plasma is a term that refers to a quasi-neutral ionized gas containing photons, free radicals, and ions as well as uncharged particles [2–4]. Traditional plasmas are categorized as thermal or non-thermal, based on the thermodynamic equilibria and non-equilibrium of electrons and other gas species. Recent advances in plasma engineering have enabled the generation of plasma-activated waters (PAW) through non-thermal atmospheric pressure plasmas (NTAPPs) [5]. The PAW can be produced by exposing water to ionized gas generated by a plasma device, either above or below the water surface. Many studies have demonstrated that the reactive species in PAW are not toxic and do not pollute the environment, paving the way for diverse applications of PAW in the life sciences [4]. The primary application areas of PAW technology are seed germination, plant growth, food preservation, antimicrobial activities, virus inactivation, and anticancer treatments, among others [6]. The disinfection effectiveness of PAW hinges upon the concentration of reactive oxygen species (ROS) and reactive nitrogen species (RNS) in the PAW. ROS generated in PAW, including hydrogen peroxide(H_2O_2), hydroxyl radical($\cdot$OH), and ozone (O_3), function as potent oxidizing agents, inducing oxidative stress on the microbial cell membranes and, consequently, bacterial damage and death [7].

RNS present in PAW, such as nitric oxide (NO), nitrite (NO_2^-), and nitrate (NO_3^-), exist primarily as peroxynitrite (ONOOH) under acidic conditions. Peroxynitrite can accumulate inside cells leading to apoptotic or necrotic cell death [8]. The production of various reactive oxygen and nitrogen species (RONS) depends on the operating parameters of plasma generation, such as the power source, treatment time, feed gases, and electrode materials [7,9]. Gao et al. reported that increasing the electrical power from 0 to 160 W led to higher concentrations of OH, H_2O_2, NO_3^-, and NH_4 in PAW [4]. Berardinelli et al. reported that longer plasma treatment duration result in elevated concentrations of NO_2^- and NO_3^- in PAW [10]. Other studies have demonstrated that the concentrations of O_3 and H_2O_2 can also be increased in PAW by manipulating how the PAW is generated [11]. Georgescu et al. reported that feed gases like helium and nitrogen can alter the reactive species in PAW by changing the electron density and surface charges compared with ambient air [12]. These parameters can be fine-tuned to create PAWs with different reactivity levels for various disinfection applications, and researchers are actively developing PAW devices to generate PAWs with different reactivities [13]. The applications of PAW have rapidly expanded to include the treatment of biomedical devices and biological materials, including foods.

Lee and Hong demonstrated that plasma discharge in tap water helps eliminate harmful microorganisms by increasing the concentrations of free residual chlorine molecules such as hypochlorous acid and hypochlorite ions [14–16]. While the exact mechanisms remain unclear, PAW has been shown to have antibacterial and cytotoxic activity, which has been attributed to the reactive species present in PAW [4,5,17,18]. Further theoretical-experimental investigations are warranted to expand our understanding of PAW and further explore its potential in biological decontamination and clinical applications [12].

In previous studies, we confirmed the bactericidal effect of PAWs on clinically abnormal vaginal microbiota in clinical practice [14,19,20]. Patients sprayed with PAW (22.3%) had fewer Gram-positive and -negative bacteria than betadine treatment (BT) patients (14.4%) [14]. A significant decrease after treatment was observed in the following pathogenic organisms: *Mycoplasma hominis* ($30 \pm 15.28\%$ decrease), *Ureaplasma urealyticum* ($25 \pm 9.93\%$ decrease), *Ureaplasma parvum* ($23 \pm 8.42\%$ decrease), and *Candida albicans* ($28 \pm 10.86\%$ decrease) [19]. Vaginitis is a common disease among women and bacterial vaginosis is the most common form of vaginitis, accounting for 22~50% of all vaginitis cases, followed by candida vulvovaginitis at 17~35%, and trichomonas vaginitis at 4~39%. Vaginosis is caused by unbalanced changes in the vaginal microbiome, which are associated with a reduction in the overall number of Lactobacilli species and a predominance of anaerobic microorganisms, including *Gardnerella vaginalis*, *Trichomonas*, and *Candida albicans* strains. Generally, the clinical symptoms include a foul-smelling vaginal discharge, fever, sexual discomfort and painful urination [21].

The excessive growth of anaerobic species, particularly *Gardnerella vaginalis*, results in a polymicrobial biofilm that adheres to the vaginal epithelium [22]. Biofilms are communities of microorganisms encased in a polymeric matrix of nucleic acids, polysaccharides, and proteins [23]. Biofilm-related infections are challenging to eradicate by both the immune system and antibiotics, leading to a high rate of relapse and recurrence in bacterial vaginosis cases [24–26].

PAW can be used to disrupt biofilms and effectively eliminate the attached bacteria. PAW created through cold atmospheric-pressure plasma discharge in water demonstrated significant antimicrobial activity against biofilms, without promoting bacterial resistance [27].

Therefore, our objective in this study was to analyze the chemical components underlying the disinfection effects of PAW with a specific focus on vaginal sterilization. Additionally, we evaluated the effects of PAW on *Lactobacillus reuteri*, a vital probiotic component of the vaginal microbiome essential for the host's health [28].

2. Materials and Methods

2.1. PAW Generating System and PAW Processing

The PAW system comprised an underwater plasma-generating device, with plasma generated using procedures established previously [20]. An illustration of the device is provided in Figure 1; the device comprises a 3 L cleaning solution container, the atmospheric plasma electrodes, and a controller.

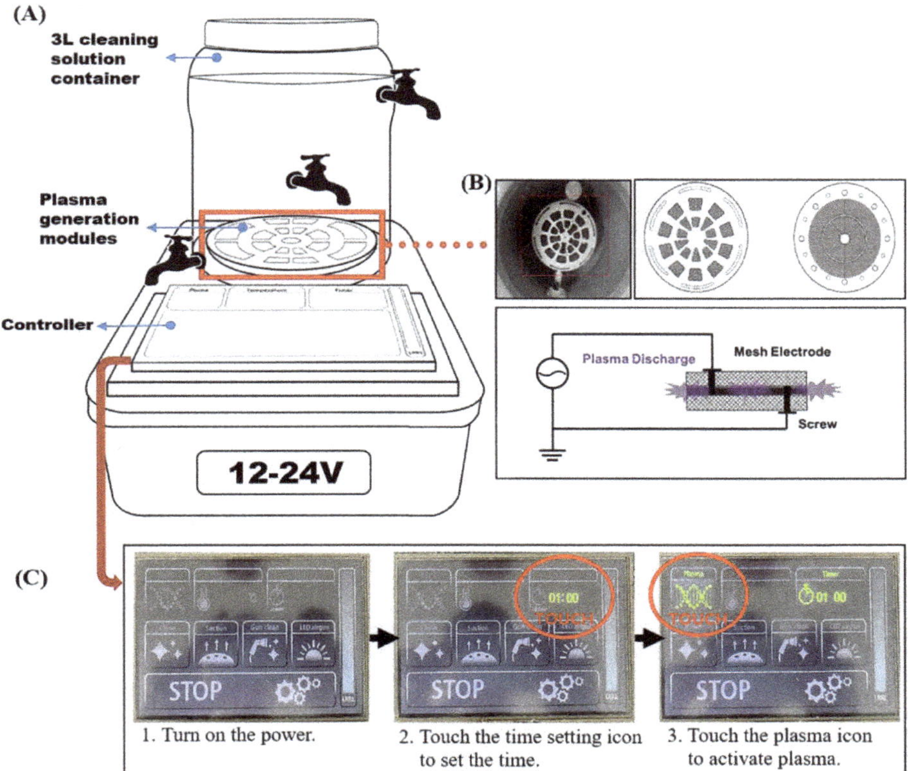

Figure 1. Plasma device and operation process. (**A**) Components of the plasma device used in the experiment. (**B**) Plasma module and schematic principle for underwater plasma discharge. Reprinted with permission from Hwang, et al., 2020 [20], and (**C**) the digital display controller and details of how to operate the plasma device.

The device facilitates plasma discharge underwater through a plasma electrode located at the container's base. The plasma generation modules consist of two plasma electrodes separated by an insulating frame, with each electrode connected to power of different polarity. These electrodes are disc-shaped grids, measuring 77 mm in diameter and 0.5 mm in thickness, with titanium used as the electrode material to prevent corrosion. The grid has a diameter of 1.07 mm and pitch of and 3.92, resulting in 50% open space. To enhance plasma generation, a 300 nm thick platinum thin film is deposited on the grid's surface using electroless plating. Plasma discharge occurs between the two electrodes with a consistent 2 mm separation maintained by the insulating frame (Figure 1B) [20]. To analyze the chemical composition of water at different positions (top, middle, bottom) within the 3 L cleaning solution container, three stopcocks were connected to the container (Figure 1A). Figure 1C shows the operation process of the plasma device.

2.2. PAW Sample Preparation Conditions

PAW samples were prepared using the following conditions: (1) electrical power of 12 or 24 V; (2) plasma treatment durations of 1, 10, or 20 min; and (3) 0, 30 or 60 min retention times after plasma exposure (Table 1). At specified time intervals, 10 mL samples of PAW were collected and immediately subjected to analysis using ion chromatography, and a residual chlorine analyzer.

Table 1. The conditions of plasma activated water.

	Position	Power	Treatment Time (min)	Retention Time (min)
Tap water	Top, middle, bottom	12 V, 24 V	1, 10, or 20	0, 30, or 60

2.3. Analytical Methods

The inorganic constituents and hypochlorous acid content of the PAW were determined as described below.

2.3.1. Ion Chromatography (IC)

IC analysis was carried out with an ion chromatograph (Dionex ICS-3000, Thermo Fisher Scientific, Waltham, MA, USA) equipped with both an anion and cation module. An Ionpac AG20 4 × 50 mm guard column (Thermo Fisher Scientific, Waltham, MA, USA) and Ionpac AS20 4 × 250 mm analytical column (Thermo Fisher Scientific, Waltham, MA, USA) were employed in the anion module while an Ionpac CG16 5 × 50 mm guard column and Ionpac CS 5 × 250 mm analytical column were used in the cation module. The column temperature was set to 30 °C for a run time of 20 min. A gradient method was used for the mobile phase of the anion module starting at 12 mM sodium hydroxide (NaOH) for the first 8 min, followed by a change to 40 mM NaOH from 8 min to 12 min, maintenance at 40 mM NaOH until 18 min, and then a decrease to 12 mM NaOH until 20 min. An isocratic method was employed in the cation module with a mobile phase consisting of 40 mM methanesulfonic acid (MSA). The flow rate was 1 mL/min and the injection volume was 25 µL. The anion module contained an ADRS 600 suppressor (Thermo Fisher Scientific, Waltham, MA, USA) while the cation module incorporated a CDRS 600 suppressor (Thermo Fisher Scientific, Waltham, MA, USA). Instrument control and data acquisition were managed through Chromeleon® chromatography management software (version 6.80) (Thermo Fisher Scientific, Waltham, MA, USA).

Working standards for seven anions were prepared (fluoride, chloride, nitrite, bromide, nitrate, phosphate, sulfate) and their calibration curves ranges were 0.1 to 9.9, 0.15 to 15, 0.5 to 50, 0.5 to 50, 0.5 to 50, 0.75 to 75, and 0.75 to 75 mg/L, respectively.

Cation standard working solutions of six cations were prepared (lithium, sodium, ammonium, magnesium, potassium, calcium) and their calibration curves ranges were 0.3 to 25, 1 to 100, 1.26 to 126, 1.27 to 127, 2.5 to 250, and 2.5 to 250 mg/L, respectively.

2.3.2. Residual Chlorine Analyzer

Hypochlorous acid concentrations in the PAW samples were determined using DPD(N,N diethyl-1,4-phenylenediamine) free chlorine reagent with a Q-CL501B analyzer (Shenzhen Sinsche Technology Co. Ltd., Shenzhen, China), following the DPD colorimetric method.

2.4. Evaluation of Antibacterial Efficacy

2.4.1. Microorganisms and Materials

To assess antibacterial activity, *Limosilactobacillus reuteri* subsp. *reuteri* (*L. reuteri*) was sourced from the Korean Collection for Type Cultures (KCTC) (Table 2). To cultivate *L. reuteri* subsp. *reuteri* (*L. reuteri*), de Man, Rogosa and Sharpe agar (MRS, Difco, Detroit, MI, USA) was employed as the growth medium.

Table 2. Microorganisms used for antibacterial activity test.

Microorganism		Strain
Gram-positive	*Lactobacillus reuteri*	KCTC 3594

2.4.2. Antibacterial Activity

The antibacterial activity of plasma-activated water (PAW) was determined using the filter paper disc method [29]. Bacterial cultures (sub-cultured before assay) were diluted with sterile water to obtain a bacterial suspension of OD_{600nm} = 0.2~0.3. Petri dishes containing 10 mL MRS media were inoculated with 0.1 mL of the bacterial suspension, dried within a sterile chamber, and incubated at 37 °C for 48 h. A filter paper disc (Ø: 6 mm, ADVANTEC, Tokyo, Japan) was impregnated with 20 µL PAW sample, placed on the medium, and then the petri dish was left at 37 °C for 15 min or 30 min. Sterile water served as the negative control. All experiments were performed in triplicate.

3. Results

3.1. Inorganic Anion and Cation Composition of PAW Samples

The inorganic anion and cation content of PAW samples generated by underwater plasma discharge for various durations (1, 10, and 20 min) at two voltage settings (12 and 24 V) was determined. Chromatography results for the standard anions and cations are shown in Figure 2. Retention times for the individual anions (fluoride, chloride, nitrite, bromide, nitrate, phosphate, sulfate) were approximately 3.74, 5.14, 6.05, 7.06, 7.87, 9.52, and 14.30 min, respectively. The retention times for the cations (lithium, sodium, ammonium, magnesium, potassium, calcium) were 4.78, 6.45, 7.94, 10.84, 11.96, and 14.77 min, respectively. The calibration curves for the standard anions and cations demonstrated linearity, indicating that anion and cation concentrations in the PAW samples could be determined accurately. Chromatographs of anions and cations within the PAW samples are shown in Figure 2C,D.

Ion chromatography analysis revealed the generation of five cations and four anions following plasma treatment of tap water (Tables 3 and 4). Table 3 presents changes in ion concentrations following 12 V plasma generation. After plasma activation for 20 min, changes in the concentrations of Ca^{2+}, Mg^{2+}, and Cl^- were noted. The baseline levels of Ca^{2+} were 10.46 mg/L (top), 15.17 mg/L (middle), and 15.00 mg/L (bottom), while those of Mg^{2+} were 3.52 mg/L (top), 3.86 mg/L (middle), and 3.89 mg/L (bottom). The baseline levels of Cl^- were 20.21 mg/L (top), 19.35 mg/L (middle), and 19.27 mg/L (bottom). The Ca^{2+} concentrations were 20.73 mg/L (top), 19.33 mg/L (middle), and 18.75 mg/L (bottom) after a 60 min treatment duration, while those of Mg^{2+} were and 4.51 mg/L (top), 4.04 mg/L (middle), and 4.01 mg/L (bottom). The Cl^- concentrations were 19.76 mg/L (top), 18.46 mg/L (middle), and 17.98 mg/L (bottom) under the same conditions. After plasma activation for 10 min, changes were observed in Ca^{2+} and Cl^-. Ca^{2+}, initially at 13.38 mg/L, changed to 17.33 (top), 17.76 (middle), 17.73 mg/L (bottom) after a 60 min treatment duration. Cl^- concentration, initially at 17.41 mg/L, changed to 16.85 (top), 16.96 (middle), and 17.17 mg/L (bottom) under the same conditions. After plasma activation for 1 min, changes were observed in Ca^{2+} and Cl^-. The baseline levels of Ca^{2+} and Cl^- ions were 13.31 mg/L and 19.08 mg/L. The Ca^{2+} concentrations were 16.33 (top), 16.20 (middle), and 16.31 mg/L (bottom) for a treatment time of 60 min. The Cl^- concentrations were 18.62 (top), 13.83 (middle), and 13.62 mg/L (bottom) under the same conditions.

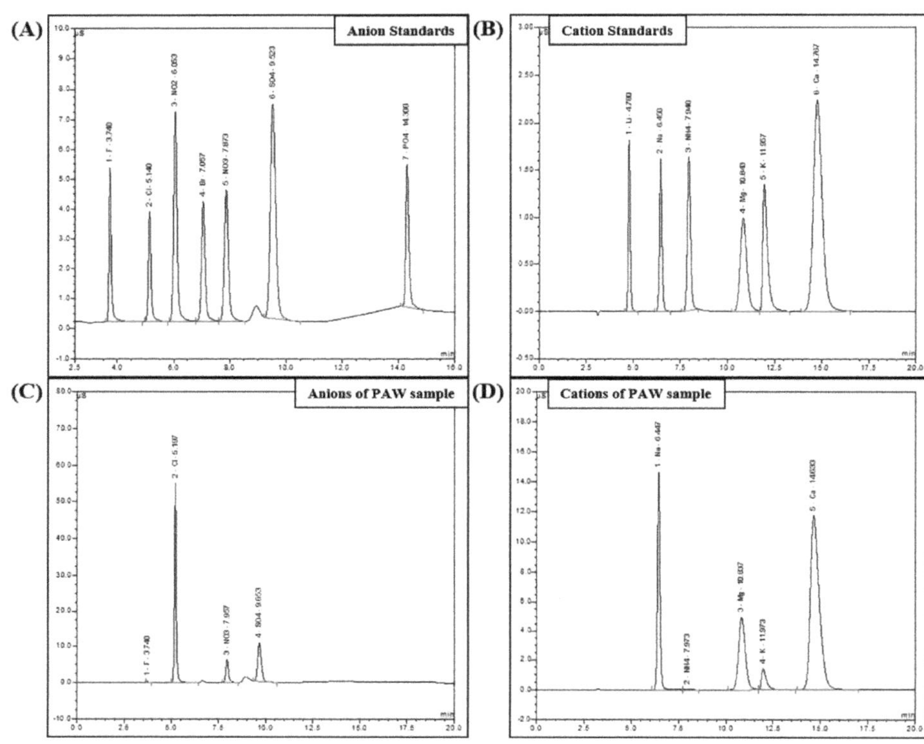

Figure 2. Ion chromatography analysis of inorganic anions and cations in PAW. Chromatograms depict the separation of inorganic anions (**A**) and cations (**B**). Determination of inorganic anions (**C**) and cations (**D**) in PAW generated using 12 V electrical power, 20 min treatment time, sample position at the top, and a 60 min retention time.

Table 3. Measurements of inorganic ions and hypochlorous acid for 12 V underwater plasma discharge.

Sample No.	Treatment Time (min)	Sampling Position	Retention Time (min)	Na$^+$ (mg/L)	NH$_4$$^+$ (mg/L)	K$^+$ (mg/L)	Mg^{2+} (mg/L)	Ca^{2+} (mg/L)	F$^-$ (mg/L)	Cl$^-$ (mg/L)	NO$_3$$^-$ (mg/L)	SO$_4$$^{2-}$ (mg/L)	HOCl (mg/L)
1			baseline	9.20	0.18	2.48	3.52	10.46	0.05	20.21	5.79	10.23	0.00
2		Sample Top	0	9.11	0.13	2.43	4.41	18.30	0.06	19.24	5.70	10.25	1.03
3		Sample Top	30	9.16	0.16	2.43	4.48	20.08	0.06	19.96	5.97	10.91	1.11
4			60	9.16	0.13	2.44	4.51	20.73	0.06	19.79	5.82	10.50	1.17
5			baseline	7.46	0.15	2.33	3.86	15.17	0.05	19.35	6.46	9.59	0.09
6	20 min	Sample Middle	0	7.45	0.16	2.32	4.02	18.37	0.05	18.54	6.44	9.67	1.21
7		Sample Middle	30	7.45	0.18	2.32	4.05	19.15	0.05	18.18	6.31	9.36	1.20
8			60	7.41	0.17	2.31	4.04	19.33	0.05	18.46	6.39	9.53	1.00
9			baseline	7.34	0.14	2.28	3.89	15.00	0.06	19.27	6.96	10.06	0.05
10		Sample Bottom	0	7.06	0.17	2.20	3.84	16.99	0.05	17.31	6.54	9.25	1.07
11		Sample Bottom	30	7.28	0.16	2.27	3.99	18.40	0.05	17.90	6.77	9.58	1.12
12			60	7.30	0.17	2.27	4.01	18.75	0.05	17.98	6.78	9.63	1.11

Table 3. *Cont.*

Sample No.	Treatment Time (min)	Sampling Position	Retention Time (min)	Na$^+$ (mg/L)	NH$_4^+$ (mg/L)	K$^+$ (mg/L)	Mg^{2+} (mg/L)	Ca^{2+} (mg/L)	F$^-$ (mg/L)	Cl$^-$ (mg/L)	NO$_3^-$ (mg/L)	SO$_4^{2-}$ (mg/L)	HOCl (mg/L)
13			baseline	7.59	0.16	2.27	3.40	13.38	0.04	17.41	7.71	9.02	0.11
14			0	7.59	0.22	2.27	3.50	16.22	0.05	17.16	7.75	9.13	0.78
15		Sample Top	30	7.62	0.20	2.28	3.51	17.02	0.05	17.23	7.74	9.09	0.66
16			60	7.57	0.26	2.30	3.50	17.33	0.05	16.85	7.65	8.95	0.60
17			0	7.65	0.20	2.28	3.55	17.66	0.05	16.92	7.69	8.98	0.71
18	10 min	Sample Middle	30	7.66	0.20	2.30	3.56	17.74	0.05	16.98	7.63	8.95	0.72
19			60	7.60	0.17	2.28	3.54	17.76	0.05	16.96	7.63	8.96	0.66
20			0	7.69	0.17	2.30	3.52	17.71	0.05	17.05	7.57	8.88	0.59
21		Sample Bottom	30	7.72	0.18	2.29	3.53	17.77	0.05	16.99	7.64	9.00	0.57
22			60	7.67	0.17	2.28	3.51	17.73	0.05	17.17	7.64	9.00	0.65
23			baseline	7.49	0.23	2.28	3.44	13.31	0.05	19.08	7.84	9.37	0.02
24			0	7.21	0.25	2.05	3.18	16.41	0.06	18.39	7.48	8.80	0.21
25		Sample Top	30	7.20	0.22	2.16	3.18	16.32	0.06	18.83	7.67	9.12	0.11
26			60	7.19	0.19	2.15	3.18	16.33	0.06	18.62	7.60	8.98	0.11
27			0	6.37	0.08	1.89	3.14	15.03	0.04	14.00	7.44	7.85	0.23
28	1 min	Sample Middle	30	6.74	0.11	1.94	3.14	15.70	0.05	13.77	7.58	8.59	0.15
29			60	6.42	0.09	1.99	3.22	16.20	0.05	13.83	7.65	7.77	0.11
30			0	6.31	0.09	1.92	3.15	16.06	0.05	13.84	7.65	7.90	0.11
31		Sample Bottom	30	6.26	0.09	1.93	3.15	16.02	0.05	13.76	7.66	7.88	0.09
32			60	6.30	0.11	1.94	3.18	16.31	0.05	13.62	7.61	7.76	0.15

Table 4. Measurements of inorganic ions and hypochlorous acid for 24 V underwater plasma discharge.

Sample No.	Treatment Time (min)	Sampling Position	Retention Time (min)	Na$^+$ (mg/L)	NH4$^+$ (mg/L)	K$^+$ (mg/L)	Mg^{2+} (mg/L)	Ca^{2+} (mg/L)	F$^-$ (mg/L)	Cl$^-$ (mg/L)	NO3$^-$ (mg/L)	SO4^{2-} (mg/L)	HOCl (mg/L)
1			baseline	6.92	0.17	2.22	3.33	14.61	0.05	18.25	7.34	8.66	0.00
2			0	6.95	0.15	2.22	3.45	16.12	0.04	15.46	7.20	8.37	2.10
3		Sample Top	30	6.94	0.13	2.22	3.49	16.59	0.05	15.67	7.35	8.63	2.56
4			60	6.93	0.13	2.20	3.50	16.78	0.05	16.17	7.42	8.81	2.19
5			baseline	6.77	0.14	2.32	3.14	14.35	0.05	17.93	7.44	8.69	0.10
6			0	6.53	0.13	2.22	3.19	15.32	0.04	14.95	7.20	8.46	2.64
7	20 min	Sample Middle	30	6.74	0.19	2.30	3.32	16.40	0.05	15.32	7.37	8.65	2.62
8			60	6.77	0.16	2.31	3.37	16.79	0.05	15.55	7.38	8.67	2.22
9			baseline	7.27	0.25	2.29	4.61	12.17	0.05	17.10	7.13	8.39	0.04
10			0	7.04	0.20	2.27	3.43	14.61	0.05	16.00	7.46	8.96	2.00
11		Sample Bottom	30	7.15	0.29	2.29	3.37	15.64	0.04	15.16	7.38	8.70	2.62
12			60	7.15	0.28	2.32	3.38	15.95	0.04	15.28	7.40	8.61	2.32
13			baseline	7.49	0.23	2.28	3.44	13.31	0.05	19.08	7.84	9.37	0.02
14			0	6.38	0.35	1.98	2.83	13.10	0.05	17.11	7.78	9.40	1.54
15		Sample Top	30	7.40	0.25	2.27	3.45	16.70	0.05	17.38	7.82	9.39	1.43
16			60	7.40	0.34	2.03	3.38	16.98	0.05	17.35	7.72	9.31	1.14
17			0	7.35	0.26	2.03	3.16	15.88	0.05	17.38	7.66	8.96	1.49
18	10 min	Sample Middle	30	7.21	0.29	2.23	3.14	15.96	0.06	17.42	7.72	8.98	1.36
19			60	7.22	0.23	2.12	3.15	16.07	0.06	17.18	7.53	8.85	1.20
20			0	7.20	0.29	2.15	3.18	16.22	0.06	17.42	7.42	8.64	1.03
21		Sample Bottom	30	7.22	0.28	2.16	3.18	16.21	0.06	17.29	7.64	9.01	1.31
22			60	7.14	0.27	2.15	3.15	16.08	0.06	17.15	7.54	8.82	1.21

Table 4. *Cont.*

Sample No.	Treatment Time (min)	Sampling Position	Retention Time (min)	Na$^+$ (mg/L)	NH4$^+$ (mg/L)	K$^+$ (mg/L)	Mg^{2+} (mg/L)	Ca^{2+} (mg/L)	F$^-$ (mg/L)	Cl$^-$ (mg/L)	NO3$^-$ (mg/L)	SO4^{2-} (mg/L)	HOCl (mg/L)
23			baseline	6.35	0.07	1.89	3.03	12.43	0.04	13.45	7.50	7.64	0.10
24		Sample Top	0	6.48	0.19	1.96	3.19	16.48	0.06	14.09	7.52	7.99	0.14
25		Sample Top	30	6.15	0.16	1.88	3.07	15.92	0.05	13.76	7.61	7.84	0.24
26			60	6.32	0.17	1.94	3.18	16.42	0.05	13.71	7.64	7.84	0.20
27	1 min	Sample Middle	0	6.32	0.12	1.93	3.15	16.33	0.05	13.84	7.55	7.78	0.23
28		Sample Middle	30	6.32	0.12	1.94	3.18	16.46	0.05	13.67	7.63	7.78	0.29
29			60	5.50	0.15	1.68	2.65	13.80	0.05	13.91	7.71	7.98	0.20
30		Sample Bottom	0	5.44	0.17	1.70	2.60	13.50	0.05	13.78	7.61	7.87	0.15
31		Sample Bottom	30	5.64	0.14	1.77	2.72	14.18	0.05	13.82	7.69	7.98	0.18
32			60	6.36	0.20	1.94	3.20	16.54	0.05	13.72	7.62	7.81	0.19

Table 4 presents changes in ion concentrations after 24 V plasma generation. After plasma activation for 20 min, changes were observed in Ca^{2+} and Cl$^-$. The baseline levels of Ca^{2+} were 14.61 mg/L (top), 14.35 mg/L (middle), and 12.17 mg/L (bottom), while those of Cl$^-$ were 18.25 mg/L (top), 17.93 mg/L (middle), and 17.10 mg/L (bottom). The Ca^{2+} concentrations reached 16.78 mg/L (top), 16.79 mg/L (middle), and 15.95 mg/L (bottom) after a 60 min treatment duration. The Cl$^-$ concentrations reached 16.17 mg/L (top), 15.55 mg/L (middle), and 15.28 mg/L (bottom) under the same conditions. After plasma activation for 10 min, the baseline concentration of Ca^{2+}, initially at 13.31 mg/L, increased to 16.98 (top), 16.07 (middle), and 16.08 mg/L (bottom) within 60 min. Cl$^-$, initially at 19.08 mg/L, decreased to 17.35 (top), 17.18 (middle), and 17.15 mg/L (bottom) under the same conditions. After plasma activation for 1 min, the baseline concentration of Ca^{2+}, initially at 12.43 mg/L, increased to 16.42 (top), 13.80 (middle), and 16.54 mg/L (bottom) after a 60 min treatment duration. To summarize, there was an increase in Ca^{2+} concentration and a decrease in Cl$^-$ concentration with plasma generation. Moreover, as the duration of plasma generation increased, the magnitude of these ion changes became more pronounced. However, there were no significant discernible effects of electrical power or retention time after plasma exposure on ion concentrations.

3.2. Hypochlorous Acid (HOCl) Concentration in PAW

Chlorine is routinely added to tap water to inhibit the growth of microorganisms, including common bacteria and *E. coli*, during the tap water supply process. As a result, residual chlorine is present in tap water in both free and combined forms. Free residual chlorine encompasses species such as HOCl, OCl$^-$, and Cl$^-$, with the chemical equation indicating the formation of hydrochloric acid and hypochlorous acid as follows:

$$Cl_2 + H_2O \rightleftharpoons HOCl + HCl \tag{1}$$

Hypochlorous acid is inherently unstable and may dissociate into the hypochlorite anion:

$$HOCl \rightleftharpoons ClO^- + H^+ \tag{2}$$

The presence of HOCl or OCl$^-$ depends on the acidity (pH 4–6) or basicity (pH 8.5–10) of the water, with HOCl being formed under acidic condition and OCl$^-$ being formed under basic conditions. Hypochlorous acid concentrations increased significantly following plasma treatment of tap water, and these concentrations were maintained for 60 min (Tables 3 and 4).

Figure 3 depicts the changes in hypochlorous acid concentration after generation of 12 and 24 V plasma. In Figure 3A, the concentrations of hypochlorous acid immediately after treatment with 24 V plasma for 20 min were 2.10 mg/L (top), 2.64 mg/L (middle), and 2.00 mg/L (bottom). For 12 V plasma, the immediately generated hypochlorous acid concentrations were 1.03 mg/L (top), 1.21 mg/L (middle), and 1.07 mg/L (bottom). As shown in Figure 3B), hypochlorous acid concentrations immediately after treatment with 24 V plasma for 10 min were 1.54 mg/L (top), 1.49 mg/L (middle), and 1.03 mg/L (bottom). For 12 V plasma, the immediately generated hypochlorous acid concentrations were 0.78 mg/L (top), 0.71 mg/L (middle), and 0.59 mg/L (bottom). Figure 3C reveals that hypochlorous acid concentrations generated after treatment with 24 V plasma or 12 V for 1 min were not significantly different. In summary, the amount of hypochlorous acid generated after plasma activation was influenced by electrical power and plasma processing time, but not retention time after plasma exposure.

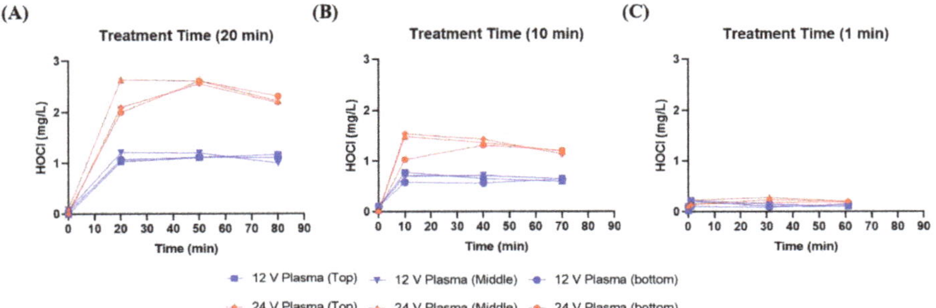

Figure 3. Variation in hypochlorous acid (HOCl) concentrations following plasma discharge for different durations ((**A**): 20 min, (**B**): 10 min, and (**C**): 1 min) at 12 and 24 V power.

3.3. Antibacterial Activity of PAW

The antibacterial activity of PAW against the Gram-positive bacterium *L. reuteri* was assessed using the filter paper disc method. PAW samples were collected immediately from the middle position after treatment with 12 V plasma for 20 min. Subsequently, *L. reuteri* cultures were exposed to filter paper treated with PAW samples for 15 min or 30 min. PAW had no significant antibacterial effect on *L. reuteri* after either a 15 min (Figure 4A) or a 30 min (Figure 4B) incubation period.

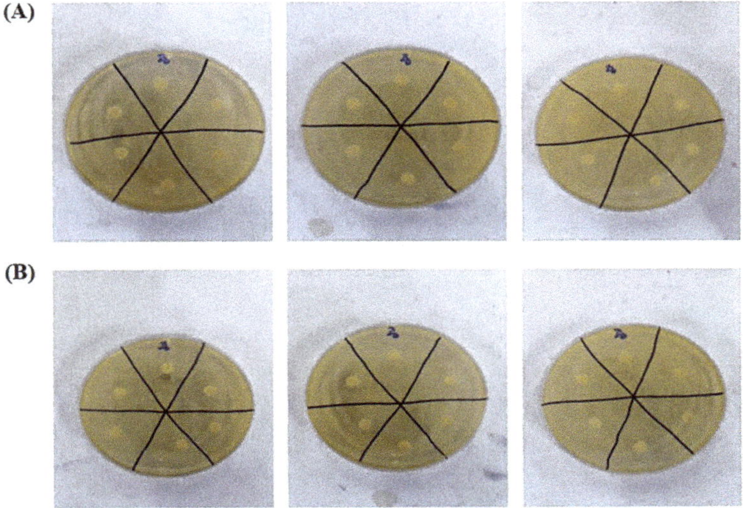

Figure 4. Antibacterial activity of PAW against Gram-positive *L. reuteri* following 15 min (**A**) or 30 min (**B**) exposure. Non-marked: PAW samples. Marked: negative control (sterile water).

4. Discussion

Our innovative vaginal cleaning device employing underwater plasma discharge generates various reactive radicals in tap water, conferring it with antibacterial activity [14,19,20]. The effectiveness of PAW in terms of antibacterial activity is contingent on its chemical composition. Thus, chemical composition analysis was undertaken to elucidate the antibacterial properties of various PAW samples.

Chemical analysis revealed an increase in Ca^{2+} and Mg^{2+} and a decrease in Cl^- as the duration of plasma generation increased. The concentration of hypochlorous acid was notably enhanced by higher electrical power input and longer plasma processing times. Previous studies have reported that magnesium and calcium ions have beneficial effects on the epidermis: Mg enhances epidermal barrier function and exhibits anti-inflammatory properties [30,31], while Ca promotes epidermal differentiation and regulates hyaluronic acid synthesis in the epidermal layer [32–35]. Various ions are constituents of the natural moisturizing factor (NMF) in the stratum corneum, the outermost layer of skin. Ionized minerals strengthen the epidermal barrier, particularly in damaged skin, and hinder the penetration of various external irritants [36]. Therefore, the generation of ions, specifically Ca^{2+} and Mg^{2+}, in PAW through plasma treatment suggests that the PAW is likely to have antibacterial and skin protection effects.

Furthermore, free chlorine, including HOCl, is a crucial component of disinfectants. Reactive chlorine species, such as HOCl, are widely employed for disinfection in industrial, hospital, and household settings [37]. As illustrated in Figure 5A, the body's innate immune system plays a pivotal role in the production of substantial amounts of oxidants, with HOCl being a key component. These oxidants, including HOCl, are generated by neutrophils as a response to the presence of invading pathogens [38]. As depicted in Figure 5B, the HOCl produced possesses a remarkable ability to breach the protective barriers of bacterial cells; HOCl launches rapid and destructive attacks. These attacks result in a cascade of effects within the bacterial cell, including the loss of adenosine triphosphate (ATP), disruption of DNA replication, and inhibition of protein synthesis [37]. This two-step process highlights the crucial role of HOCl in the body's immune defense mechanisms. HOCl is produced by neutrophils in response to infection and is followed by the infiltration of bacterial cells and the initiation of multiple processes that lead to the destruction of invading pathogens. The increased Ca^{2+} and Mg^{2+} concentrations in PAW generated using our plasma device and the augmentation of hypochlorous acid production indicate that the PAW samples produced using this device are likely to have antibacterial efficacy.

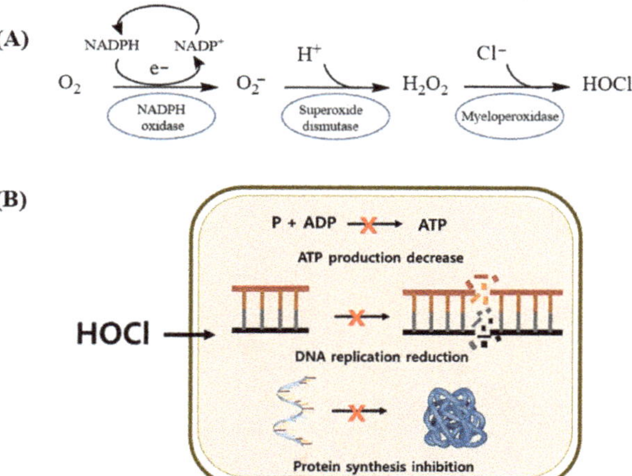

Figure 5. Schematic of HOCl production mechanisms in neutrophils (**A**) and HOCl targets in Gram negative bacterial cells (**B**).

Importantly, PAW did not exhibit antibacterial effects against *L. reuteri*, a probiotic crucial for vaginal health that produces lactic acid and helps maintain an acidic environment in the vagina to inhibit the growth of pathogens [39].

5. Conclusions

In this study, we conducted a comprehensive analysis of the chemical composition of PAW to determine the potential applications of PAW for vaginal sterilization and mucosal protection. The major findings of this research can be summarized as follows: There were notable changes in ion composition within PAW following plasma treatment. Specifically, levels of Ca^{2+} and Mg^{2+} increased while those of Cl^- decreased, with these changes becoming more pronounced over longer plasma generation times. This suggests that PAW generated by underwater plasma discharge can have different ion compositions depending on the conditions under which the PAW is generated. We found a significant increase in hypochlorous acid within PAW samples immediately after plasma treatment. Moreover, this increased level of hypochlorous acid was sustained for at least 60 min, indicating the potential disinfection capabilities of PAW in comparison with untreated water. Through in-depth exploration of the ion content of PAW, we demonstrated the presence of HOCl, a key component of the body's innate immune defense mechanisms. HOCl is known for its ability to eliminate invading pathogens efficiently. Intriguingly, PAW had no significant antibacterial effects against *L. reuteri*, a probiotic that plays a crucial role in maintaining vaginal health by producing lactic acid, and thereby creating an acidic environment that hinders pathogen growth. This selective antibacterial activity of PAW is encouraging for its potential use in female intimate hygiene products.

In summary, by providing insights into the ion composition of PAW, we have opened avenues for further exploration and development of feminine hygiene solutions that can harness these unique characteristics while respecting the delicate balance of the vaginal environment. The potential of PAW to offer both protection and cleansing within this context represents a promising direction for future research and applications.

Author Contributions: Conceptualization, J.-S.K., H.-J.K. and H.-A.S.; investigation: H.-J.K., H.-A.S., W.-K.C., A.J., E.-K.K. and W.A.; supervision: J.-S.K.; writing—original draft, H.-J.K.; writing—review and editing, J.-S.K., A.-S.O. and H.-A.S. All authors have read and agreed to the published version of the manuscript.

Funding: This research received no external funding.

Institutional Review Board Statement: Not applicable.

Informed Consent Statement: Not applicable.

Data Availability Statement: Data are contained within the article.

Conflicts of Interest: The authors have no conflict of interest to declare.

References

1. Cortese, E.; Settimi, A.G.; Pettenuzzo, S.; Cappellin, L.; Galenda, A.; Famengo, A.; Dabala, M.; Antoni, V.; Navazio, L. Plasma-Activated Water Triggers Rapid and Sustained Cytosolic $Ca^{(2+)}$ Elevations in Arabidopsis thaliana. *Plants* **2021**, *10*, 2516. [CrossRef]
2. Pankaj, S.K.; Keener, K.M. Cold plasma: Background, applications and current trends. *Curr. Opin. Food Sci.* **2017**, *16*, 49–52. [CrossRef]
3. Misra, N.N.; Pankaj, S.K.; Segat, A.; Ishikawa, K. Cold plasma interactions with enzymes in foods and model systems. *Trends Food Sci. Technol.* **2016**, *55*, 39–47. [CrossRef]
4. Gao, Y.; Francis, K.; Zhang, X. Review on formation of cold plasma activated water (PAW) and the applications in food and agriculture. *Food Res. Int.* **2022**, *157*, 111246. [CrossRef] [PubMed]
5. Sampaio, A.D.G.; Chiappim, W.; Milhan, N.V.M.; Botan Neto, B.; Pessoa, R.; Koga-Ito, C.Y. Effect of the pH on the Antibacterial Potential and Cytotoxicity of Different Plasma-Activated Liquids. *Int. J. Mol. Sci.* **2022**, *23*, 13893. [CrossRef]
6. Rathore, V.; Nema, S.K. A comparative study of dielectric barrier discharge plasma device and plasma jet to generate plasma activated water and post-discharge trapping of reactive species. *Phys. Plasmas* **2022**, *29*, 033510. [CrossRef]

7. Wong, K.S.; Chew, N.S.L.; Low, M.; Tan, M.K. Plasma-Activated Water: Physicochemical Properties, Generation Techniques, and Applications. *Processes* **2023**, *11*, 2213. [CrossRef]

8. Zhou, R.; Zhou, R.; Prasad, K.; Fang, Z.; Speight, R.; Bazaka, K.; Ostrikov, K. Cold atmospheric plasma activated water as a prospective disinfectant: The crucial role of peroxynitrite. *Green Chem.* **2018**, *20*, 5276–5284. [CrossRef]

9. Gott, R.P.; Engeling, K.W.; Olson, J.; Franco, C. Plasma activated water: A study of gas type, electrode material, and power supply selection and the impact on the final frontier. *Phys. Chem. Chem. Phys.* **2023**, *25*, 5130–5145. [CrossRef]

10. Berardinelli, A.; Pasquali, F.; Cevoli, C.; Trevisani, M.; Ragni, L.; Mancusi, R.; Manfreda, G. Sanitisation of fresh-cut celery and radicchio by gas plasma treatments in water medium. *Postharvest. Biol. Technol.* **2016**, *111*, 297–304. [CrossRef]

11. Neretti, G.; Taglioli, M.; Colonna, G.; Borghi, C.A. Characterization of a dielectric barrier discharge in contact with liquid and producing a plasma activated water. *Plasma Sources Sci. Technol.* **2016**, *26*, 015013. [CrossRef]

12. Georgescu, N.; Apostol, L.; Gherendi, F. Inactivation of Salmonella enterica serovar Typhimurium on egg surface, by direct and indirect treatments with cold atmospheric plasma. *Food Control* **2017**, *76*, 52–61. [CrossRef]

13. Rathore, V.; Patel, D.; Butani, S.; Nema, S.K. Investigation of physicochemical properties of plasma activated water and its bactericidal efficacy. *Plasma Chem. Plasma Process.* **2021**, *41*, 871–902. [CrossRef]

14. Jang, Y.; Bok, J.; Ahn, D.K.; Kim, C.K.; Kang, J.S. Human Trial for the Effect of Plasma-Activated Water Spray on Vaginal Cleaning in Patients with Bacterial Vaginosis. *Med. Sci.* **2022**, *10*, 33. [CrossRef]

15. Lee, S.J.; Ma, S.-H.; Hong, Y.C.; Choi, M.C. Effects of pulsed and continuous wave discharges of underwater plasma on Escherichia coli. *Sep. Purif. Technol.* **2018**, *193*, 351–357. [CrossRef]

16. Hong, Y.C.; Park, H.J.; Lee, B.J.; Kang, W.-S.; Uhm, H.S. Plasma formation using a capillary discharge in water and its application to the sterilization of E. coli. *Phys. Plasmas* **2010**, *17*, 053502. [CrossRef]

17. Lee, H.R.; Lee, Y.S.; You, Y.S.; Huh, J.Y.; Kim, K.; Hong, Y.C.; Kim, C.H. Antimicrobial effects of microwave plasma-activated water with skin protective effect for novel disinfectants in pandemic era. *Sci. Rep.* **2022**, *12*, 5968. [CrossRef]

18. Perinban, S.; Orsat, V.; Lyew, D.; Raghavan, V. Effect of plasma activated water on Escherichia coli disinfection and quality of kale and spinach. *Food Chem.* **2022**, *397*, 133793. [CrossRef]

19. Kang, J.S.; Kang, E.K.; Jeon, A.; An, W.; Shin, H.A.; Kim, Y.J.; Om, A.S. Vaginal cleansing effect using plasma-activated water(PAW) spray method in patients with vaginitis (suspected). *J. Rehabil. Welf. Eng. Assisitve Technol.* **2023**, *17*, 18–28.

20. Hwang, Y.; Jeon, H.; Wang, G.Y.; Kim, H.K.; Kim, J.-H.; Ahn, D.K.; Choi, J.S.; Jang, Y. Design and Medical Effects of a Vaginal Cleaning Device Generating Plasma-Activated Water with Antimicrobial Activity on Bacterial Vaginosis. *Plasma* **2020**, *3*, 204–213. [CrossRef]

21. Marnach, M.L.; Wygant, J.N.; Casey, P.M. Evaluation and Management of Vaginitis. *Mayo Clin. Proc.* **2022**, *97*, 347–358. [CrossRef] [PubMed]

22. Swidsinski, A.; Mendling, W.; Loening-Baucke, V.; Ladhoff, A.; Swidsinski, S.; Hale, L.P.; Lochs, H. Adherent biofilms in bacterial vaginosis. *Obstet. Gynecol.* **2005**, *106*, 1013–1023. [CrossRef]

23. Høiby, N.; Ciofu, O.; Johansen, H.K.; Song, Z.j.; Moser, C.; Jensen, P.Ø.; Molin, S.; Givskov, M.; Tolker-Nielsen, T.; Bjarnsholt, T. The clinical impact of bacterial biofilms. *Int. J. Oral Sci.* **2011**, *3*, 55–65. [CrossRef] [PubMed]

24. Machado, D.; Castro, J.; Palmeira-de-Oliveira, A.; Martinez-de-Oliveira, J.; Cerca, N. Bacterial Vaginosis Biofilms: Challenges to Current Therapies and Emerging Solutions. *Front. Microbiol.* **2015**, *6*, 1528. [CrossRef] [PubMed]

25. Cerca, N.; Jefferson, K.K.; Oliveira, R.; Pier, G.B.; Azeredo, J. Comparative antibody-mediated phagocytosis of Staphylococcus epidermidis cells grown in a biofilm or in the planktonic state. *Infect. Immun.* **2006**, *74*, 4849–4855. [CrossRef] [PubMed]

26. Xie, Z.; Thompson, A.; Sobue, T.; Kashleva, H.; Xu, H.; Vasilakos, J.; Dongari-Bagtzoglou, A. Candida albicans biofilms do not trigger reactive oxygen species and evade neutrophil killing. *J. Infect. Dis.* **2012**, *206*, 1936–1945. [CrossRef] [PubMed]

27. Mai-Prochnow, A.; Zhou, R.; Zhang, T.; Ostrikov, K.; Mugunthan, S.; Rice, S.A.; Cullen, P.J. Interactions of plasma-activated water with biofilms: Inactivation, dispersal effects and mechanisms of action. *Npj Biofilms Microbiomes* **2021**, *7*, 11. [CrossRef]

28. Mu, Q.; Tavella, V.J.; Luo, X.M. Role of Lactobacillus reuteri in Human Health and Diseases. *Front. Microbiol.* **2018**, *9*, 757. [CrossRef] [PubMed]

29. Qadrie, Z.L.; Jacob, B.; Anandan, R.; Rajkapoor, B.; Ulla, M.R. Anti-bacterial activity of ethanolic extract of *Indoneesiella echioides* (L) nees. evaluated by the filter paper disc method. *Pak. J. Pharm. Sci.* **2009**, *22*, 123–125.

30. Denda, M.; Katagiri, C.; Hirao, T.; Maruyama, N.; Takahashi, M. Some magnesium salts and a mixture of magnesium and calcium salts accelerate skin barrier recovery. *Arch. Dermatol. Res.* **1999**, *291*, 560–563. [CrossRef]

31. Schempp, C.M.; Dittmar, H.C.; Hummler, D.; Simon-Haarhaus, B.; Schöpf, E.; Simon, J.C.; Schulte-Mönting, J. Magnesium ions inhibit the antigen-presenting function of human epidermal Langerhans cells in vivo and in vitro. Involvement of ATPase, HLA-DR, B7 molecules, and cytokines. *J. Investig. Dermatol.* **2000**, *115*, 680–686. [CrossRef]

32. Hennings, H.; Michael, D.; Cheng, C.; Steinert, P.; Holbrook, K.; Yuspa, S.H. Calcium regulation of growth and differentiation of mouse epidermal cells in culture. *Cell* **1980**, *19*, 245–254. [CrossRef] [PubMed]

33. Pillai, S.; Bikle, D.D.; Hincenbergs, M.; Elias, P.M. Biochemical and morphological characterization of growth and differentiation of normal human neonatal keratinocytes in a serum-free medium. *J. Cell. Physiol.* **1988**, *134*, 229–237. [CrossRef] [PubMed]

34. Yuspa, S.H.; Kilkenny, A.E.; Steinert, P.M.; Roop, D.R. Expression of murine epidermal differentiation markers is tightly regulated by restricted extracellular calcium concentrations in vitro. *J. Cell Biol.* **1989**, *109*, 1207–1217. [CrossRef] [PubMed]

35. Lee, S.E.; Jun, J.E.; Choi, E.H.; Ahn, S.K.; Lee, S.H. Stimulation of epidermal calcium gradient loss increases the expression of hyaluronan and CD44 in mouse skin. *Clin. Exp. Dermatol.* **2010**, *35*, 650–657. [CrossRef]

36. Tsukui, K.; Kakiuchi, T.; Suzuki, M.; Sakurai, H.; Tokudome, Y. The ion balance of Shotokuseki extract promotes filaggrin fragmentation and increases amino acid production and pyrrolidone carboxylic acid content in three-dimensional cultured human epidermis. *Nat. Prod. Bioprospecting* **2022**, *12*, 37. [CrossRef]

37. Da Cruz Nizer, W.S.; Inkovskiy, V.; Overhage, J. Surviving Reactive Chlorine Stress: Responses of Gram-Negative Bacteria to Hypochlorous Acid. *Microorganisms* **2020**, *8*, 1220. [CrossRef]

38. Gray, M.J.; Wholey, W.Y.; Jakob, U. Bacterial responses to reactive chlorine species. *Annu. Rev. Microbiol.* **2013**, *67*, 141–160. [CrossRef]

39. Reid, G. Probiotic and Prebiotic Applications for Vaginal Health. *J. AOAC Int.* **2019**, *95*, 31–34. [CrossRef] [PubMed]

Article

Synergistic Antimicrobial Effect of Cold Atmospheric Plasma and Redox-Active Nanoparticles

Artem M. Ermakov [1,2,3], Vera A. Afanasyeva [1,2], Alexander V. Lazukin [4], Yuri M. Shlyapnikov [2], Elizaveta S. Zhdanova [1,2], Anastasia A. Kolotova [2], Artem S. Blagodatski [2], Olga N. Ermakova [2], Nikita N. Chukavin [2,5], Vladimir K. Ivanov [6] and Anton L. Popov [2,5,*]

[1] Hospital of the Pushchino Scientific Center of the Russian Academy of Sciences, 142290 Pushchino, Russia; beoluchi@yandex.ru (A.M.E.); va_vera_afanaseva@mail.ru (V.A.A.); dla_lisa@mail.ru (E.S.Z.)
[2] Institute of Theoretical and Experimental Biophysics of the Russian Academy of Sciences, 142290 Pushchino, Russia; yuri.shlyapnikov@gmail.com (Y.M.S.); anas.kolotowa2010@ya.ru (A.A.K.); bswin2000@gmail.com (A.S.B.); knopsik-svetik@yandex.ru (O.N.E.); chukavinnik@gmail.com (N.N.C.)
[3] ANO Engineering Physics Institute, 142210 Serpukhov, Russia
[4] Troitsk Institute of Innovative and Thermonuclear Research (JSC "SSC RF TRINITY"), 108840 Moscow, Russia; lazukin_av@mail.ru
[5] Scientific and Educational Center, State University of Education, 105005 Moscow, Russia
[6] Kurnakov Institute of General and Inorganic Chemistry, Russian Academy of Sciences, 119991 Moscow, Russia; van@igic.ras.ru
* Correspondence: antonpopovleonid@gmail.com

Abstract: Cold argon plasma (CAP) and metal oxide nanoparticles are well known antimicrobial agents. In the current study, on an example of *Escherichia coli*, a series of analyses was performed to assess the antibacterial action of the combination of these agents and to evaluate the possibility of using cerium oxide and cerium fluoride nanoparticles for a combined treatment of bacterial diseases. The joint effect of the combination of cold argon plasma and several metal oxide and fluoride nanoparticles (CeO_2, CeF_3, WO_3) was investigated on a model of *E. coli* colony growth on agar plates. The mutagenic effect of different CAP and nanoparticle combinations on bacterial DNA was investigated, by means of a blue–white colony assay and RAPD-PCR. The effect on cell wall damage, using atomic force microscopy, was also studied. The results obtained demonstrate that the combination of CAP and redox-active metal oxide nanoparticles (RAMON) effectively inhibits bacterial growth, providing a synergistic antimicrobial effect exceeding that of any of the agents alone. The combination of CAP and CeF_3 was shown to be the most effective mutagen against plasmid DNA, and the combination of CAP and WO_3 was the most effective against bacterial genomic DNA. The analysis of direct cell wall damage by atomic force microscopy showed the combination of CAP and CeF_3 to be the most effective antimicrobial agent. The combination of CAP and redox-active metal oxide or metal fluoride nanoparticles has a strong synergistic antimicrobial effect on bacterial growth, resulting in plasmid and genomic DNA damage and cell wall damage. For the first time, a strong antimicrobial and DNA-damaging effect of CeF_3 nanoparticles has been demonstrated.

Keywords: cold argon plasma; cerium oxide; cerium fluoride; tungsten oxide; nanoparticles; bacteria; antimicrobial effect; combined treatment

Citation: Ermakov, A.M.; Afanasyeva, V.A.; Lazukin, A.V.; Shlyapnikov, Y.M.; Zhdanova, E.S.; Kolotova, A.A.; Blagodatski, A.S.; Ermakova, O.N.; Chukavin, N.N.; Ivanov, V.K.; et al. Synergistic Antimicrobial Effect of Cold Atmospheric Plasma and Redox-Active Nanoparticles. *Biomedicines* **2023**, *11*, 2780. https://doi.org/10.3390/biomedicines11102780

Academic Editor: Christoph Viktor Suschek

Received: 5 September 2023
Revised: 3 October 2023
Accepted: 6 October 2023
Published: 13 October 2023

1. Introduction

Cold argon plasma (CAP) is ionised gas which temperature is comparable to a physiological value (30–40 °C), whereas so-called thermal plasma used in industrial applications has a temperature of 3000–5000 °C, having a devastating effect on biological objects [1]. In 1996, Laroussi et al. first demonstrated that glow discharge plasma generated at atmospheric pressure can be a good alternative for a sterilisation process [2]. Its microbiocidal effect is achieved via damaging and killing bacteria by reactive oxygen species (ROS), reactive nitrogen species (RNS), charged particles, UV photons, generated in the plasma flow [3,4].

In vitro and in vivo studies have shown that CAP has no allergic, toxic or mutagenic effects and is safe for medical applications [5]. The most common applications of cold plasma are regenerative medicine (wound healing) and cancer treatment [6]. The presence of conditionally pathogenic or pathogenic microorganisms in a wound often leads to prolongation of its healing period. CAP is effective in the sterilisation of gram-positive and gram-negative bacteria, biofilms, viruses and fungi [1]. Thus, cold argon plasma is used in medicine for reducing wound healing duration due to the destruction of microorganisms, thus improving postoperative quality of life [7,8].

Metal oxide nanoparticles have been repeatedly tested for their antimicrobial activity. It has been shown that they are able to inactivate most pathogens, in various environments. Among metal oxides, cerium oxide nanoparticles (CeO_2 NPs) are uniquely able to act as antioxidants, due to the transition between Ce^{3+} and Ce^{4+} states [9]. CeO_2 NPs also show an ability to increase the efficiency of existing antimicrobials, and thus they could be used as antibiotic adjuvants against drug-resistant pathogens. The synergistic antibacterial effect is the result of the interaction of CeO_2 NPs with bacterial outer membranes [10]. It has been demonstrated that CeO_2 NPs can destroy bacterial biofilms [11]. CeO_2 NP-modified surfaces can prevent the adhesion, proliferation and spread of *Pseudomonas aeruginosa* [12]. Additionally, it has been shown that CeO_2 nanorods have an antimicrobial action through haloperoxidase activity. Enzyme-like activities of cerium-based nanorods vary with differences in Ce^{3+} concentration, oxygen vacancy concentration and band gap energy [13].

Given that the presence and concentration of Ce^{3+} on the surface of nanoparticles determines its enzyme-like activities, the use of other cerium-containing nanoparticles (for example, cerium fluoride, CeF_3 NPs) should be considered. Previously, a comprehensive analysis of the biological activity of CeF_3 NPs, in various models, confirmed the promise of this nanomaterial in various biomedical applications. For example, it has been demonstrated that CeF_3 NPs protect living cells from exogenous H_2O_2 [14], and also protect freshwater planarian flatworms from the detrimental effects of X-ray irradiation as effective radioprotective agents [15]. These unique properties suggest that cerium-based or cerium-doped nanomaterials might provide the basis for new effective antimicrobial agents.

Another promising biologically active nanomaterial is photochromic tungsten oxide nanoparticles (WO_3 NPs). WO_3 NPs also demonstrate good antimicrobial properties, especially when illuminated, due to their high photocatalytic activity [16]. The antibacterial activity of tungsten oxide particles increases as their size decreases, and significantly depends on the degree of surface hydration. WO_3 NPs have already been used successfully as a photocatalyst of the visible spectrum in the treatment and disinfection of wastewater. These particles have high biocompatibility and can be considered as antibacterial agents that are safe for human cells [17].

Combination therapies are an advanced approach whereby several therapeutic agents are used to effectively treat a condition, using synergistic or additive effects or reducing undesired effects. The principles of combination therapy are widely used in various fields of medicine, including HIV antiviral therapies [18], oncology [19] and the prevention of the development of drug resistance [20], e.g., bacterial resistance to antibiotics [21]. In this study, we combined the treatment of a model bacterium (*Escherichia coli*) with both CAP and various types of metal oxide/metal fluoride nanoparticles, to investigate the possible effect of two agents, for the development of advanced antimicrobial combination therapy.

2. Materials and Methods

2.1. Cold Atmospheric Argon Plasma Source

Low-temperature atmospheric pressure argon plasma was generated using a specially designed high-frequency current generator (frequency 50 kHz) with an amplitude of 4 kV

(Figure S1a). A plasma torch was created in an argon flow in a quartz tube 8 mm in diameter, using central and ring copper electrodes (Figure S1b). The flow rate of argon was 2 L per min.

2.2. Model Organism

E. coli K-12 strain M61655 was used as a model organism. The bacterial species was cultured on Petri dishes (Merck, Darmstadt, Germany) with Luria–Bertani (LB) agar in super optimal medium with catabolic repressor (SOC) medium.

2.3. Nanoparticle Synthesis and Characterisation

CeO_2 nanoparticles were synthesised using the precipitation method, according to a previously described protocol [22]. Briefly, 2.0 g of citric acid (Sigma-Aldrich, St. Louis, MO, USA, #251275) were mixed with 25 mL of 0.4 M aqueous cerium(III) chloride (Aldrich, #228931). The resulting solution was quickly poured, with stirring, into 100 mL of a 3 M ammonia solution (Khimmed, Moscow, Russia), kept for 2 h at ambient conditions and then boiled for 4 h. Then, the solution was cooled to room temperature and purified from precursors and by-products by precipitation and further redispersion.

CeF_3 nanoparticles were synthesised by precipitation in alcohol media [23]. Briefly, 1.86 g of cerium(III) chloride heptahydrate (5 mM) (Aldrich, #228931) was dissolved in 15 mL of distilled water and added to 150 mL of isopropyl alcohol (Aldrich, #W292907). Hydrofluoric acid (20 mM) (Sigma-Aldrich, #30107) dissolved in 50 mL of isopropyl alcohol was added dropwise to the cerium salt solution, with vigorous stirring. The resulting white precipitate was filtered off and washed thoroughly several times with pure isopropyl alcohol. Then, the suspension was slightly dried, to form a pasty substance, and dispersed in 110 mL of distilled water using an ultrasonic bath. The resulting transparent colloidal solution was boiled for 5 min to remove alcohol residues. UV–visible absorption spectra of colloidal solutions were recorded in standard quartz cuvettes for liquid samples, using a UV5 Nano spectrophotometer (Mettler Toledo, Columbus, OH, USA). Transmission electron microscopy was performed using a Leo 912 AB Omega electron microscope (Carl Zeiss, Oberkochen, Germany) operating at an accelerating voltage of 100 kV. The particle size and zeta potential of nanoparticles were measured on a Benano 90 Zeta particle size analyser (Bettersize, Dandong, China).

Ultra-small hydrated WO_3 nanoparticles were synthesised from tungstic acid in the presence of polyvinylpyrrolidone (PVP K-30, average mol. wt. 40,000) as a template, stabiliser and growth regulator, according to a method described earlier [24]. The electron microphotographs, UV/visible spectra and hydrodynamic radii distribution of the nanomaterials used in this study are presented in Figure S2. Zeta-potential values of the nanoparticles in distilled water were as follows: CeO_2——28.73 mV, CeF_3—+39.22 mV, WO_3——8.11 mV (Figure S4). The following concentrations of the nanoparticles were used: CeO_2—10^{-4} M, CeF_3—10^{-4} M, WO_3—2×10^{-3} M.

2.4. Inhibition Zone Test

The *E. coli* cell suspension (OD_{625} of 0.1) mixed with NP solution and the control sample sprayed onto LB agar plates were exposed to the plasma treatment. The plasma-emitting jet outlet was set at 1.0 cm above the agar layer containing the bacterial suspension. CAP was performed for two treatment periods (3 and 6 min). Cells were allowed to grow for 16–20 h at 37 °C and then inhibition zones were measured. The degree of inhibition was measured using ImageJ software (Version 1.53t).

2.5. Mutation Assay

The pal2-T plasmid was introduced into competent *E. coli* XL1-Blue cells (Evrogen, Moscow, Russia), according to the manufacturer's instructions for chemical transformation: 0.5 µL of plasmid DNA was added to competent cells in 50 µL aliquots, left for 30 min on ice, subjected to heat shock (42 °C) for 60 s and transferred to ice for 10 min. Then, 300 µL of SOC was added to cells and incubated for 40 min at 37 °C, with shaking. The suspension with transformed cells was spread onto LB agar containing 1 mM isopropyl-β-D-thiogalactopyranoside (IPTG), 0.2 μg/mL 5-bromo-4-chloro-3-indolyl β-D-galactopyranoside (X-gal) (SibEnzyme LLC, Novosibirsk, Russia) and 100 μg/mL of ampicillin (SibEnzyme LLC, Novosibirsk, Russia) and incubated at 37 °C for 16–18 h. Blue-coloured cells were selected and placed in a tube with PBS to an OD_{625} of ~0.3. PBS was used instead of culture media during plasma treatment to eliminate the protective effects of media components. Then, 500 μL of the suspension was mixed with NP solutions and placed in 4-well plates for plasma treatment. The treatment procedure was performed as described previously. The exposure time was 3 min. The cultures were allowed to recover for 30 min at 37 °C, after which dilutions of 1:100 and 1:1000 were prepared and plated onto LB plates containing ampicillin, X-gal and IPTG, and incubated for 16–18 h at 37 °C and 1 h at 4 °C. Surviving colonies were then counted and photographs were taken. The results are represented as colony-forming units (CFUs).

2.6. RAPD PCR

The DNA isolation procedure was performed using an ExtractDNA Blood & Cells Kit (Evrogen, Moscow, Russia). DNA concentration was measured using a Qubit 4 Fluorometer (Thermo Fisher Scientific, Waltham, MA, USA); DNA samples were stored at −20 °C until use. A total of 20 random DNA oligonucleotide primers, synthesised by Evrogen (Russia), were used in the PCR reaction. Only six primers generated reproducible polymorphic DNA products. PCR amplification was performed in a 10 μL volume containing 20 ng of genomic DNA, 0.5 μL of primer (0.2 μM), 0.2 μL of dNTPs (2.5 mM), 1.0 μL of 10× buffer, 0.2 μL of Tersus polymerase (Tersus Plus PCR kit, Evrogen, Moscow, Russia) and sterile ddH_2O. Amplifications were implemented in a DNA thermocycler (Bio-Rad, Hercules, CA, USA) and programmed to one cycle at 95 °C for 5 min, followed by 40 cycles at 94 °C for 1 min, 37 °C for 1 min and 72 °C for 2 min, with a final cycle of 72 °C for 5 min. PCR products were analysed by electrophoresis at 100 V for 45 min on 1.5% agarose gels in 1× TBE buffer. After electrophoresis, the RAPD patterns were visualised using a UV transilluminator (Vilber, Eberhardzell, Germany). RAPD markers were scored from the gels as DNA fragments that were present or absent in all lanes. The electropherograms were photographed.

Genomic template stability (GTS) was calculated for each primer (Table 1), using the formula: GTS (%) = (1 − a/n) × 100, where "a" is the number of polymorphic bands detected in each treated sample and "n" is the total number of bands in the control. Polymorphism observed in the RAPD profile included the disappearance of a normal band and the appearance of a new band in comparison with the control profile. To compare the sensitivity of the GTS parameter, changes in the values of this parameter were calculated as a percentage of the control (which was set to 100%).

Table 1. Set of primers used for RAPD-PCR analysis.

Primer	Nucleotide Sequence (5′→3′)
OPB5	TGCGCCCTTC
OPB7	GGTGACGCAG
OPV14	GGTCGATCTG
OPX7	GAGCGAGGCT
OP-B04	GATGACCGCC
OP-Q18	GGGAGCGAGT

2.7. Atomic Force Microscopy (AFM) Study of Bacteria after CAP Irradiation

Samples for the AFM characterisation were prepared as follows. A volume of 500 μL of *E. coli* suspension in PBS (OD_{625} of 0.1) with added NPs was placed in a 4-well culture plate, which was then exposed to plasma treatment. The plasma-emitting jet outlet was located 1.0 cm above the wells. The treatment duration was 3 and 6 min. After the treatment, the bacterial solution was centrifuged at 3000 rpm for 5 min, the supernatant was discarded, and bacterial sediment was resuspended in 50 µL of H_2O. A 3 µL aliquot of the suspension was transferred onto a 1×1 cm piece of freshly-cleaved mica (TipsNano, Tallinn, Estonia) pretreated with a 1% polyethylenimine solution. AFM characterisation was performed using a SmartSPM-1000 atomic force microscope (AIST-NT, Co., Moscow, Russia). The tapping mode with a resonance frequency of 150–300 kHz was used in all scanning experiments.

2.8. Statistical Analysis

Each experiment included three parallel runs. The data obtained were processed using the SPSS software package (version 21.0 for Windows), featuring one-way analysis of variance (ANOVA). The results are expressed as the mean and standard deviation.

3. Results

3.1. The Inactivation Effect of Plasma and NPs on E. coli

Photographs showing the enhanced antimicrobial effect of NPs in combination with plasma treatment are presented in Figure 1. ImageJ software (Version 1.53t) was used to quantify inactivation efficiency. The results are shown in Table 2 and demonstrate differences from irradiated control without NPs. Here, two treatment modes (3 and 6 min) and two concentrations of NPs (for CeO_2 and CeF_3—10^{-5} M and 10^{-4} M; for WO_3—0.1 and 0.5 mg/mL) were used. A more pronounced effect was observed when using higher concentrations of all NPs investigated. The greatest effect was achieved in the case of 6 min of cold plasma treatment in combination with WO_3 NPs. The use of WO_3 NPs resulted in an additional 37% suppression of bacterial growth, in comparison with CAP treatment only. In the case of 3 min of CAP treatment, different types of NPs had the same effect. The plate surface without bacterial growth increased by an average of 24.89% when NPs were added to the bacterial suspension before treatment.

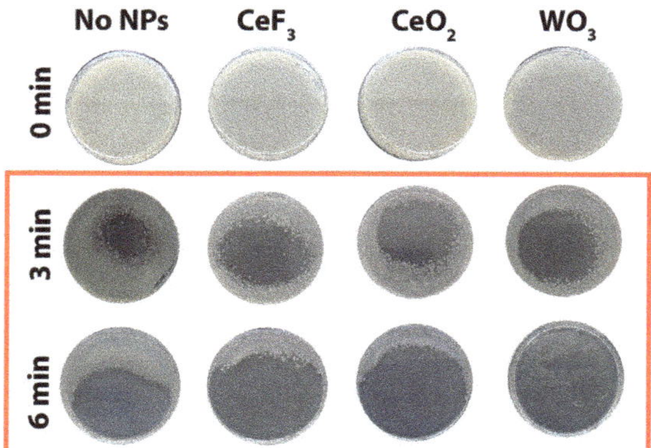

Figure 1. *E. coli* growth inhibition zones after CAP treatment. Control plate (0 min)—no CAP treatment, no NPs.

Table 2. The effect of CeF_3, CeO_2 and WO_3 nanoparticles on the size of the no-growth area of microorganisms.

NP Type	Bacterial Growth Suppression with the Combination of CAP and NPs (in Comparison with the Control Treated with CAP Only)	
	3 min treatment	6 min treatment
CeF_3	28.12% ± 1.41	17.05% ± 0.85
CeO_2	24.35% ± 1.12	24.73% ± 1.19
WO_3	22.21% ± 0.91	37.04% ± 1.84

3.2. The Effect of NPs on Mutation in E. coli

To investigate the effects of NP pretreatment before plasma treatment on DNA in bacterial cells, a mutation assay was used. The impact on the DNA structure was tested using the pAL2-T plasmid. This plasmid carries the genetic information for two enzymes, β-lactamase, which confers ampicillin resistance, and b-galactosidase (lacZ), which can convert colourless X-Gal to a blue dye. When competent *E. coli* cells are transformed with the plasmid, the growth of colonies on ampicillin-containing LB agar indicates successful transformation. Blue colonies indicate the presence of functional b-galactosidase, whereas in white colonies, a functional b-galactosidase is absent (Figure 2). An experimental group without treatment (negative control group) did not contain white colonies and is not represented on the graph, for clarity. Figure 2 shows an overall decrease in the number of colonies after CAP treatment and an even greater reduction after pretreatment with NPs. It should be noted that NP pretreatment itself, without further CAP treatment, did not influence the number of colonies (Figure 2a). The CFU-counting method was used to quantify bacterial inactivation efficiency by plasma treatment. The CFU number of bacteria after the treatment decreased by, on average, 69.26%; in the case of combined treatment (NPs + CAP), the reduction was by 94.14%. The number of blue and white colonies determined after incubation is shown in Figure 2. The most pronounced effect on mutation appearance was caused by the combination of CAP and CeF_3 NPs, resulting in approximately 20% colourless colonies. For CeO_2, colourless colonies accounted for 10.53%; for WO_3, the figure was 9.09%.

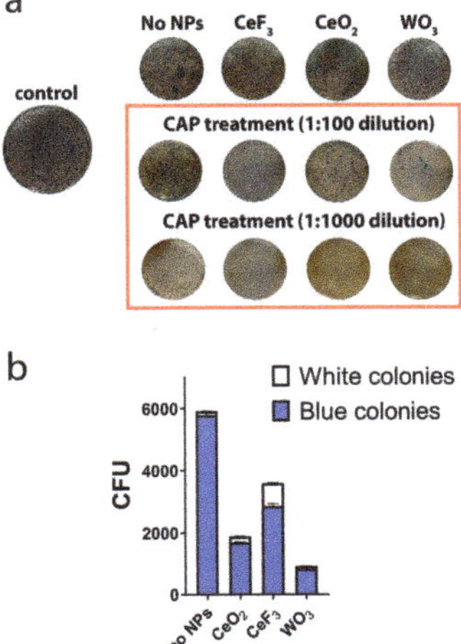

Figure 2. Antimicrobial and mutagenic effects of CAP in combination with CeO_2, CeF_3 and WO_3 NPs, as assessed using blue/white colouring assay on *E. coli* (**a**). Blue/white colony ratio after CAP treatment (**b**).

3.3. RAPD PCR

The RAPD-PCR profile of *E. coli* showed a total of 43 reproducible bands amplified with six different primers (Table 3). Genomic template stability (GTS) is a percentage value that reflects the PCR amplification profile changes of a test sample relative to a control sample. The total GTS value was calculated in each of the RAPD profiles generated using six selected screening primers. GTS% was calculated as follows [25]: $\text{GTS\%} = (1 - a/n) \times 100$, where (a) is the DNA polymorphism profile in each polluted site and (n) is the total number of bands in the control. Polymorphism included the disappearance of a normal band and the appearance of a new band in comparison with the control profiles. The percentage of polymorphism ranged from 100% to 41.86%, as shown in Figure 3.

Table 3. Changes in the total number of bands in the control and treated samples of *E. coli*. The first value reflects the appearance of new bands; the second the disappearance of control bands. n—total number of bands in the control.

Primer	Control	Group										
		No NPs		n	CeF$_3$		n	CeO$_2$		n	WO$_3$	
		3	6		3	6		3	6		3	6
OPB5	8	0; 4	0; 6	0; 0	0; 3	0; 0	0; 0	0; 4	0; 1	0; 0	0; 6	0; 3
OPB7	8	0; 0	0; 0	0; 0	0; 3	0; 2	0; 0	0; 3	0; 2	0; 0	0; 4	0; 3
OPV14	7	0; 3	0; 1	0; 0	0; 0	0; 1	0; 0	0; 0	0; 4	0; 0	0; 5	0; 3
OPX7	6	0; 5	2; 1	0; 0	1; 0	0; 0	0; 0	1; 0	0; 1	0; 0	0; 2	0; 0
OPB4	7	0; 1	0; 0	0; 0	0; 1	0; 1	0; 0	0; 1	0; 1	0; 0	0; 3	0; 0
OPQ18	7	0; 0	0; 0	0; 0	0; 2	0; 2	0; 0	0; 2	0; 2	0; 0	0; 5	0; 2
Total bands	43											
Polymorphic bands		13	5	0	10	6	0	11	11	0	25	11

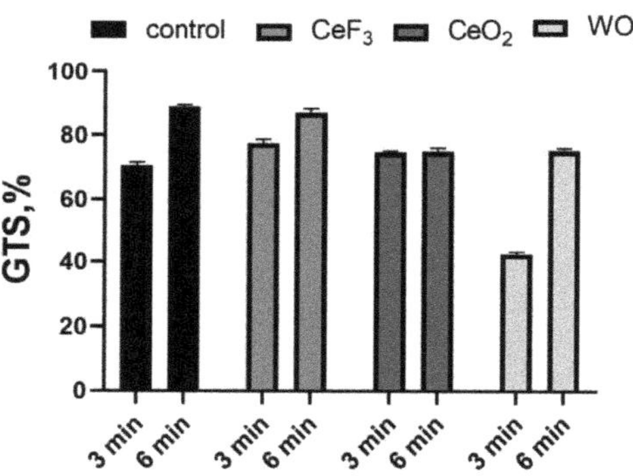

Figure 3. Genomic DNA template stability (GTS) of *E. coli*.

Groups treated with NPs only showed no differences from the control. The GTS of CAP-treated groups without NPs was 69.77% after 3 min of treatment and 88.37% after 6 min of treatment. Treatment with both NPs and plasma, in the case of CeO$_2$ and CeF$_3$ NPs, had virtually no effect on GTS, but a significant difference was observed after 3 min of CAP treatment with WO$_3$-NPs. Here, the GTS value was 41.86%.

3.4. Atomic Force Microscopy

Plasma affects the bacterial cell walls in a destructive way, leading to their deformation (Figure 4). This effect is enhanced with increasing exposure time. The destruction of the cell walls and the aggregation of bacterial cells can be clearly observed. The addition of cerium- or tungsten-containing oxide nanoparticles enhanced the effect of the plasma. The most obvious changes and the strongest effect were observed when using CeF_3 NPs with CAP: the cell wall was completely destroyed, and the cell content was released. CeO_2 NPs acted in the same way, but the effect was less pronounced. WO_3 NPs did not influence the bacterial cell structure in such a destructive way.

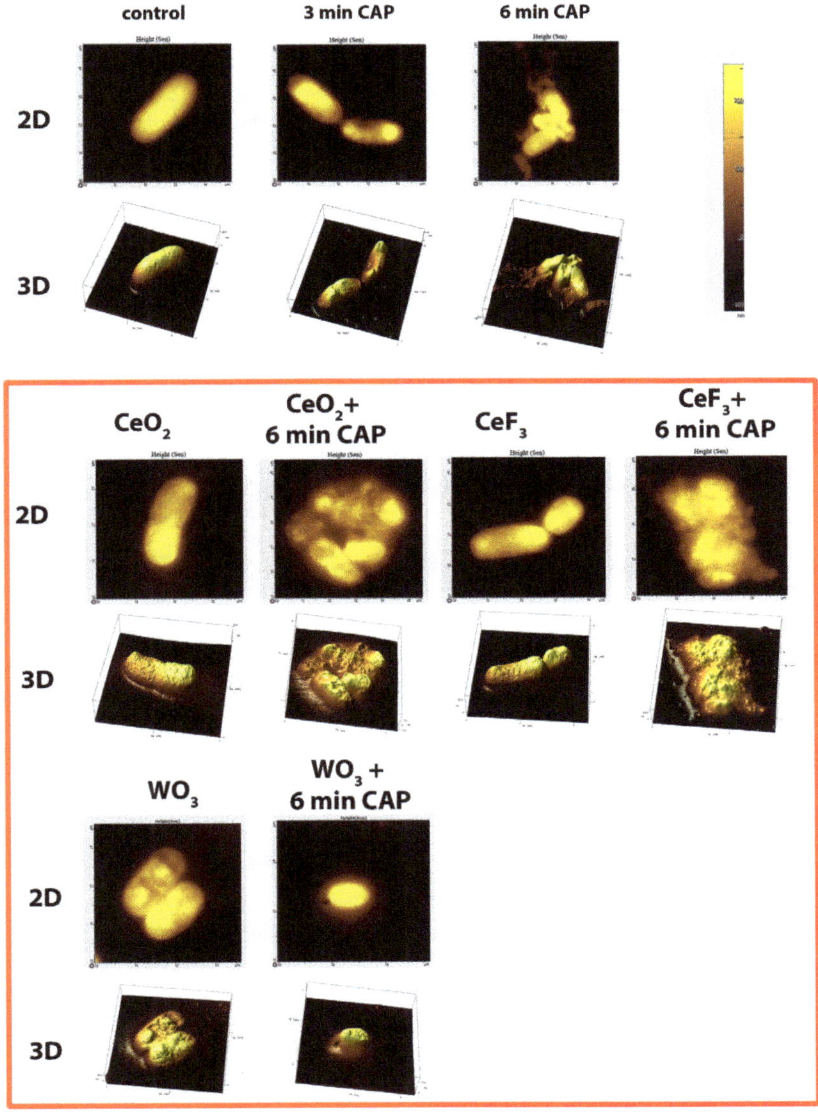

Figure 4. Representative AFM images of bacteria after treatment with CAP, nanoparticles or combinations of these. The scan area was 30–100 µm^2; a typical z-axis scale is 200 nm.

4. Discussion and Conclusions

The study showed that a combined treatment of bacterial colonies with cold argon plasma and nanoparticles had a pronounced antimicrobial effect, exceeding that of any of the components alone. Metal oxide nanoparticles are known to possess antimicrobial activity [26], and WO_3 was previously shown to inhibit the growth of gram-positive and gram-negative bacteria and viruses, and to have a damaging effect on cell membranes [27,28]. CeO_2 has also been reported previously to be an effective agent against a broad spectrum of pathogens [29]. Information on the antibacterial properties of CeF_3 NPs, however, continues to be absent. The enhancement of NPs' antimicrobial properties by CAP treatment paves the way for the effective use of this combination to heal chronic wounds, where CAP is already being used efficiently [30].

In this study, different combinations of the argon plasma treatment with inorganic redox-active NPs were tested. Bacterial growth inhibition tests showed a synergistic effect for the combination of CAP with all types of nanoparticles tested, exceeding the effect of CAP alone. The blue–white colony test showed that the CAP-WO_3 combination was the most effective against colony growth. The CAP-CeF_3 combination, however, had the most pronounced mutagenic effect on plasmid DNA, producing the largest percentage of white colonies. CAP-CeF_3 and, to a lesser degree, CAP-CeO_2 were responsible for the most pronounced cell wall damage in AFM studies, while CAP-WO_3 had a more moderate impact. For the first time, the antimicrobial and DNA-damaging effect of CeF_3 nanoparticles has been shown; previously, these NPs were reported as being antioxidant agents only [23]. Finally, the CAP-WO_3 combination demonstrated a strong mutagenic effect on bacterial genomic DNA, while the impact of the other combinations did not significantly differ from that of CAP alone.

The key factor determining the greater impact of CAP-CeF_3 on cell walls and plasmid DNA was possibly the positive charge of NPs. Surface charge has been shown previously to be crucial for the cytotoxic effect of CeO_2 nanoparticles [31]. The zeta potential measurements presented above show that CeF_3 NPs possess a substantial positive surface charge, whereas CeO_2 and WO_3 are charged negatively. Thus, CeF_3 nanoparticles can be electrostatically adsorbed on a negatively charged bacterial cell wall, providing a higher cytotoxicity for the CAP-CeF_3 combination. It is assumed that the mutagenic effect is responsible for the penetration of NPs into bacterial cells. While genomic DNA is supercoiled and protected from interactions with NPs by HU and other nucleoid-associated proteins [32], plasmid DNA can be more accessible for electrostatic interactions with positively charged CeF_3. At the same time, the mutagenic effect of negatively charged CeO_2 and WO_3 NPs in combination with CAP on genomic DNA is governed by other factors, which may be more prominent for WO_3 because of its higher concentration in the experiment.

In general, the combination of cold argon plasma treatment and metal oxide/metal fluoride nanoparticles has demonstrated a pronounced synergistic antimicrobial effect, and it can be recommended as the subject of further studies, to develop novel wound-healing therapies and disinfection methods.

Supplementary Materials: The following supporting information can be downloaded at: https://www.mdpi.com/article/10.3390/biomedicines11102780/s1, Figure S1. Irradiation scheme with the CAP irradiator (a) and demonstration of the nozzle with the outgoing plasma jet (b). Figure S2. Characterisation of CeF_3, CeO_2 and WO_3 nanoparticles: transmission electron microscopy (a), UV/visible spectra (b) and hydrodynamic diameters after dilution in distilled water (c). Figure S3. Results of RAPD PCR. (a) Samples treated with NPs, without CAP exposition. M—marker, C—control without NPs, 1—CeO_2, 2—CeF_3, 3—WO_3. (b) Samples exposed to 3 min of CAP. M—marker, C—control, 1—no NPs, 2—CeF_3, 3—CeO_2, 4—WO_3. (c) Samples exposed to 6 min of CAP. M—marker, C—control without NPs, 1—no NPs, 2—CeF_3, 3—CeO_2, 4—WO_3. Figure S4. Zeta-potential values of the nanoparticles in distilled water: CeF_3 (a), CeO_2 (b), WO_3 (c).

Author Contributions: Conceptualisation, A.M.E. and A.L.P.; methodology, Y.M.S., V.A.A. and E.S.Z.; investigation, V.A.A., Y.M.S., E.S.Z., A.A.K., N.N.C. and A.V.L.; writing—original draft preparation,

A.S.B., A.M.E., A.L.P. and V.K.I.; writing—review and editing, O.N.E., A.S.B., A.M.E. and A.L.P.; visualisation, A.A.K.; supervision, V.K.I. and A.L.P. All authors have read and agreed to the published version of the manuscript.

Funding: The work was supported by the State Assignment of the Ministry of Science and Higher Education of the Russian Federation No. FNSZ-2023-0008.

Institutional Review Board Statement: Not applicable.

Informed Consent Statement: Not applicable.

Data Availability Statement: Data is contained within the article or Supplementary Materials.

Conflicts of Interest: The authors declare no conflict of interest.

References

1. Martusevich, A.K.; Surovegina, A.V.; Bocharin, I.V.; Nazarov, V.V.; Minenko, I.A.; Artamonov, M.Y. Cold Argon Athmospheric Plasma for Biomedicine: Biological Effects, Applications and Possibilities. *Antioxidants* **2022**, *11*, 1262. [CrossRef] [PubMed]
2. Laroussi, M. Sterilization of contaminated matter with an atmospheric pressure plasma. *IEEE Trans. Plasma Sci.* **1996**, *24*, 1188–1191. [CrossRef]
3. Moisan, M.; Barbeau, J.; Moreau, S.; Pelletier, J.; Tabrizian, M.; Yahia, L.H. Low-temperature sterilization using gas plasmas: A review of the experiments and an analysis of the inactivation mechanisms. *Int. J. Pharm.* **2001**, *226*, 1–21. [CrossRef]
4. Vatansever, F.; de Melo, W.C.; Avci, P.; Vecchio, D.; Sadasivam, M.; Gupta, A.; Chandran, R.; Karimi, M.; Parizotto, N.A.; Yin, R.; et al. Antimicrobial strategies centered around reactive oxygen species—Bactericidal antibiotics, photodynamic therapy, and beyond. *FEMS Microbiol. Rev.* **2013**, *37*, 955–989.
5. Maisch, T.; Bosserhoff, A.K.; Unger, P.; Heider, J.; Shimizu, T.; Zimmermann, J.L.; Morfill, G.E.; Landthaler, M.; Karrer, S. Investigation of toxicity and mutagenicity of cold atmospheric argon plasma. *Environ. Mol. Mutagen.* **2017**, *58*, 172–177. [CrossRef]
6. Boeckmann, L.; Schäfer, M.; Bernhardt, T.; Semmler, M.L.; Jung, O.; Ojak, G.; Fischer, T.; Peters, K.; Nebe, B.; Müller-Hilke, B.; et al. Cold Atmospheric Pressure Plasma in Wound Healing and Cancer Treatment. *Appl. Sci.* **2020**, *10*, 6898. [CrossRef]
7. Bernhardt, T.; Semmler, M.L.; Schäfer, M.; Bekeschus, S.; Emmert, S.; Boeckmann, L. Plasma Medicine: Applications of Cold Atmospheric Pressure Plasma in Dermatology. *Oxid. Med. Cell. Longev.* **2019**, *2019*, 3873928. [CrossRef]
8. Plattfaut, I.; Besser, M.; Severing, A.-L.; Stürmer, E.K.; Opländer, C. Plasma medicine and wound management: Evaluation of the antibacterial efficacy of a medically certified cold atmospheric argon plasma jet. *Int. J. Antimicrob. Agents* **2021**, *57*, 106319. [CrossRef]
9. Corral-Diaz, B.; Peralta-Videa, J.R.; Alvarez-Parrilla, E.; Rodrigo-García, J.; Morales, M.I.; Osuna-Avila, P.; Niu, G.; Hernandez-Viezcas, J.A.; Gardea-Torresdey, J.L. Cerium oxide nanoparticles alter the antioxidant capacity but do not impact tuber ionome in *Raphanus sativus* (L). *Plant Physiol. Biochem.* **2014**, *84*, 277–285. [CrossRef]
10. Bellio, P.; Luzi, C.; Mancini, A.; Cracchiolo, S.; Passacantando, M.; Di Pietro, L.; Perilli, M.; Amicosante, G.; Santucci, S.; Celenza, G. Cerium oxide nanoparticles as potential antibiotic adjuvant. Effects of CeO_2 nanoparticles on bacterial outer membrane permeability. *Biochim. Biophys. Acta Biomembr.* **2018**, *1860*, 2428–2435. [CrossRef]
11. Xu, Y.; Wang, C.; Hou, J.; Wang, P.; You, G.; Miao, L. Effects of cerium oxide nanoparticles on bacterial growth and behaviors: Induction of biofilm formation and stress response. *Environ. Sci. Pollut. Res. Int.* **2019**, *26*, 9293–9304. [CrossRef]
12. Sarif, M.; Jegel, O.; Gazanis, A.; Hartmann, J.; Plana-Ruiz, S.; Hilgert, J.; Frerichs, H.; Viel, M.; Panthöfer, M.; Kolb, U.; et al. High-throughput synthesis of CeO_2 nanoparticles for transparent nanocomposites repelling *Pseudomonas aeruginosa* biofilms. *Sci. Rep.* **2022**, *12*, 3935. [CrossRef] [PubMed]
13. He, X.; Tian, F.; Chang, J.; Bai, X.; Yuan, C.; Wang, C.; Neville, A. Haloperoxidase Mimicry by CeO_{2-x} Nanorods of Different Aspect Ratios for Antibacterial Performance. *ACS Sustain. Chem. Eng.* **2020**, *8*, 6744–6752.
14. Shcherbakov, A.B.; Zholobak, N.M.; Baranchikov, A.E.; Ryabova, A.V.; Ivanov, V.K. Cerium fluoride nanoparticles protect cells against oxidative stress. *Mater. Sci. Eng. C* **2015**, *50*, 151–159. [CrossRef] [PubMed]
15. Filippova, K.O.; Ermakov, A.M.; Popov, A.L.; Ermakova, O.N.; Blagodatsky, A.S.; Chukavin, N.N.; Shcherbakov, A.B.; Baranchikov, A.E.; Ivanov, V.K. Mitogen-like Cerium-Based Nanoparticles Protect Schmidtea mediterranea against Severe Doses of X-rays. *Int. J. Mol. Sci.* **2023**, *24*, 1241. [CrossRef] [PubMed]
16. Popov, A.L.; Zholobak, N.M.; Balko, O.I.; Balko, O.B.; Shcherbakov, A.B.; Popova, N.R.; Ivanova, O.S.; Baranchikov, A.E.; Ivanov, V.K. Photo-induced toxicity of tungsten oxide photochromic nanoparticles. *J. Photochem. Photobiol. B Biol.* **2018**, *178*, 395–403. [CrossRef] [PubMed]
17. Han, B.; Popov, A.; Shekunova, T.; Kozlov, D.; Ivanova, O.; Rumyantsev, A.; Shcherbakov, A.; Popova, N.; Baranchikov, A.Y.; Ivanov, V. Highly crystalline WO_3 nanoparticles are non-toxic to stem cells and cancer cells. *J. Nanomater.* **2019**, *2019*, 5384132. [CrossRef]
18. The Antiretroviral Therapy Cohort Collaboration. Survival of HIV-positive patients starting antiretroviral therapy between 1996 and 2013: A collaborative analysis of cohort studies. *Lancet HIV* **2017**, *4*, e349–e356. [CrossRef]

19. Bayat Mokhtari, R.; Homayouni, T.S.; Baluch, N.; Morgatskaya, E.; Kumar, S.; Das, B.; Yeger, H. Combination therapy in combating cancer. *Oncotarget* **2017**, *8*, 38022–38043. [CrossRef]
20. Chen, S.H.; Lahav, G. Two is better than one; toward a rational design of combinatorial therapy. *Curr. Opin. Struct. Biol.* **2016**, *41*, 145–150. [CrossRef]
21. Tängdén, T. Combination antibiotic therapy for multidrug-resistant Gram-negative bacteria. *Upsala J. Med. Sci.* **2014**, *119*, 149–153. [CrossRef]
22. Ivanova, O.S.; Shekunova, T.O.; Ivanov, V.K.; Shcherbakov, A.B.; Popov, A.L.; Davydova, G.A.; Selezneva, I.I.; Kopitsa, G.P.; Tret'yakov, Y.D. One-stage synthesis of ceria colloid solutions for biomedical use. *Dokl. Chem.* **2011**, *437*, 103–106. [CrossRef]
23. Popov, A.L.; Zholobak, N.M.; Shcherbakov, A.B.; Kozlova, T.O.; Kolmanovich, D.D.; Ermakov, A.M.; Popova, N.R.; Chukavin, N.N.; Bazikyan, E.A.; Ivanov, V.K. The strong protective action of Ce^{3+}/F^- combined treatment on tooth enamel and epithelial cells. *Nanomaterials* **2022**, *12*, 3034. [CrossRef] [PubMed]
24. Popov, A.L.; Han, B.; Ermakov, A.M.; Savintseva, I.V.; Ermakova, O.N.; Popova, N.R.; Shcherbakov, A.B.; Shekunova, T.O.; Ivanova, O.S.; Kozlov, D.A.; et al. PVP-stabilized tungsten oxide nanoparticles: pH sensitive anti-cancer platform with high cytotoxicity. *Mater. Sci. Eng. C* **2020**, *108*, 110494. [CrossRef] [PubMed]
25. Atienzar, F.A.; Evenden, A.J.; Jha, A.N.; Depledge, M.H. Use of the random amplified polymorphic DNA (RAPD) assay for the detection of DNA damage and mutations: Possible implications of confounding factors. *Biomarkers* **2002**, *7*, 94–101. [CrossRef] [PubMed]
26. Azam, A.; Ahmed, A.S.; Oves, M.; Khan, M.S.; Habib, S.S.; Memic, A. Antimicrobial activity of metal oxide nanoparticles against Gram-positive and Gram-negative bacteria: A comparative study. *Int. J. Nanomed.* **2012**, *7*, 6003–6009. [CrossRef]
27. Duan, G.; Chen, L.; Jing, Z.; De Luna, P.; Wen, L.; Zhang, L.; Zhao, L.; Xu, J.; Li, Z.; Yang, Z.; et al. Robust Antibacterial Activity of Tungsten Oxide (WO_{3-x}) Nanodots. *Chem. Res. Toxicol.* **2019**, *32*, 1357–1366. [CrossRef]
28. Matharu, R.K.; Ciric, L.; Ren, G.; Edirisinghe, M. Comparative Study of the Antimicrobial Effects of Tungsten Nanoparticles and Tungsten Nanocomposite Fibres on Hospital Acquired Bacterial and Viral Pathogens. *Nanomaterials* **2020**, *10*, 1017. [CrossRef]
29. Dar, M.A.; Gul, R.; Karuppiah, P.; Al-Dhabi, N.A.; Alfadda, A.A. Antibacterial Activity of Cerium Oxide Nanoparticles against ESKAPE Pathogens. *Crystals* **2022**, *12*, 179. [CrossRef]
30. Isbary, G.; Heinlin, J.; Shimizu, T.; Zimmermann, J.L.; Morfill, G.; Schmidt, H.U.; Monetti, R.; Steffes, B.; Bunk, W.; Li, Y.; et al. Successful and safe use of 2 min cold atmospheric argon plasma in chronic wounds: Results of a randomized controlled trial. *Br. J. Dermatol.* **2012**, *167*, 404–410. [CrossRef]
31. He, X.; Kuang, Y.; Li, Y.; Zhang, H.; Ma, Y.; Bai, W.; Zhang, Z.; Wu, Z.; Zhao, Y.; Chai, Z. Changing exposure media can reverse the cytotoxicity of ceria nanoparticles for *Escherichia coli*. *Nanotoxicology* **2012**, *6*, 233240. [CrossRef] [PubMed]
32. Verma, S.C.; Qian, Z.; Adhya, S.L. Architecture of the *Escherichia coli* nucleoid. *PLoS Genet.* **2019**, *15*, e1008456, Erratum in *PLoS Genet.* **2020**, *16*, e1009148. [CrossRef] [PubMed]

 *biomedicines*

Article

Gas Flow-Dependent Modification of Plasma Chemistry in µAPP Jet-Generated Cold Atmospheric Plasma and Its Impact on Human Skin Fibroblasts

Dennis Feibel [1], Judith Golda [2], Julian Held [3], Peter Awakowicz [4], Volker Schulz-von der Gathen [3], Christoph V. Suschek [1], Christian Opländer [5,*] and Florian Jansen [1]

[1] Department of Orthopedics Trauma Surgery, Medical Faculty of the Heinrich Heine University, 40225 Düsseldorf, Germany
[2] Plasma Interface Physics, Ruhr University Bochum, 44801 Bochum, Germany
[3] Experimental Physics II, Ruhr University Bochum, 44801 Bochum, Germany
[4] Institute for Electrical Engineering and Plasma Technology, Ruhr University Bochum, 44801 Bochum, Germany
[5] Institute for Research in Operative Medicine (IFOM), Witten/Herdecke University, 51109 Cologne, Germany
* Correspondence: christian.oplaender@uni-wh.de; Tel.: +49-221-989570

Abstract: The micro-scaled Atmospheric Pressure Plasma Jet (µAPPJ) is operated with low carrier gas flows (0.25–1.4 slm), preventing excessive dehydration and osmotic effects in the exposed area. A higher yield of reactive oxygen or nitrogen species (ROS or RNS) in the µAAPJ-generated plasmas (CAP) was achieved, due to atmospheric impurities in the working gas. With CAPs generated at different gas flows, we characterized their impact on physical/chemical changes of buffers and on biological parameters of human skin fibroblasts (hsFB). CAP treatments of buffer at 0.25 slm led to increased concentrations of nitrate (~352 µM), hydrogen peroxide (H_2O_2; ~124 µM) and nitrite (~161 µM). With 1.40 slm, significantly lower concentrations of nitrate (~10 µM) and nitrite (~44 µM) but a strongly increased H_2O_2 concentration (~1265 µM) was achieved. CAP-induced toxicity of hsFB cultures correlated with the accumulated H_2O_2 concentrations (20% at 0.25 slm vs. ~49% at 1.40 slm). Adverse biological consequences of CAP exposure could be reversed by exogenously applied catalase. Due to the possibility of being able to influence the plasma chemistry solely by modulating the gas flow, the therapeutic use of the µAPPJ represents an interesting option for clinical use.

Keywords: cold atmospheric plasma; hydrogen peroxide; nitrite; nitrate; nitric oxide; inhibition of proliferation

Citation: Feibel, D.; Golda, J.; Held, J.; Awakowicz, P.; Schulz-von der Gathen, V.; Suschek, C.V.; Opländer, C.; Jansen, F. Gas Flow-Dependent Modification of Plasma Chemistry in µAPP Jet-Generated Cold Atmospheric Plasma and Its Impact on Human Skin Fibroblasts. *Biomedicines* **2023**, *11*, 1242. https://doi.org/10.3390/biomedicines11051242

Academic Editor: Eugenia Pechkova

Received: 24 March 2023
Revised: 17 April 2023
Accepted: 19 April 2023
Published: 22 April 2023

1. Introduction

It is generally recognized that non-thermal "cold" atmospheric pressure plasma (CAP) has great potential for numerous technological as well as medical-therapeutic applications [1,2]. A common feature of such plasma is the reduced spatial dimension (from a few microns up to a few millimeters) of the confining structures, e.g., electrodes, stabilizing the discharge and preventing the transition to a "thermal" arc discharge. These types of generated discharges are often summarized as "microplasmas" and contain high concentrations of radicals at low gas temperatures [3–5]. Therefore, microplasma-producing jets as the source are suitable for many applications, in particular, the modifications of sensitive surfaces and plasma medicinal applications, including sterilization, the treatment of cancer and impaired wound healing [6–8]. A microplasma-producing jet often uses helium as the carrier gas, and numerous investigations of RF plasma jets and their potential clinical use have been conducted [9–18].

Many research groups around the world use technically diverse and mostly self-designed and developed atmospheric pressure plasma jets to establish biomedical applications [19–22]. This complicates comparisons between studies conducted with the different plasma sources and greatly delays exploration and insight into the fundamental understanding of CAPs and their interaction with biological structures [23]. This in turn slows down the national and international approval procedures for this technology for medical applications [24]. As part of the European Cooperation in Science and Technology (COST)-Action MP1011, the microscale atmospheric pressure plasma jet (μAPPJ) developed by Schulz-von der Gathen and colleagues was selected as the basis for the development of a reference source to solve these problems [25,26], the COST reference microplasma jet [27–29]. For the μAPPJ, small admixtures of oxygen and additional nitrogen contributions from surrounding air result in the generation of reactive oxygen nitrogen species (RONS), such as ozone (O_3), hyperoxide (O_2^-), hydroxyl radicals (-OH) as well as nitric oxide (NO) and nitrogen dioxide (NO_2) [2,15,26,30,31]. In particular, NO regulates many processes in human skin physiology [32–34] and many pathological skin conditions, including psoriasis, impaired wound healing microcirculation, and skin tumor formation, are also associated with an imbalance in NO-biosynthesis [35–38]. Thus, the μAPPJ device could be a tool in treating different skin and wound conditions with cold atmospheric-pressure plasma (CAP) through the delivery of NO or bioactive NO compounds and the induction of NO-dependent pathways, as already shown for other plasma sources by several studies [39–41]. It has been shown in numerous studies that CAP can induce severe damage up to cell death in a large number of mammalian cells, including skin and blood cells, which is a disadvantage per se when using plasma to treat chronic and acute wounds or inflammatory wounds, which can represent skin diseases [42–48].

In cancer, the induction of apoptotic (programmed) or necrotic cell death by CAP could be an excellent and attractive therapeutic tool [49–57]. Many different plasma devices were used in these and other studies, but the observed CAP-induced cell toxicity correlated positively with plasma treatment time or plasma dose independent of the CAP device used [43,44,50,52,56,58–61].

Lower doses or shorter treatment time showed lesser toxicity as well as many interesting biological effects, such as enhanced proliferation of porcine endothelial cells by short CAP exposure (30 s) using a dielectric barrier discharge (DBD) device, correlated with a significant release of FGF-2 [43]. By using a KINpen plasma jet as the CAP source, a treatment of HaCaT (30 s) reduced the cell number and down-regulated E-cadherin and EGF receptor expression, whereas no effects were observed after shorter treatment time (10 s) or after changing the media after treatment [59]. Using a microwave plasma torch as the CAP source, Arnd et al., 2013, described a long-term inhibition of the proliferation of melanoma cells after 1 min and toxic effects after 2 min of treatment [50]. It is believed that the oxidative effects of reactive oxygen/nitrogen species (ROS/RNS) are responsible for the observed plasma-induced effects, since the presence of antioxidants can positively influence the toxic plasma effects [55,62]. However, the survival and functionality of cells depends, among other things, on a humid milieu, with constant physiological pH value, temperature and osmolarity, and the sufficient supply of nutrients and oxygen. In this context, plasma jets can lead to dehydration effects in vitro and in vivo due to the relatively high gas flows of up to 6 slm, which, for example, can also severely limit the treatment times with such devices. In comparison, the μAPPJ operates with gas flows between 0.25 and 1.4 slm, which allows longer treatment times and limits potential desiccation effects.

In addition, plasma treatment can acidify cell culture media and other exposed aqueous solutions, might increase nitrite and nitrate concentrations [52,57,58,60,63], and can lead to the formation of hydrogen peroxide (H_2O_2) [58,64], all of which are known to mediate cellular responses and may be responsible for many of the observed CAP-induced effects.

Human skin fibroblasts (hsFB) play a central role in the regulation and maintenance of wound healing. In the process of wound healing, hsFB-relevant cytokines proliferate and synthesize and form a provisional extracellular matrix (ECM) by generating collagen and fibronectin [65,66]. Regarding the small volume of liquids in wounds, CAP sources may induce many biological responses via chemical and physical modifications of the liquid microenvironment of cells, in addition to the generation of reactive oxygen/nitrogen species. Therefore, the current study aims to investigate the effects of µAPPJ treatment on the viability and proliferation of human dermal fibroblasts, elucidating the role of possible CAP-induced chemical and physical changes in the treatment medium, such as an accumulation of NO-related compounds and H_2O_2, as well as acidification and an increase in osmolarity.

2. Materials and Methods

2.1. Plasma/CAP Source

The µAPPJ is a capacitively coupled microplasma jet consisting of two stainless steel electrodes (length 30 mm and width 1 mm) with a gap of 1 mm. The plasma is ignited in this gap, filling a volume of $1 \times 1 \times 30$ mm^3. The complete device consists of the electrode stack with 1.5 mm thick quartz panes enclosing the plasma volume (identical to the COST reference electrode head [28]), the jet holder containing the electric connection, the gas connector and the gas tubing. One electrode is connected to a power supply (13.56 MHz, <1 W) and the other one is grounded. The µAPPJ can be operated for helium gas flows from 0.25 slm up to several slm. Furthermore, small admixtures of molecular oxygen (>0.6%) are possible [15]. The helium flow is controlled by a mass flow controller (FC280S, Millipore-Tylan, Bedford, MA, USA). For all measurements, the µAPPJ was operated at 13.56 MHz and at a helium gas flow of either 0.25 slm or 1.40 slm (helium purity 99.999%, Linde, Munich, Germany).

In anticipation of atmospheric impurities in the process gas, e.g., in the form of traces of water, oxygen and/or nitrogen from the ambient air, which should lead to a higher yield of reactive oxygen species and nitrogen oxide species of the produced CAPs, we exchanged the metal gas lines of the processing gas supply with silicone gas lines. With this procedure, we were fully aware that such a procedure could not generate a previously defined degree of the aforementioned contamination of the processing gas. However, a detailed control of the admixtures, especially with very small gas flows, was unfeasible. The method we used was intended to demonstrate the principle involved. N_2, O_2 and H_2O would always be involved in small amounts from the ambient air in any treatment.

Optical emission spectroscopy (OES, HR 4000, Ocean Optics, Duvien, The Netherlands) using an optical fiber was applied to characterize the plasma under experimental conditions at about 1 mm from the tip of the jet electrode (Figure 1).

2.2. Determination of Evaporation and Temperature

As preliminary work, the evaporation rate was estimated by measuring the liquid volumes (250, 500, 750, 1000 µL) in cell culture plates (24-well) before and after µAPPJ treatments or gas control (flow rates 0.25 slm or 1.4 slm) by a pipette (Eppendorf, Wesseling, Germany). In the preliminary experiments, the temperature of treated buffer was measured by using a digital thermometer (GMH3230, Greisinger Electronic, Regenstauf, Germany) after different treatment intervals (0, 5, 10 min).

2.3. Determination of pH Values

The pH values were determined before and after the plasma treatment using a calibrated pH meter (Calimatic 766, Knick, Berlin, Germany) and a corresponding pH electrode from Mettler-Toledo (Giessen, Germany).

2.4. Measurement of Dissolved Oxygen

The oxygen saturation/concentration of the treated and untreated buffer (500 µL/24-well plate, 0–10 min) at different gas flows (0.25 and 1.4 slm) was measured by using a multiparameter pH-Meter (HI2020-edge, Hanna Instruments, Carrollton, TX, USA) and a digital dissolved oxygen/temperature electrode (HI764080, Hanna Instruments, Carrollton, TX, USA).

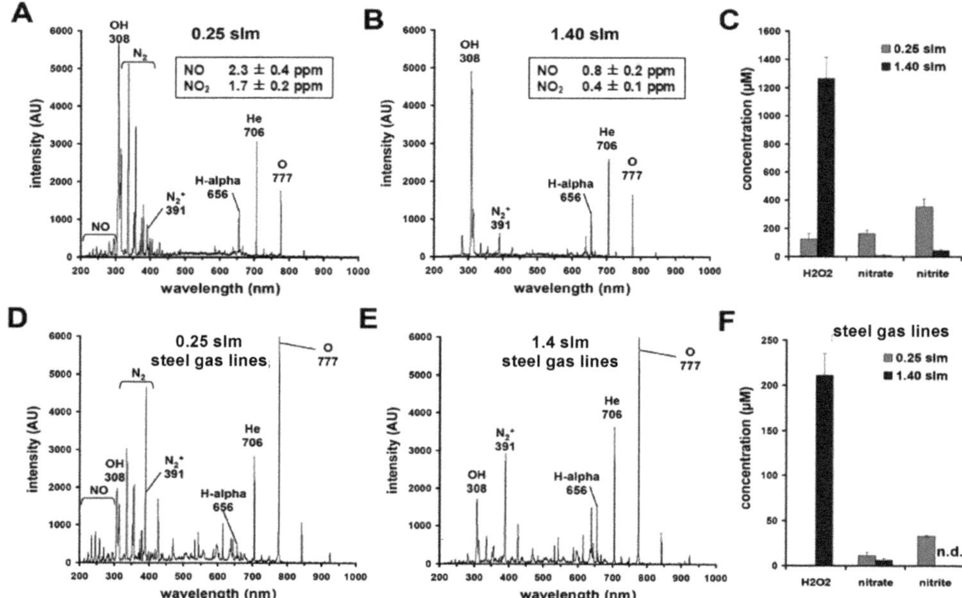

Figure 1. Gas line material affects plasma composition and the accumulation rate of H_2O_2, nitrate and nitrite. OES-spectra of µAPPJ plasma generated at different gas flows (0.25 slm (**A**); 1.4 slm (**B**)) using silicon gas lines for gas delivery. (**C**) The plasma-induced generation of hydrogen peroxide (H_2O_2), nitrate and nitrite found in PBS (500 µL) after a single plasma treatment (10 min) under the same conditions. By replacing the silicon gas lines by steel gas lines, different OES spectra were obtained (**D,E**). The H_2O_2, nitrate and nitrite concentrations found in PBS (500 µL) after CAP treatment (10 min) using steel gas lines are shown in (**F**). Bars shown in (**C,F**) represent the mean $\pm$ SD of three individual experiments.

2.5. Detection of Nitrite and Nitrate

The nitrite concentrations were quantified by an iodine/iodide-based assay using the NO analyzer CLD 88 from Ecophysics (Munich, Germany), and the determination of the total concentration of nitrite and nitrate was carried out using the vanadium (III) chloride method as previously described [67–69].

2.6. Measurement of H_2O_2

The concentration of H_2O_2 in CAP-treated PBS was determined by the titanium oxide oxalate method as previously described [70].

2.7. Measurement of Nitric Oxide and Nitrogen Dioxide

The determination of nitrogen monoxide (NO) and nitrogen oxides (NO_2) in the gas phase at the outlet of the µAPPJ was measured using the NO/NO_x analyzer CLD 822r (Ecophysics, Munich, Germany).

2.8. Cell Culture

Primary cultures of hsFB were isolated from abdominoplasty skin specimens obtained with donor consent and approval of the Ethics Commission of Düsseldorf University (Study No. 3634) from 7 female and 1 male patients (39–75 years old, mean 52.4 $\pm$ 15.3 years).

The hsFB cultures were isolated, cultivated and cryopreserved as previously described [71]. For the experiments, cryopreserved primary hsFB cultures were thawed and cultivated in T75 cell culture flasks (Cellstar, Greiner Bio-One, Frickenhausen, Germany) at 5% CO_2 and 37 °C. For seeding, the cells were detached from the surface by adding 0.05% trypsin/0.02% EDTA/0.9%, and the remaining trypsin activity was neutralized by adding 1 mL fetal calf serum (FCS). hsFB cultures were grown in Dulbecco's modified Eagle's medium (DMEM, Gibco-Invitrogen, Karlsruhe, Germany), plus 10% FCS (SeraPlus, Pan-Biotech, Aidenbach, Germany), 100 U/mL penicillin and 100 µg/mL streptomycin (PAA, Pasching, Austria) cultivated. Experiments were carried out in 24-well cell culture plates with a cell density of 2.5×10^4 per well. All measurements were performed with hsFB from passages 4–6.

2.9. Plasma Treatments of Fibroblasts

Prior to plasma treatment in PBS (500–1000 µL), the hsFB were carefully washed (PBS, 500 µL), and the culture plate was placed under the µAPPJ. The distance between the electrode end and the culture plate well's bottom in all experiments was kept at 7 mm (Figure 2A) for direct treatments. For indirect treatments, the buffer or media was separately treated from the hsFB and immediately transferred to the hsFB-containing cell culture plate. The directly and indirectly treated hsFB were incubated as indicated (0–5 min) before buffer was removed and fresh media was added. In addition, the hsFB were incubated with a buffer containing bovine catalase (1000 U/mL, Sigma Aldrich, St. Louis, MO, USA).

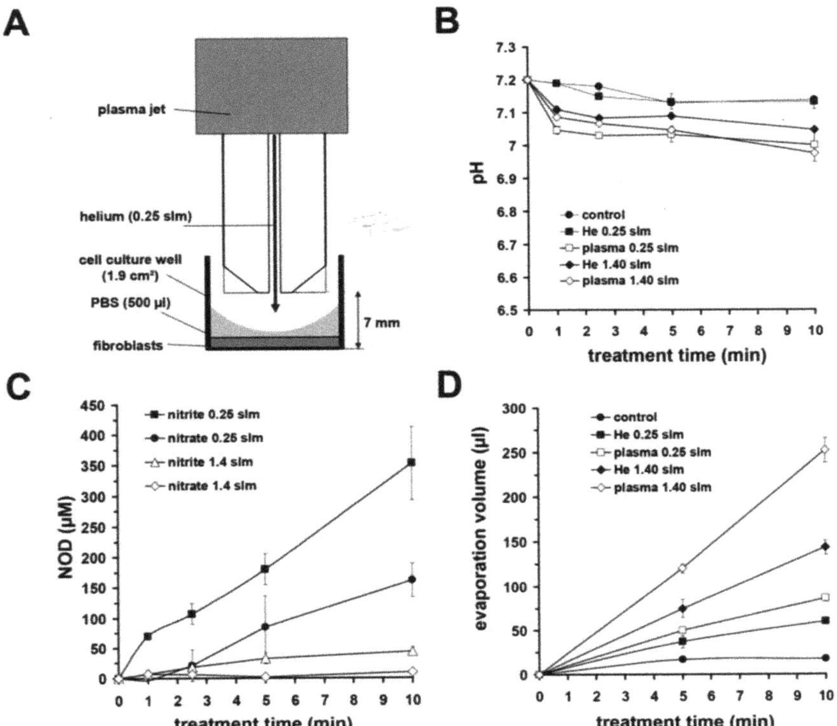

Figure 2. Experimental set-up and plasma-induced modifications of buffer. (**A**) Experimental set-up for µAPPJ treatments. (**B**) Evaluated plasma-induced pH changes. (**C**) Accumulation of nitrite and nitrate by plasma treatment obtained with different gas flows. (**D**) Loss of buffer volume after plasma/gas application using different gas flows. Values represent the means ± SD of five individual experiments.

2.10. Toxicity, Viability and Proliferation

We used three vital dyes to detect and quantify cytotoxic events as described previously [72,73]. Vital cells were detected using fluorescein diacetate (FDA), apoptotic cells were visualized and quantified using Hoechst 33342 dye and we used propidium iodide to detect and quantify necrotic events. The fluorescent dyes were each used in a concentration of 0.5 µg/mL. The experiments were evaluated using a fluorescence microscope (Zeiss, Wetzlar, Germany). In addition, we characterized the vitality and cell number in the respective differently treated cell cultures using a resazurin-based assay (CellTiter-Blue, Promega, Madison, WI, USA) and using a fluorescence spectrometer (VICTOR II Plate Reader, PerkinElmer, Waltham, MA, USA) at an excitation wavelength of 540 nm and an emission wavelength of 590 nm, as described previously [72,73].

2.11. Statistical Analysis

Significant differences were evaluated by using either paired two-tailed Student's *t*-test or ANOVA followed by appropriate post hoc multiple comparison tests (Tukey method). A value of $p < 0.05$ was considered as significant.

3. Results

3.1. Plasma Characterization

The comparison of OES spectra taken with different gas flows and supply lines is shown in Figure 1.

In Figure 1A,B, we show the OES spectra of µAPPJ plasma at gas flows of 0.25 slm and 1.4 slm using silicon gas lines for gas supply, while Figure 1D,E shows the spectra after replacing silicon lines with steel lines. Irrespective of the chosen gas flow values, after change of the gas lines, apart from the expected signals of helium, other dominant signals appeared, which were identified as impurities. At 0.25 slm (Figure 1A), in addition to signals from atomic hydrogen, neutral/ionic molecular nitrogen and nitric oxide, a marked peak at 308 nm can be observed, indicating the occurrence of OH in the plasma. This is most probably the result of enhanced humidity in the system's gas line of silicon due to stored moisture as well as greater diffusion from the immediate environment. At a gas flow of 1.40 slm (Figure 1B), the signals of NO and neutral nitrogen molecules are much lower.

Additionally, Figure 1C,F shows the concentration of H_2O_2, nitrate and nitrite, which were obtained after a plasma treatment (10 min) of PBS (500 µL) under the respective gas flow conditions.

Using the silicone gas lines, the 10 min treatment of the aqueous solutions, regardless of the gas flow used, resulted in a significant and many times higher accumulation of H_2O_2, nitrite and nitrate (Figure 1C) than when using the µAPPJ with the steel gas lines (Figure 1F). Furthermore, as we show in the inserted tables in Figure 1A, B, we were able to detect considerable amounts of nitric oxide (NO) and nitrogen dioxide (NO_2) in the µ-APPJ plasmas using the CLD technique. The measured NO and NO_2 concentrations at a lower gas flow (0.25 slm) were 2.3 ± 0.4 ppm and 1.7 ± 0.2 ppm, respectively, whereas at a gas flow of 1.40 slm, NO achieved only $0.8 \pm 0\ 0.2$ ppm and NO_2 0.4 ± 0.1 ppm. Based on these results, all of the following experiments were conducted using silicone gas supply lines.

3.2. Impact of CAP on Physicochemical Modifications of Aqueous Solutions

Using an experimental setup, as shown in Figure 2A, we found that a treatment of buffer with µAPPJ plasma led to a slight, albeit significant, reduction in pH (Figure 2B) and to a partly strong evaporation, particularly at the higher gas flow (Figure 2D). We observed that after plasma treatment (10 min), the volume loss of the treated buffer at a 1.4 slm gas flow was almost 50%. In contrast, using 0.25 slm, the loss of volume was about 17% of the initial volume. Without plasma ignition, the helium gas flow resulted in volume losses of ~29% (at 1.40 slm) and ~12% (at 0.25 slm) without plasma (Figure 2D).

As shown in Figure 2C, the chosen gas flow also had a modulative effect on the nitrite and nitrate concentrations in µAPPJ-plasma exposed buffers. At the gas flow of 0.25 slm, we were able to detect nitrite (161.4 ± 27.7 µM) and nitrate (352.7 ± 59.9 µM) after plasma treatment (10 min), whereas the nitrite concentration in the treated buffer was 9.7 ± 2.4 µM, and the nitrate concentration was 44.1 ± 6.6 µM at 1.40 slm (Figure 2C).

We also evaluated the influence of the gas flow rate on the oxygen content of the He-exposed (10 min) solution and found, as expected, a significantly greater oxygen depletion of the treated solution with the treatment with 1.40 slm compared to the treatment with 0.25 slm (Figure 3A).

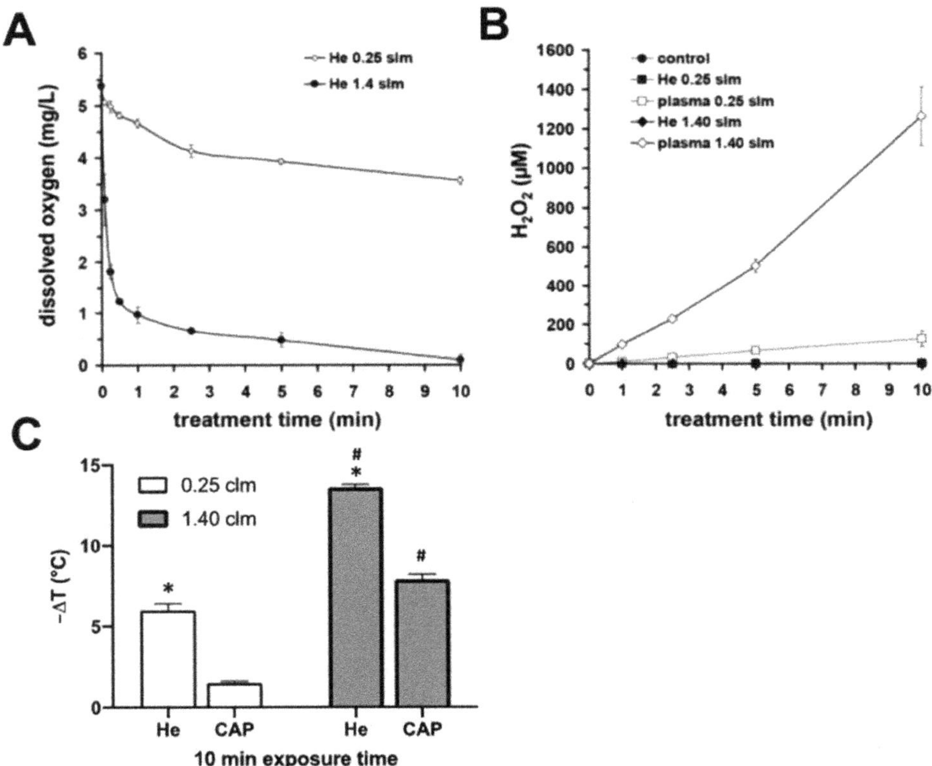

Figure 3. Plasma-induced modifications of buffer. (**A**) The gas flow-induced changes of dissolved oxygen in the buffer. (**B**) Accumulation of hydrogen peroxide by plasma treatment using different gas flows. Values represent the means ± SD of five individual experiments. (**C**) Decrease in temperature (−ΔT) of gas (He) or plasma (CAP) exposed aqueous solutions using gas flows of 0.25 slm or 1.40 slm. (**A–C**) Values represent the mean ± SD of four individual experiments. * $p < 0.05$ as compared to CAP-treatment; # $p < 0.05$ as compared to respective values obtained with 0.25 slm.

Characterizing the ability of µAPPJ to generate H_2O_2 in exposed buffer solutions, we observed a time-dependent linear H_2O_2 production rate in the treated solutions (Figure 3B). With a gas flow of 1.4 slm, we observed an H_2O_2 concentration of 1265 ± 148 µM after 10 min of exposure and at 0.25 slm, we quantified an approximately 10-fold lower H_2O_2 concentration of 124 ± 40 µM.

The gas flow we used also had a significant effects on the temperature of the CAP-treated buffer. After treatment (10 min, 1.40 slm) with helium (without plasma ignition), the temperature of the buffer dropped by 13.6 ± 0.2 °C and by 7.9 ± 0.3 °C with the helium plasma ignited. The cooling effects mentioned were −6.0 ± 0.4 °C in the case of helium gas exposure alone (without plasma ignition) and $-1.5 \pm 0.1\ °C$ after CAP exposure (10 min) using 0.25 slm (Figure 3C).

3.3. Impact of Helium Stream Exposure on Viability of Primary Human Skin Fibroblasts

Based on some of the phenomena mentioned above, one can assume that just a physical modification of a solution by the gas flow used, e.g., a lowering of the oxygen content, the reduction of the medium volume and the resulting increased osmotic stress, has a negative effect on the vitality of the exposed cell cultures. To verify this assumption, we overlaid human fibroblast cultures in cell culture dishes with medium (100–1000 μL) and exposed them to a 1.40 or 0.25 slm flow of helium for 10 min via the μAPPJ, but without plasma ignition. As shown in Figure 4A, treatment with 1.40 slm led to a significant increase in toxicity, which correlated with decreasing medium volumes of the respective cell culture. This observed toxic effect of the helium flow was significantly reduced when using the lower 0.25 slm gas flow (Figure 4B).

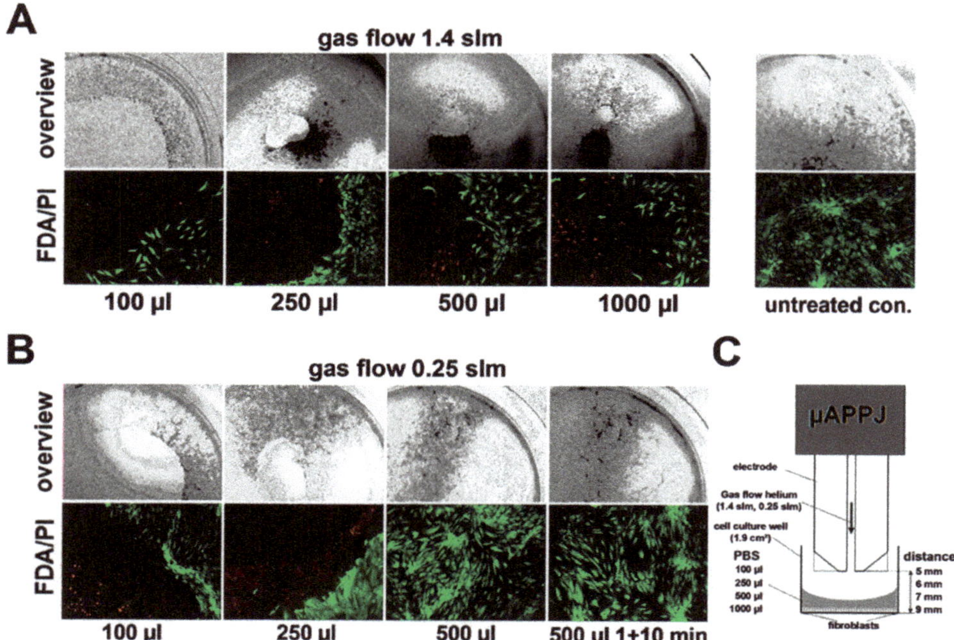

Figure 4. Helium treatment induced toxicity. Representative microscopy images of human skin fibroblasts cultures and their respective fluorescence images (fluorescein diacetate, FDA/propidium iodide, PI) directly after helium treatment (10 min) at different gas flows (1.40 slm (**A**); 0.25 slm (**B**)) applied by the μAPPJ without plasma ignition (**C**) and treatment volumes (as indicated) of buffered phosphate saline (PBS).

3.4. Impact of CAP on Viability of Primary Human Skin Fibroblasts

In addition to direct CAP exposure, as shown in Sections 3.2 and 3.3, to avoid the subsequent dehydration effects caused by the carrier gas flow, the fibroblast cultures were "indirectly" treated with the plasma. In this case, the buffer was treated with CAP separately and immediately added to the fibroblast cultures for further incubation (0, 1, 3, 5 min). Alternatively, in order to estimate the possible osmotic effects, the CAP-induced loss of buffer volume as a result of the treatment was compensated by adding the respective evaporated amount of water.

In Figure 5A, we show the results of indirect treatment of the fibroblast cultures with the µAPPJ-generated CAP at 1.40 slm without compensating for the volume loss. Here, it can be seen that the toxic effect of the treatment correlates with the length of the treatment time of the cell culture medium and the exposure time of the cells to this medium.

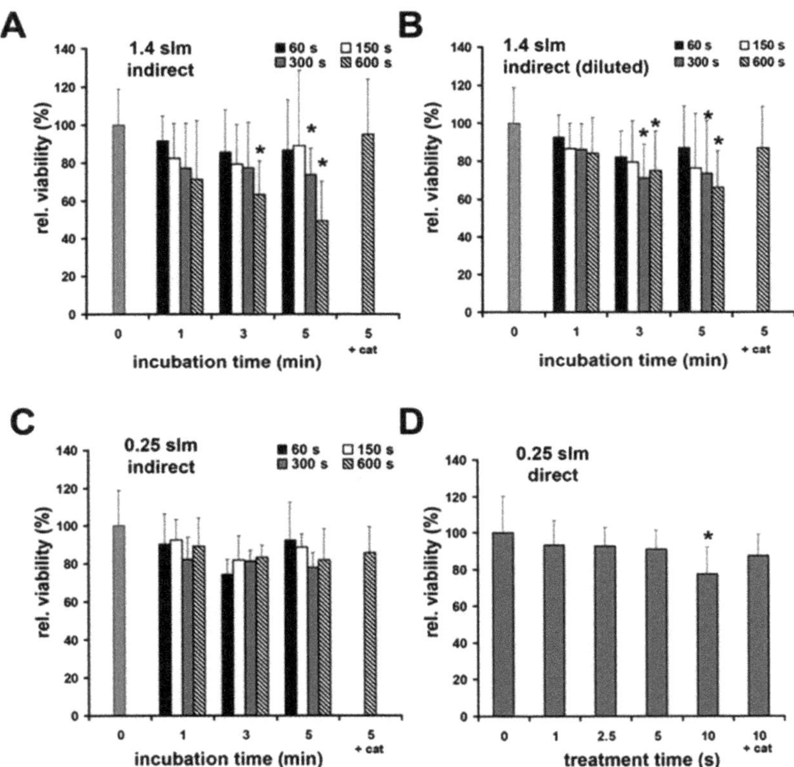

Figure 5. Plasma effects on cell viability. (**A**) Cell viability of human skin fibroblasts measured by a resazurin-based assay 24 h after indirect plasma treatment (gas flow 1.4 slm) followed by short incubation as indicated, (**B**) 24 h after indirect plasma treatment (gas flow 1.4 slm) followed by short incubation as indicated with prior compensation of plasma-induced loss of buffer, (**C**) 24 h after indirect plasma treatment (gas flow 0.25 slm) followed by short incubation as indicated, and (**D**) 24 h after direct plasma treatment (gas flow 0.25 slm). Catalase (1000 U; +cat) was added for a treatment time of 600 s followed by 5 min incubation. Catalase and gas treatments alone did not show significant effects on cell viability (not shown). Values represent the means ± SD of 7–8 independent experiments. * $p < 0.05$ as compared to the control values (0 min incubation time).

We were able to determine a significant decrease in cell vitality after treatment of cell cultures with CAP-exposed (10 min at 1.40 slm) buffers to 49 ± 29% of the vitality of the original culture. However, the toxic effect of indirect treatment of cell cultures was significantly weaker when the volume loss of CAP-treated buffer was replenished to the initial volume by the addition of water before distribution to cell cultures was performed (Figure 5B). It should be kept in mind that the final H_2O_2 concentration (600 s treatment, 5 min incubation) documented in Figure 5A was again reduced approximately two-fold after volume compensation (Figure 5B). The addition of catalase (+cat) resulted in a significant mitigation of the toxic effect of indirect CAP treatment of cell cultures, further underscoring the dominant role of H_2O_2 in the experiment described above.

After indirect plasma treatments using 0.25 slm gas flow, no significant effects on cell viability were observed, regardless of the presence of catalase in the CAP-exposed buffer (Figure 5C). After direct plasma-treatment (0.25 slm), we observed a slight, albeit significant, reduction in cell viability (down to $77 \pm 15\%$) only after a 10 min treatment, which was significantly reduced by the addition of catalase (Figure 5D). Treatment with helium as a control did not show significant differences to the untreated control (96 ± 14, 0.25 slm; $95 \pm 13\%$, 1.4 slm).

3.5. Impact of CAP on Proliferation Capacity of Primary Human Skin Fibroblasts

The results documented in Figure 6A,B also show that the proliferation of cell cultures treated indirectly with CAP (1.4 slm) followed by a 5 min incubation was significantly inhibited when volume compensation was omitted. Here, too, this effect could be significantly reduced by the addition of catalase (+cat), which in turn points to a CAP-induced H_2O_2 accumulation as the cause of the reduction in the proliferation rate.

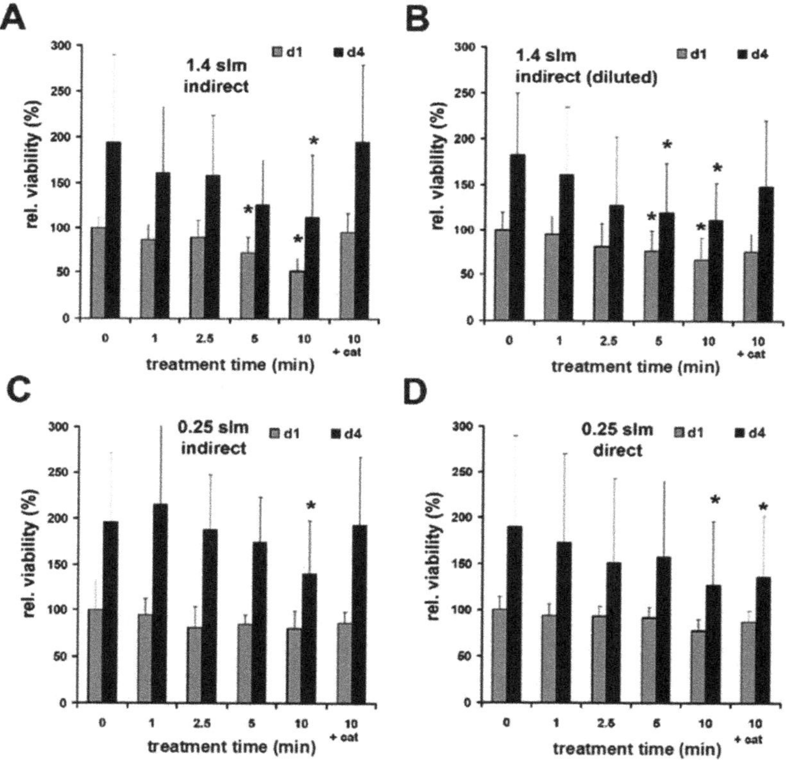

Figure 6. Plasma effects on cell proliferation. The cell numbers of human skin fibroblasts normalized to the untreated control were measured on d1 and d4 after plasma treatment by a resazurin-based assay. (**A**) Indirect plasma treatment (gas flow 1.4 slm) followed by 5 min incubation. (**B**) Indirect plasma treatment (gas flow 1.4 slm) followed by 5 min incubation with prior compensation for buffer's plasma-induced loss. (**C**) Indirect plasma treatment (gas flow 0.25 slm) followed by 5 min incubation. (**D**) Direct plasma treatment (gas flow 0.25 slm). Catalase (1000 U; cat) was added for 600 s treatment followed by 5 min incubation. Catalase and gas treatments alone did not show significant effects on cell numbers (not shown). Values represent the means $\pm$ SD of 7–8 independent experiments; * $p < 0.05$ as compared to the control values (0 min incubation time).

Direct and indirect treatment (10 min, 0.25 slm) reduced the cell numbers on day one and day four as compared to the controls (Figure 6C,D). By the addition of catalase, the observed inhibition of proliferation could be reserved only for indirect treatment (Figure 6C).

4. Discussion

For more than two decades, modern plasma technology has allowed the generation of "cold" atmospheric plasmas (CAP), even under atmospheric pressure conditions and low process temperatures, which enables interactions with biological tissues. Based on good tolerability, there are a variety of plasma-based therapy options for the treatment of different diseases in humans and animals [74,75]. Depending on the technology used and the atmospheric environment, CAPs contain reactive nitrogen species (RNS), such as nitric oxide (NO), nitrogen dioxide (NO_2), reactive oxygen species (ROS and ozone (O_3)), superoxide radicals ($.O_2^-$), hydroxyl radicals (.OH) and many other radical products in different concentrations and compositions [76,77]. The components contained in CAPs, individually or in combination, can affect the biological functions of tissues and cells [41]. Nitric oxide and some of its derivatives have an important physiological function in the regulation of inflammatory response, vasodilation, angiogenesis, thrombogenesis, immune response, cell proliferation/differentiation, antibacterial defense, collagen metabolism, apoptosis and necrosis [78]. The pivotal importance of NO in the regulation of tissue homeostasis becomes particularly clear in situations characterized by insufficient NO production or NO availability. A relative and absolute NO deficiency correlates with the corresponding chronic, bacterially infected and poorly healing wounds seen in the clinic [79].

In this respect, it is not surprising that the positive effect of NO-based therapies with exogenously applied NO gas, NO donors or NO-containing plasmas in the therapy of chronic wounds have been shown in various studies [80–82]. In addition, the bacterial infection of wounds is a driving factor in delayed wound healing, and all measures that lead to a reduction of the bacterial burden may improve the wound healing status. Therefore, plasma compositions, in particular those characterized by dominant H_2O_2 production, were shown to be very effective in combating the bacterial load on the wound and support wound healing in a particularly positive way [83]. It is therefore expedient to generate plasmas with different properties adapted to the desired therapy goals by modulating the plasma chemistry. A particular bacteriotoxic effect could be aimed at using H_2O_2-enriched or H_2O_2-generating plasmas, whereas the modulation of NO-dependent physiological parameters could be achieved most likely with strongly NOD-generating plasmas.

With a DBD device, we very recently observed that by increasing the power dissipation in the discharge, it was possible to shift from a plasma chemistry characterized by oxygen radicals towards a nitrogen oxide–dominated chemistry [84]. Under the conditions given here, i.e., in the presence of the atmospheric impurities mentioned, such a shift in plasma chemistry could be achieved with the μAPP-Jet simply by selecting the flow rate of the operating gas (Figure 1C,E). With the "regular" use of the plasma source, i.e., the operating gas supplied via stainless steel lines [85], oxygen radical-dominated plasmas could be generated at higher flow velocities (1.40 slm), and at lower flow velocities (0.25 slm) plasmas predominantly dominated by nitrogen oxides are generated. Establishing a reference jet based on μAPPJ technology (COST jet; see introduction), Schulz-von der Gathen and colleagues found that the material of the gas lines can strongly influence the reproducibility and the purity of the generated plasmas. Here, even the slightest contamination of the process gas with air, oxygen and/or H_2O could lead to a significant change in the composition of the resulting plasmas [28,29].

With this knowledge, in our experiments we used silicone tubes instead of metal tubes. The intention was to provoke contamination and thereby achieve a higher yield of reactive species in the plasma. Our technical equipment did not allow us to actively and specifically control the degree of contamination of the process gas, but the optical emission spectroscopy (OES) analysis revealed a defined time-constant level of specific impurities with the use of silicone tubes. In the OES diagram, in addition to NO, neutral/ionic molecular nitrogen and atomic hydrogen also the presence of OH radicals in the form of a clear peak at 308 nm could be seen, most likely representing the result of increased humidity in the tube atmosphere. However, this supposedly low level of contamination had serious consequences for the composition and chemical properties of the plasma. Using the silicone tubes, plasma exposure of aqueous solutions led to the accumulation of six times the H_2O_2 concentration and an almost 16-fold higher NOD concentration compared to the µAPPJ plasmas driven with metal tubes. Additionally, regardless of the significantly higher plasma concentrations of the reactive species when using the silicone gas lines, we observed up to 20-fold higher production rates of NO derivatives in the plasma-exposed solutions when using the lower gas flow of 0.25 slm than with 1.40 slm. In contrast, up to 14 times higher H_2O_2 concentrations were obtained in the CAP-treated solutions using 1.40 slm than in those using 0.25 slm. A possible and the most reasonable explanation for this phenomenon is related to the very short lifetime of the generated oxygen atoms due to their large reaction potential. It can be assumed that with a low gas flow, the generated reactive oxygen species may not even reach the exposed object and react quickly to the corresponding NOX if N is present. At high flux levels, one would accordingly expect a greater influence of the ROS chemistry.

The simple option of the µAPPJ, which involves modulating the flow rate of the processing gas to decisively change the plasma chemistry in the direction of a ROS- or RNS-directed therapy option, has a major advantage over other plasma-device technologies, such as DBD technology. For example, the preference for plasma-induced nitrite accumulation and acidification is an essential aspect supporting wound healing and other NO-dependent physiological processes [86] but also an important reinforcing mechanism in the antibacterial effect of NOD-containing or NOD-generating plasma [63]. On the other hand, an effective therapeutic cytotoxic effect against cells and bacteria could be achieved through a preferred plasma-induced formation of the strong oxidizing agent H_2O_2 [58,64]. Irrespective of the injurious H_2O_2-induced effects on different cell types described in the literature [87,88], we were only able to observe very low cytotoxic effects, even with high H_2O_2 concentrations, with the human skin fibroblasts investigated here. These cell-damaging effects of the µAPPJ-induced H_2O_2-generating plasma could also be completely avoided by the exogenous addition of catalase. In addition to the change in plasma chemistry described above by modulating the gas flow, it is technically possible to change the plasma chemistry in a targeted and defined manner by adding the appropriate gas species in a controlled manner.

Another relevant aspect in the therapeutic use of plasma jet technology is the flow rate of the processing gas required for optimal plasma generation. Every cell requires a certain moist environment with a controlled physiological pH and physiological osmolar pressure, a suitable temperature and an adequate supply of oxygen and nutrients for its survival and the exercise of physiological functions. In particular, plasma jets with very high gas flow rates of up to 6 slm could increase the osmolarity and greatly reduce the temperature due to dehydration effects and reduce the oxygen supply to the cells due to the relatively high concentration of the operating gases. Loss of volume through evaporation and simultaneous accumulation of the NOD and/or H_2O_2 concentration can lead to hyperosmolarity and all the associated negative effects [89,90], including cell death. Since the µAPPJ can also be operated with incomparably low flow rates compared to other plasma jet devices using different plasma generation technologies, the described negative influences of a high gas flow could be completely neglected.

In summary, the µAPP-Jet can be operated with a comparably wide variation in the flow rate of the processing gas. In particular, a possible operation of the device at low gas flow rates takes into account the problem of dehydration of the exposed areas or samples, as can be observed when using plasma devices operated with very high gas flow rates. With the µAPP-Jet, a low operating gas flow rate correlates with a dominant NO chemistry of the generated plasma, whereas a dominant oxygen radical chemistry can be controlled at high flow rates of the processing gas. Through the targeted and defined introduction of traces of air gases or water, the concentration of ROS and RNS in the generated plasma can be increased many times over, with the above-mentioned effects of the gas flow remaining intact. The properties mentioned make the µAPPJ technology very attractive for novel therapy options in the treatment, e.g., bacteria-infested, poorly healing wounds. By choosing the flow rate of the processing gas, NOD-based support of the physiological processes of wound healing can be addressed and ROS-based therapy goals of bacterial disinfection of a wound can be promoted.

Author Contributions: Conceptualization, C.V.S. and C.O.; methodology, D.F., J.G., J.H., V.S.-v.d.G. and C.O.; validation, J.G., J.H. and C.O.; formal analysis, D.F., J.G., J.H., C.V.S. and C.O.; investigation, D.F., J.G., J.H. and F.J.; resources, P.A., V.S.-v.d.G. and C.V.S.; data curation, C.V.S. and C.O.; writing—original draft preparation, D.F., C.O. and F.J.; writing—review and editing, P.A., V.S.-v.d.G. and C.V.S.; visualization, D.F. and F.J.; supervision, V.S.-v.d.G., C.V.S. and C.O.; project administration, P.A. and C.V.S.; funding acquisition, C.O. All authors have read and agreed to the published version of the manuscript.

Funding: This research received funding from the German Research Foundation DFG (OP 207/11-1) and from the DFG PlasNOW project 430219886.

Institutional Review Board Statement: The study was conducted in accordance with the Declaration of Helsinki and approved by the Ethics Commission of Düsseldorf University (Study No. 3634).

Informed Consent Statement: Informed consent was obtained from all subjects involved in the study.

Data Availability Statement: The data that support the findings of this study are available from the corresponding author upon reasonable request.

Acknowledgments: We thank Samira Seghrouchni, Christa-Maria-Wilkens, and Jutta Schneider for technical assistance.

Conflicts of Interest: The authors declare no conflict of interest.

References

1. Becker, K.; Schoenbach, K.; Eden, J. Microplasmas and applications. *J. Phys. D Appl. Phys.* **2006**, *39*, R55. [CrossRef]
2. Laroussi, M.; Akan, T. Arc-Free Atmospheric Pressure Cold Plasma Jets: A Review. *Plasma Process. Polym.* **2007**, *4*, 777–788. [CrossRef]
3. Schoenbach, K.H.; El-Habachi, A.; Shi, W.; Ciocca, M. High-pressure hollow cathode discharges. *Plasma Sources Sci. Technol.* **1997**, *6*, 468–477. [CrossRef]
4. Eden, J.G.; Park, S.-J.; Ostrom, N.P.; McCain, S.T.; Wagner, C.J.; Vojak, B.A.; Chen, J.; Liu, C.; Von Allmen, P.; Zenhausern, F.; et al. Microplasma devices fabricated in silicon, ceramic, and metal/polymer structures: Arrays, emitters and photodetectors. *J. Phys. D Appl. Phys.* **2003**, *36*, 2869–2877. [CrossRef]
5. Baars-Hibbe, L.; Sichler, P.; Schrader, C.; Lucas, N.; Gericke, K.-H.; Büttgenbach, S. High frequency glow discharges at atmospheric pressure with micro-structured electrode arrays. *J. Phys. D Appl. Phys.* **2005**, *38*, 510–517. [CrossRef]
6. Kong, M.; Kroesen, G.; Morfill, G.; Nosenko, T.; Shimizu, T.; Van Dijk, J.; Zimmermann, J.L. Plasma medicine: An introductory review. *New J. Phys.* **2009**, *11*, 115012. [CrossRef]
7. Graves, D.B. Reactive Species from Cold Atmospheric Plasma: Implications for Cancer Therapy. *Plasma Process. Polym.* **2014**, *11*, 1120–1127. [CrossRef]
8. Weltmann, K.D.; Kindel, E.; von Woedtke, T.; Hähnel, M.; Stieber, M.; Brandenburg, R. Atmospheric-pressure plasma sources: Prospective tools for plasma medicine. *Pure Appl. Chem.* **2010**, *82*, 1223–1237. [CrossRef]
9. Hemke, T.; Wollny, A.; Gebhardt, M.; Brinkmann, R.P.; Mussenbrock, T. Spatially resolved simulation of a radio-frequency driven micro-atmospheric pressure plasma jet and its effluent. *J. Phys. D Appl. Phys.* **2011**, *44*, 285206. [CrossRef]
10. McKay, K.; Liu, D.X.; Rong, M.Z.; Iza, F.; Kong, M.G. Dynamics and particle fluxes in atmospheric-pressure electronegative radio frequency microplasmas. *Appl. Phys. Lett.* **2011**, *99*, 091501. [CrossRef]
11. Niemi, K.; Reuter, S.; Graham, L.M.; Waskoenig, J.; Gans, T. Diagnostic based modeling for determining absolute atomic oxygen densities in atmospheric pressure helium-oxygen plasmas. *Appl. Phys. Lett.* **2009**, *95*, 151504. [CrossRef]
12. Niemi, K.; Waskoenig, J.; Sadeghi, N.; Gans, T.; O'Connell, D. The role of helium metastable states in radio-frequency driven helium–oxygen atmospheric pressure plasma jets: Measurement and numerical simulation. *Plasma Sources Sci. Technol.* **2011**, *20*, 055005. [CrossRef]

13. Shi, J.J.; Kong, M.G. Mechanisms of the α and γ modes in radio-frequency atmospheric glow discharges. *J. Appl. Phys.* **2005**, *97*, 023306. [CrossRef]

14. Yang, A.; Wang, X.; Rong, M.; Liu, D.; Iza, F.; Kong, M.G. 1-D fluid model of atmospheric-pressure rf He+O_2 cold plasmas: Parametric study and critical evaluation. *Phys. Plasmas* **2011**, *18*, 113503. [CrossRef]

15. Ellerweg, D.; Benedikt, J.; von Keudell, A.; Knake, N.; Schulz-von der Gathen, V. Characterization of the effluent of a He/O_2 microscale atmospheric pressure plasma jet by quantitative molecular beam mass spectrometry. *New J. Phys.* **2010**, *12*, 013021. [CrossRef]

16. Herrmann, H.W.; Henins, I.; Park, J.; Selwyn, G.S. Decontamination of chemical and biological warfare (CBW) agents using an atmospheric pressure plasma jet (APPJ). *Phys. Plasmas* **1999**, *6*, 2284–2289. [CrossRef]

17. Gibson, A.R.; McCarthy, H.O.; Ali, A.A.; O'Connell, D.; Graham, W.G. Interactions of a Non-Thermal Atmospheric Pressure Plasma Effluent with PC-3 Prostate Cancer Cells. *Plasma Process. Polym.* **2014**, *11*, 1142–1149. [CrossRef]

18. O'connell, D.; Cox, L.J.; Hyland, W.B.; McMahon, S.J.; Reuter, S.; Graham, W.G.; Gans, T.; Currell, F.J. Cold atmospheric pressure plasma jet interactions with plasmid DNA. *Appl. Phys. Lett.* **2011**, *98*, 043701. [CrossRef]

19. Coulombe, S.; Léveillé, V.; Yonson, S.; Leask, R.L. Miniature atmospheric pressure glow discharge torch (APGD-t) for local biomedical applications. *Pure Appl. Chem.* **2006**, *78*, 1147–1156. [CrossRef]

20. Foest, R.; Kindel, E.; Lange, H.; Ohl, A.; Stieber, M.; Weltmann, K.-D. RF Capillary Jet—A Tool for Localized Surface Treatment. *Contrib. Plasma Phys.* **2007**, *47*, 119–128. [CrossRef]

21. Voráč, J.; Dvořák, P.; Procházka, V.; Ehlbeck, J.; Reuter, S. Measurement of hydroxyl radical (OH) concentration in an argon RF plasma jet by laser-induced fluorescence. *Plasma Sources Sci. Technol.* **2013**, *22*, 025016. [CrossRef]

22. Robert, E.; Barbosa, E.; Dozias, S.; Vandamme, M.; Cachoncinlle, C.; Viladrosa, R.; Pouvesle, J.M. Experimental Study of a Compact Nanosecond Plasma Gun. *Plasma Process. Polym.* **2009**, *6*, 795–802. [CrossRef]

23. Ehlbeck, J.; Schnabel, U.; Polak, M.; Winter, J.; Von Woedtke, T.; Brandenburg, R.; von dem Hagen, T.; Weltmann, K.-D. Low temperature atmospheric pressure plasma sources for microbial decontamination. *J. Phys. D Appl. Phys.* **2011**, *44*, 013002. [CrossRef]

24. von Woedtke, T.; Metelmann, H.-R.; Weltmann, K.-D. Clinical Plasma Medicine: State and Perspectives of *in Vivo* Application of Cold Atmospheric Plasma. *Contrib. Plasma Phys.* **2014**, *54*, 104–117. [CrossRef]

25. der Gathen, V.S.-V.; Buck, V.; Gans, T.; Knake, N.; Niemi, K.; Reuter, S.; Schaper, L.; Winter, J. Optical Diagnostics of Micro Discharge Jets. *Contrib. Plasma Phys.* **2007**, *47*, 510–519. [CrossRef]

26. der Gathen, V.S.-V.; Schaper, L.; Knake, N.; Reuter, S.; Niemi, K.; Gans, T.; Winter, J. Spatially resolved diagnostics on a microscale atmospheric pressure plasma jet. *J. Phys. D Appl. Phys.* **2008**, *41*, 194004. [CrossRef]

27. Golda, J.; Kogelheide, F.; Awakowicz, P.; der Gathen, V.S.-V. Dissipated electrical power and electron density in an RF atmospheric pressure helium plasma jet. *Plasma Sources Sci. Technol.* **2019**, *28*, 095023. [CrossRef]

28. Golda, J.; Held, J.; Redeker, B.; Konkowski, M.; Beijer, P.; Sobota, A.; Kroesen, G.; Braithwaite, N.S.J.; Reuter, S.; Turner, M.M.; et al. Concepts and characteristics of the 'COST Reference Microplasma Jet'. *J. Phys. D Appl. Phys.* **2016**, *49*, 084003. [CrossRef]

29. Kelly, S.; Golda, J.; Turner, M.M.; der Gathen, V.S.-V. Gas and heat dynamics of a micro-scaled atmospheric pressure plasma reference jet. *J. Phys. D Appl. Phys.* **2015**, *48*, 444002. [CrossRef]

30. Preissing, P.; Korolov, I.; Schulze, J.; der Gathen, V.S.-V.; Böke, M. Three-dimensional density distributions of NO in the effluent of the COST reference microplasma jet operated in He/N_2/O_2. *Plasma Sources Sci. Technol.* **2020**, *29*, 125001. [CrossRef]

31. Steuer, D.; Korolov, I.; Chur, S.; Schulze, J.; der Gathen, V.S.-V.; Golda, J.; Böke, M. 2D spatially resolved O atom density profiles in an atmospheric pressure plasma jet: From the active plasma volume to the effluent. *J. Phys. D Appl. Phys.* **2021**, *54*, 355204. [CrossRef]

32. Frank, S.; Kämpfer, H.; Wetzler, C.; Pfeilschifter, J. Nitric oxide drives skin repair: Novel functions of an established mediator. *Kidney Int.* **2002**, *61*, 882–888. [CrossRef]

33. Suschek, C.V.; Schewe, T.; Sies, H.; Kroncke, K.D. Nitrite, a naturally occurring precursor of nitric oxide that acts like a 'prodrug'. *Biol. Chem.* **2006**, *387*, 499–506. [CrossRef]

34. Liebmann, J.; Born, M.; Kolb-Bachofen, V. Blue-Light Irradiation Regulates Proliferation and Differentiation in Human Skin Cells. *J. Investig. Dermatol.* **2010**, *130*, 259–269. [CrossRef] [PubMed]

35. Bruch-Gerharz, D.; Schnorr, O.; Suschek, C.; Beck, K.-F.; Pfeilschifter, J.; Ruzicka, T.; Kolb-Bachofen, V. Arginase 1 Overexpression in Psoriasis: Limitation of inducible nitric oxide synthase activity as a molecular mechanism for keratinocyte hy-perproliferation. *Am. J. Pathol.* **2003**, *162*, 203–211. [CrossRef] [PubMed]

36. Seabra, A.; Pankotai, E.; Fehér, M.; Somlai, A.; Kiss, L.; Bíró, L.; Szabó, C.; Kollai, M.; de Oliveira, M.; Lacza, Z. S-nitrosoglutathione-containing hydrogel increases dermal blood flow in streptozotocin-induced diabetic rats. *Br. J. Dermatol.* **2007**, *156*, 814–818. [CrossRef] [PubMed]

37. Lee, P.C.; Salyapongse, A.N.; Bragdon, G.A.; Shears, L.L.; Watkins, S.C.; Edington, H.D.; Billiar, T.R. Impaired wound healing and angiogenesis in eNOS-deficient mice. *Am. J. Physiol.* **1999**, *277*, H1600–H1608. [CrossRef] [PubMed]

38. Joshi, M.; Strandhoy, J.; White, W.L. Nitric oxide synthase activity is up-regulated in melanoma cell lines: A potential mechanism for metastases formation. *Melanoma Res.* **1996**, *6*, 121–126. [CrossRef]

39. Shekhter, A.B.; Kabisov, R.K.; Pekshev, A.V.; Kozlov, N.P.; Perov Iu, L. Experimental clinical substantiation of plasma dy-namic therapy of wounds with nitric oxide. *Biulleten' Eksperimental'noi Biol. I Meditsiny* **1998**, *126*, 210–215.

40. Shekhter, A.B.; Serezhenkov, V.A.; Rudenko, T.G.; Pekshev, A.V.; Vanin, A.F. Beneficial effect of gaseous nitric oxide on the healing of skin wounds. *Nitric Oxide* **2005**, *12*, 210–219. [CrossRef]
41. Fridman, G.; Friedman, G.; Gutsol, A.; Shekhter, A.B.; Vasilets, V.N.; Fridman, A. Applied Plasma Medicine. *Plasma Process. Polym.* **2008**, *5*, 503–533. [CrossRef]
42. Liebmann, J.; Scherer, J.; Bibinov, N.; Rajasekaran, P.; Kovacs, R.; Gesche, R.; Awakowicz, P.; Kolb-Bachofen, V. Biological effects of nitric oxide generated by an atmospheric pressure gas-plasma on human skin cells. *Nitric Oxide* **2010**, *24*, 8–16. [CrossRef]
43. Kalghatgi, S.; Friedman, G.; Fridman, A.; Clyne, A.M. Endothelial Cell Proliferation is Enhanced by Low Dose Non-Thermal Plasma Through Fibroblast Growth Factor-2 Release. *Ann. Biomed. Eng.* **2010**, *38*, 748–757. [CrossRef] [PubMed]
44. Haertel, B.; Hähnel, M.; Blackert, S.; Wende, K.; von Woedtke, T.; Lindequist, U. Surface molecules on HaCaT keratinocytes after interaction with non-thermal atmospheric pressure plasma. *Cell Biol. Int.* **2012**, *36*, 1217–1222. [CrossRef]
45. Haertel, B.; Straßenburg, S.; Oehmigen, K.; Wende, K.; von Woedtke, T.; Lindequist, U. Differential Influence of Components Resulting from Atmospheric-Pressure Plasma on Integrin Expression of Human HaCaT Keratinocytes. *BioMed Res. Int.* **2013**, *2013*, 1–9. [CrossRef] [PubMed]
46. Blackert, S.; Haertel, B.; Wende, K.; von Woedtke, T.; Lindequist, U. Influence of non-thermal atmospheric pressure plasma on cellular structures and processes in human keratinocytes (HaCaT). *J. Dermatol. Sci.* **2013**, *70*, 173–181. [CrossRef]
47. Shi, X.-M.; Zhang, G.-J.; Yuan, Y.-K.; Ma, Y.; Xu, G.-M.; Yang, Y. Effects of Low-Temperature Atmospheric Air Plasmas on the Activity and Function of Human Lymphocytes. *Plasma Process. Polym.* **2008**, *5*, 482–488. [CrossRef]
48. Haertel, B.; Volkmann, F.; von Woedtke, T.; Lindequist, U. Differential sensitivity of lymphocyte subpopulations to non-thermal atmospheric-pressure plasma. *Immunobiology* **2012**, *217*, 628–633. [CrossRef] [PubMed]
49. Barekzi, N.; Laroussi, M. Effects of Low Temperature Plasmas on Cancer Cells. *Plasma Process. Polym.* **2013**, *10*, 1039–1050. [CrossRef]
50. Arndt, S.; Wacker, E.; Li, Y.-F.; Shimizu, T.; Thomas, H.M.; Morfill, G.E.; Karrer, S.; Zimmermann, J.L.; Bosserhoff, A.-K. Cold atmospheric plasma, a new strategy to induce senescence in melanoma cells. *Exp. Dermatol.* **2013**, *22*, 284–289. [CrossRef]
51. Ahn, H.J.; Kim, K.I.; Hoan, N.N.; Kim, C.H.; Moon, E.; Choi, K.S.; Yang, S.S.; Lee, J.-S. Targeting Cancer Cells with Reactive Oxygen and Nitrogen Species Generated by Atmospheric-Pressure Air Plasma. *PLoS ONE* **2014**, *9*, e86173. [CrossRef] [PubMed]
52. Fridman, G.; Shereshevsky, A.; Jost, M.M.; Brooks, A.D.; Fridman, A.; Gutsol, A.; Vasilets, V.; Friedman, G. Floating Electrode Dielectric Barrier Discharge Plasma in Air Promoting Apoptotic Behavior in Melanoma Skin Cancer Cell Lines. *Plasma Chem. Plasma Process.* **2007**, *27*, 163–176. [CrossRef]
53. Kang, S.U.; Cho, J.-H.; Chang, J.W.; Shin, Y.S.; Kim, K.I.; Park, J.K.; Yang, S.S.; Lee, J.-S.; Moon, E.; Lee, K.; et al. Nonthermal plasma induces head and neck cancer cell death: The potential involvement of mitogen-activated protein kinase-dependent mitochondrial reactive oxygen species. *Cell Death Dis.* **2014**, *5*, e1056. [CrossRef] [PubMed]
54. Kim, J.Y.; Ballato, J.; Foy, P.; Hawkins, T.; Wei, Y.; Li, J.; Kim, S.-O. Apoptosis of lung carcinoma cells induced by a flexible optical fiber-based cold microplasma. *Biosens. Bioelectron.* **2011**, *28*, 333–338. [CrossRef]
55. Lee, S.Y.; Kang, S.U.; Kim, K.I.; Kang, S.; Shin, Y.S.; Chang, J.W.; Yang, S.S.; Lee, K.; Lee, J.-S.; Moon, E.; et al. Nonthermal Plasma Induces Apoptosis in ATC Cells: Involvement of JNK and p38 MAPK-Dependent ROS. *Yonsei Med. J.* **2014**, *55*, 1640–1647. [CrossRef]
56. Panngom, K.; Baik, K.Y.; Nam, M.K.; Han, J.H.; Rhim, H.; Choi, E.H. Preferential killing of human lung cancer cell lines with mitochondrial dysfunction by nonthermal dielectric barrier discharge plasma. *Cell Death Dis.* **2013**, *4*, e642. [CrossRef] [PubMed]
57. Yan, X.; Xiong, Z.; Zou, F.; Zhao, S.; Lu, X.; Yang, G.; He, G.; Ostrikov, K.K. Plasma-Induced Death of HepG2 Cancer Cells: Intracellular Effects of Reactive Species. *Plasma Process. Polym.* **2012**, *9*, 59–66. [CrossRef]
58. Bekeschus, S.; Kolata, J.; Winterbourn, C.; Kramer, A.; Turner, R.; Weltmann, K.D.; Bröker, B.; Masur, K. Hydrogen peroxide: A central player in physical plasma-induced oxidative stress in human blood cells. *Free. Radic. Res.* **2014**, *48*, 542–549. [CrossRef]
59. Haertel, B.; Wende, K.; Von Woedtke, T.; Weltmann, K.D.; Lindequist, U. Non-thermal atmospheric-pressure plasma can influence cell adhesion molecules on HaCaT-keratinocytes. *Exp. Dermatol.* **2011**, *20*, 282–284. [CrossRef] [PubMed]
60. Arjunan, K.P.; Friedman, G.; Fridman, A.; Clyne, A.M. Non-thermal dielectric barrier discharge plasma induces angiogenesis through reactive oxygen species. *J. R. Soc. Interface* **2012**, *9*, 147–157. [CrossRef]
61. Kalghatgi, S.; Kelly, C.M.; Cerchar, E.; Torabi, B.; Alekseev, O.; Fridman, A.; Friedman, G.; Azizkhan-Clifford, J. Effects of Non-Thermal Plasma on Mammalian Cells. *PLoS ONE* **2011**, *6*, e16270. [CrossRef] [PubMed]
62. Sensenig, R.; Kalghatgi, S.; Cerchar, E.; Fridman, G.; Shereshevsky, A.; Torabi, B.; Arjunan, K.P.; Podolsky, E.; Fridman, A.; Friedman, G.; et al. Non-thermal Plasma Induces Apoptosis in Melanoma Cells via Production of Intracellular Reactive Oxygen Species. *Ann. Biomed. Eng.* **2011**, *39*, 674–687. [CrossRef] [PubMed]
63. Oehmigen, K.; Hähnel, M.; Brandenburg, R.; Wilke, C.; Weltmann, K.-D.; von Woedtke, T. The Role of Acidification for Antimicrobial Activity of Atmospheric Pressure Plasma in Liquids. *Plasma Process. Polym.* **2010**, *7*, 250–257. [CrossRef]
64. Chen, L.C.; Suzuki, H.; Mori, K.; Ariyada, O.; Hiraoka, K. Mass Spectrometric Detection of Gaseous Hydrogen Peroxide in Ambient Air Using Dielectric Barrier Discharge as an Excitation Source. *Chem. Lett.* **2009**, *38*, 520–521. [CrossRef]
65. Pieraggi, M.T.; Bouissou, H.; Angelier, C.; Uhart, D.; Magnol, J.P.; Kokolo, J. The fibroblast. *Ann. Pathol.* **1985**, *5*, 65–76.
66. Baum, C.L.; Arpey, C.J. Normal cutaneous wound healing: Clinical correlation with cellular and molecular events. *Dermatol. Surg. Off. Publ. Am. Soc. Dermatol. Surg.* **2005**, *31*, 674–686. [CrossRef]

67. Feelisch, M.; Rassaf, T.; Mnaimneh, S.; Singh, N.; Bryan, N.S.; Jourd'Heuil, D.; Kelm, M. Concomitant S-, N-, and heme-nitros(yl)ation in biological tissues and fluids: Implications for the fate of NO in vivo. *FASEB J.* **2002**, *16*, 1775–1785. [CrossRef]
68. Suschek, C.V.; Paunel, A.; Kolb-Bachofen, V. Nonenzymatic Nitric Oxide Formation during UVA Irradiation of Human Skin: Experimental Setups and Ways to Measure. *Methods Enzymol.* **2005**, *396*, 568–578. [CrossRef]
69. Yang, F.; Troncy, E.; Francœur, M.; Vinet, B.; Vinay, P.; Czaika, G.; Blaise, G. Effects of reducing reagents and temperature on conversion of nitrite and nitrate to nitric oxide and detection of NO by chemiluminescence. *Clin. Chem.* **1997**, *43*, 657–662. [CrossRef]
70. Sellers, R.M. Spectrophotometric determination of hydrogen peroxide using potassium titanium(IV) oxalate. *Anal.* **1980**, *105*, 950–954. [CrossRef]
71. Opländer, C.; Hidding, S.; Werners, F.B.; Born, M.; Pallua, N.; Suschek, C.V. Effects of blue light irradiation on human dermal fibroblasts. *J. Photochem. Photobiol. B Biol.* **2011**, *103*, 118–125. [CrossRef] [PubMed]
72. Balzer, J.; Heuer, K.; Demir, E.; Hoffmanns, M.A.; Baldus, S.; Fuchs, P.C.; Awakowicz, P.; Suschek, C.V.; Opländer, C. Non-Thermal Dielectric Barrier Discharge (DBD) Effects on Proliferation and Differentiation of Human Fibroblasts Are Primary Mediated by Hydrogen Peroxide. *PLoS ONE* **2015**, *10*, e0144968. [CrossRef] [PubMed]
73. Hegmann, L.; Sturm, S.; Niegisch, G.; Windolf, J.; Suschek, C.V. Enhancement of human bladder carcinoma cell chemosensitivity to Mitomycin C through quasi-monochromatic blue light ($\lambda = 453 \pm 10$ nm). *J. Photochem. Photobiol. B Biol.* **2022**, *236*, 112582. [CrossRef]
74. Ermolaeva, S.A.; Sysolyatina, E.V.; Gintsburg, A.L. Atmospheric pressure nonthermal plasmas for bacterial biofilm prevention and eradication. *Biointerphases* **2015**, *10*, 029404. [CrossRef] [PubMed]
75. Yousfi, M.; Merbahi, N.; Pathak, A.; Eichwald, O. Low-temperature plasmas at atmospheric pressure: Toward new pharmaceutical treatments in medicine. *Fundam. Clin. Pharmacol.* **2014**, *28*, 123–135. [CrossRef] [PubMed]
76. Laroussi, M.; Leipold, F. Evaluation of the roles of reactive species, heat, and UV radiation in the inactivation of bacterial cells by air plasmas at atmospheric pressure. *Int. J. Mass Spectrom.* **2004**, *233*, 81–86. [CrossRef]
77. Nosenko, T.; Shimizu, T.; Steffes, B.; Zimmermann, J.; Stolz, W.; Schmidt, H.U.; Isbary, G.; Pompl, R.; Bunk, W.; Fujii, S.; et al. Low-Temperature Atmospheric-Pressure Plasmas as a Source of Reactive Oxygen and Nitrogen Species for Chronic Wound Disinfection. *Free Radic. Bio. Med.* **2009**, *47*, S128.
78. Bruch-Gerharz, D.; Ruzicka, T.; Kolb-Bachofen, V. Nitric oxide and its implications in skin homeostasis and disease—a review. *Arch. Dermatol. Res.* **1998**, *290*, 643–651. [CrossRef]
79. Cals-Grierson, M.-M.; Ormerod, A.D. Nitric oxide function in the skin. *Nitric Oxide* **2004**, *10*, 179–193. [CrossRef]
80. Kandhwal, M.; Behl, T.; Kumar, A.; Arora, S. Understanding the Potential Role and Delivery Approaches of Nitric Oxide in Chronic Wound Healing Management. *Curr. Pharm. Des.* **2021**, *27*, 1999–2014. [CrossRef]
81. Soneja, A.; Drews, M.; Malinski, T. Role of nitric oxide, nitroxidative and oxidative stress in wound healing. *Pharmacol. Rep.* **2005**, *57*, 108–119.
82. Isenberg, J.S.; Ridnour, L.A.; Espey, M.G.; Wink, D.A.; Roberts, D.A. Nitric oxide in wound-healing. *Microsurgery* **2005**, *25*, 442–451. [CrossRef]
83. Isbary, G.; Morfill, G.; Schmidt, H.; Georgi, M.; Ramrath, K.; Heinlin, J.; Karrer, S.; Landthaler, M.; Shimizu, T.; Steffes, B.; et al. A first prospective randomized controlled trial to decrease bacterial load using cold atmospheric argon plasma on chronic wounds in patients. *Br. J. Dermatol.* **2010**, *163*, 78–82. [CrossRef]
84. Feibel, D.; Kwiatkowski, A.; Opländer, C.; Grieb, G.; Windolf, J.; Suschek, C.V. Enrichment of Bone Tissue with Antibacterially Effective Amounts of Nitric Oxide Derivatives by Treatment with Dielectric Barrier Discharge Plasmas Optimized for Nitrogen Oxide Chemistry. *Biomedicines* **2023**, *11*, 244. [CrossRef]
85. Golda, J.; Sgonina, K.; Held, J.; Benedikt, J.; der Gathen, V.S.-V. Treating Surfaces with a Cold Atmospheric Pressure Plasma using the COST-Jet. *J. Vis. Exp.* **2020**, *165*, e61801. [CrossRef]
86. Suschek, C.V.; Feibel, D.; von Kohout, M.; Opländer, C. Enhancement of Nitric Oxide Bioavailability by Modulation of Cutaneous Nitric Oxide Stores. *Biomedicines* **2022**, *10*, 2124. [CrossRef]
87. de Bono, D.; Yang, W. Exposure to low concentrations of hydrogen peroxide causes delayed endothelial cell death and inhibits proliferation of surviving cells. *Atherosclerosis* **1995**, *114*, 235–245. [CrossRef] [PubMed]
88. Chen, Q.; Ames, B.N. Senescence-like growth arrest induced by hydrogen peroxide in human diploid fibroblast F65 cells. *Proc. Natl. Acad. Sci. USA* **1994**, *91*, 4130–4134. [CrossRef] [PubMed]
89. Mammone, T.; Ingrassia, M.; Goyarts, E. Osmotic stress induces terminal differentiation in cultured normal human epidermal keratinocytes. *Vitr. Cell. Dev. Biol. Anim.* **2008**, *44*, 135–139. [CrossRef] [PubMed]
90. Hashimoto, S.; Matsumoto, K.; Gon, Y.; Nakayama, T.; Takeshita, I.; Horie, T. Hyperosmolarity-induced Interleukin-8 Expression in Human Bronchial Epithelial Cells through p38 Mitogen-activated Protein Kinase. *Am. J. Respir. Crit. Care Med.* **1999**, *159*, 634–640. [CrossRef] [PubMed]

 *biomedicines*

Article

Can Cold Atmospheric Plasma Be Used for Infection Control in Burns? A Preclinical Evaluation

Mahsa Bagheri [1,2], Maria von Kohout [1,2], Andreas Zoric [2], Paul C. Fuchs [1], Jennifer L. Schiefer [1] and Christian Oplander [2,*]

[1] Plastic Surgery, Hand Surgery, Burn Center, Cologne-Merheim Hospital, Witten/Herdecke University, Ostmerheimer Str. 200, 51109 Cologne, Germany; mahsa.bagheri@posteo.de (M.B.); maria.vonkohout@uni-wh.de (M.v.K.); fuchsp@kliniken-koeln.de (P.C.F.); schieferj@kliniken-koeln.de (J.L.S.)

[2] Institute for Research in Operative Medicine (IFOM), Cologne-Merheim Hospital, Witten/Herdecke University, Ostmerheimer Str. 200, 51109 Cologne, Germany; andreas.zoric@googlemail.com

* Correspondence: christian.oplaender@uni-wh.de; Tel.: +49-221-989570

Abstract: Wound infection with Pseudomonas aeruginosa (PA) is a serious complication and is responsible for higher rates of mortality in burn patients. Because of the resistance of PA to many antibiotics and antiseptics, an effective treatment is difficult. As a possible alternative, cold atmospheric plasma (CAP) can be considered for treatment, as antibacterial effects are known from some types of CAP. Hence, we preclinically tested the CAP device PlasmaOne and found that CAP was effective against PA in various test systems. CAP induced an accumulation of nitrite, nitrate, and hydrogen peroxide, combined with a decrease in pH in agar and solutions, which could be responsible for the antibacterial effects. In an ex vivo contamination wound model using human skin, a reduction in microbial load of about 1 $\log_{10}$ level was observed after 5 min of CAP treatment as well as an inhibition of biofilm formation. However, the efficacy of CAP was significantly lower when compared with commonly used antibacterial wound irrigation solutions. Nevertheless, a clinical use of CAP in the treatment of burn wounds is conceivable on account of the potential resistance of PA to common wound irrigation solutions and the possible wound healing-promoting effects of CAP.

Keywords: burns; wound infection; cold atmospheric plasma hydrogen peroxide; Pseudomonas aeruginosa; biofilm; nitric oxide

Citation: Bagheri, M.; von Kohout, M.; Zoric, A.; Fuchs, P.C.; Schiefer, J.L.; Oplander, C. Can Cold Atmospheric Plasma Be Used for Infection Control in Burns? A Preclinical Evaluation. *Biomedicines* **2023**, *11*, 1239. https://doi.org/10.3390/biomedicines11051239

Academic Editor: Francois Niyonsaba

Received: 13 March 2023
Revised: 16 April 2023
Accepted: 19 April 2023
Published: 22 April 2023

1. Introduction

The skin is the largest organ of the human body and serves as protection from the environment. It is not only a physical barrier against pathogens and mechanical injuries, but also serves to protect against unregulated loss of water and solutes. Furthermore, the skin provides a chemical/biochemical barrier with antimicrobial activity [1]. The skin of the human body is colonized by a diverse milieu of microorganisms, most of which are harmless, or even beneficial to the host. Colonization is determined by the nature and characteristics of the skin surface, which can vary widely, depending on topographic location, endogenous host factors, and exogenous environmental factors [2].

A burn occurs when the skin is exposed to a heat source. Here, the longer the exposure to heat and the higher the temperature, the greater and deeper the tissue damage. In addition, burn trauma can also result from freezing, electricity, chemicals, radiation, or friction [3]. After a burn, the skin and its barrier function are damaged or destroyed, sometimes over a large area. Deep second- or third-degree skin burns, if spontaneous healing occurs, heal slowly. The regeneration is dependent on the migration of keratinocytes from the surrounding uninjured skin toward the wound surface [3]. This process is also the basis for standard plastic surgery therapy for the rapid, permanent closure of burn wounds, in which donor split-thickness skin grafts are used to fill the wound [4].

Bacterial infections of burn wounds are among the most important and potentially serious complications that can occur in the acute period following injury [5,6]. Thermal destruction of the skin barrier and the reduced local and systemic immune responses of the body are critical factors that contribute to infectious complications in patients with severe burns [7–9]. Although the burned wound surface is primarily sterile, immediately after thermal injury, colonization often occurs within a few days [10,11]. Here, the second- or third-degree burn wound is a protein-rich environment consisting of non-vascularized necrotic tissue that provides a favorable niche for microorganism colonization and proliferation [12]. Most nosocomial infections of burn wounds result from multidrug-resistant Gram-negative bacteria [13]; *Pseudomonas* spp., *Staphylococcus aureus*, *Klebsiella* spp., *Proteus* spp., *Enterococcus* spp., and *Escherichia coli* are the pathogens that can be isolated from infected burn wounds [14,15]. In particular, *Pseudomonas* spp. was detected here most frequently. Sepsis induced by wound infection is associated with high mortality and poses a threat to patient survival [13].

In particular, infections with *Pseudomonas aeruginosa* (PA) significantly increase mortality in burn patients, especially when acquired nosocomially. Here, MDR (multi-drug resistant) PA has increasing importance as a main cause of death in burn patients, due to the high occurrence of MDR bacteria in burn centers [16,17]. In addition, burn wounds trigger PA to produce pathogenic factors and biofilms, which, in turn, can cause delays in healing [18]. Hence, wound infections with PA significantly increase the length of hospital stays, the number of days on mechanical ventilation, the number of surgical procedures, and the amounts of blood products used [19].

PA is able to survive in almost any environment on account of its low nutrient requirements. It is named for its characteristic green coloration in purulent wound infections, which is caused by the blue-green pigment pyocyanin and the dye fluorescein. Infections can be easily identified by the characteristic, penetrating sweet aromatic odor, because of the formation of 2-aminoacetophenone. In addition, for the identification of PA and differentiation from other pathogens, its fluorescent property can be used [20–22]. PA, like all Gram-negative bacteria, has an outer membrane that is a diffusion barrier for most antibiotics. Antibiotics can only enter the bacterium through narrow and impermeable porins, which is why most antibiotics cannot penetrate the bacterium. Hence, PA is naturally resistant to many antibiotics [20]. In addition, PA possesses other intrinsic resistance mechanisms to antibiotics, such as the expression of efflux pumps that expel antibiotics from the cell and the production of antibiotic-inactivating enzymes like β-lactamases [23–25].

Infections of burn wounds caused by *P. aeruginosa* are treated systemically with antibiotics and locally with the antiseptics mafenide acetate and citric acid [26–28]. However, mafenide acetate has—apart from an antimicrobial effect—cytotoxic properties, and can cause metabolic acidosis and delayed wound healing [29–31]. As an addition to and/or alternative for topical treatment, 3% citric acid is used to treat burn wounds infected with PA [32]. Very often, conservative therapy attempts are unsuccessful, so that a distressing and aggressive surgical debridement is necessary as a last therapy option [33,34].

Cold atmospheric plasma (CAP) is a physical low-temperature plasma generated at atmospheric pressure, containing a highly reactive mixture of UV radiation, reactive oxygen, and reactive nitrogen species (ROS/RNS), such as O_3, O_2^-, OH^-, NO_2, and NO, and shows, in general, a broad-spectrum antimicrobial efficacy [35]. There are many devices that produce different types of CAP. A distinction is made between direct, indirect, and hybrid CAPs. Direct CAP is generated between two electrodes by energizing the surrounding air, where the body or tissue can act as a counter-electrode. The CAP is, therefore, created between the electrode and skin/wound and no carrier gas is required. An example of a direct CAP source is dielectric barrier discharge (DBD). Here, the electrodes are enclosed by a non-conductive layer, a dielectric, and the discharges take the form of many small micro-discharges. Indirect CAPs are generated by plasma jets, which form the plasma between two electrodes within the device and are transported to the target via a carrier gas, normally with the noble gases argon or helium. Hybrid CAPs represent a combination of direct and indirect plasma. These are generated in the same way as direct plasma, but have current-free properties on account of a grounded grid electrode. An example of a hybrid plasma source is the corona discharge. Unlike DBD, there is no air space between the barrier and the counter electrode [35–38].

As a kind of direct plasma source, the PlasmaOne cold plasma device does not require any gas supply with noble gases to generate therapeutically effective CAPs, only the ambient air. The control center supplies the converter with a direct current, which is converted into high frequencies. Inside the treatment probe made of glass—which is filled with helium—these high frequencies are conducted via the ionized helium molecules to the tip of the probe and an electric field is now formed between the patient's skin and the treatment probe, leading to the ionization of the atoms and molecules of the ambient air. This generates the therapeutically effective CAP between the skin/wound and the probe surface. The predominant reactive species found in the generated CAP are NO, NO_2, and O_3, according to information provided by the company [39].

Normally, CAPs do not cause thermal damage to the tissue, which significantly expands the possible applications of the plasma. In medical applications, CAP therapies are sometimes used in dermatology and dentistry, with distinct CAP devices for treating acute and chronic wounds [4]. However, the evidence base on the efficacy of cold plasma is still narrow and difficult to assess. The individual studies with different generation methods and CAP devices are difficult to compare. In addition, the effectiveness of the therapy also depends on the dose and duration of application [40]. Furthermore, although CAP shows good antibacterial efficacy and promising effects on biofilms on more or less dry biomaterials in vitro, it seems that the antibacterial effects in vivo or under wet conditions are much less pronounced and possibly not clinically relevant [41,42].

Therefore, in this present study, we have examined the antimicrobial efficacy against PA of a direct CAP, generated by the PlasmaOne® device, as an alternative option for the treatment of (burn) wound infections, using a quantifiable human skin wound contamination model, which is closer to the clinical reality than standard microbiology assays [43,44].

2. Materials and Methods

2.1. Cold Atmospheric Plasma (CAP) Source and CAP Treatment

For CAP treatments, the medically accredited device Plasma One (MEDICAL SYSTEMS GmbH, Nassau, Germany) was used. This device is battery-powered; CAP is generated by a floating electrode dielectric barrier discharge (FE-DBD). To generate CAP, the device does not require any admixture of noble gases like argon or neon, only the ambient air. The energy control center supplies the converter with a direct current and this is converted into high frequencies (Hf), which are conducted to the tip of the treatment probe via the noble gases within it. Through an electrode, the patient/sample has contact with the energy control during treatment and an electric field is formed between the patient's skin and the treatment glass probe, which leads to the ionization of the atoms and molecules of the ambient air, generating a direct CAP in this space between probe and tissue/sample.

For our experiments, the device was conducted in Mode 5, which is the maximal power level, and the treatment glass probe PS30 with an outer tip diameter of 34 mm (power: 5 W, repetition frequency: 1220 Hz, converter pulse width: 10 µs). The CAP was applied at a distance of 1–2 mm between the tip of the treatment glass probe and the sample (agar plate, skin wound). At this distance, the activity monitor of the device is activated and the CAP generation is continued [45].

2.2. CAP-Induced Accumulation of Nitrite and Nitrate

On casein/soy peptone agar plates (diameter 10 cm), 100 µL TSB was spread and incubated for 20 min at 37 °C prior to CAP treatment. The CAP treatment glass probe was mounted on a stand and an agar plate was placed centrally on the laboratory lifting platform, which was used to slowly lift the agar plate to the probe tip until stable ignition and CAP generation were formed. This was normally the case with a distance of 1–2 mm between the probe tip and the agar surface. This approach avoided contact of the probe with the agar surface. The agar was grounded by a wire and treated with CAP for 0, 30, 60, 150, 300, or 600 s.

After treatment, certain areas of the agar plates were punched out (diameter 4 mm), as seen in Figure 1A, dissolved in 950 µL phosphate-buffered saline (PBS, pH 7.4), homogenized for 1 s by an Ultra turrex T8 disperser (IKA, Staufen, Germany), and centrifuged at $1000 \times g$ for 1 min. The supernatants obtained were used for nitrite/nitrate measurements by an iodine/iodide-based and vanadium (III) chloride/hydrochloric acid-based assay using a NO analyzer (CLD 88, Ecophysics, Munich, Germany), as described elsewhere [46]. These experiments were performed four times independently.

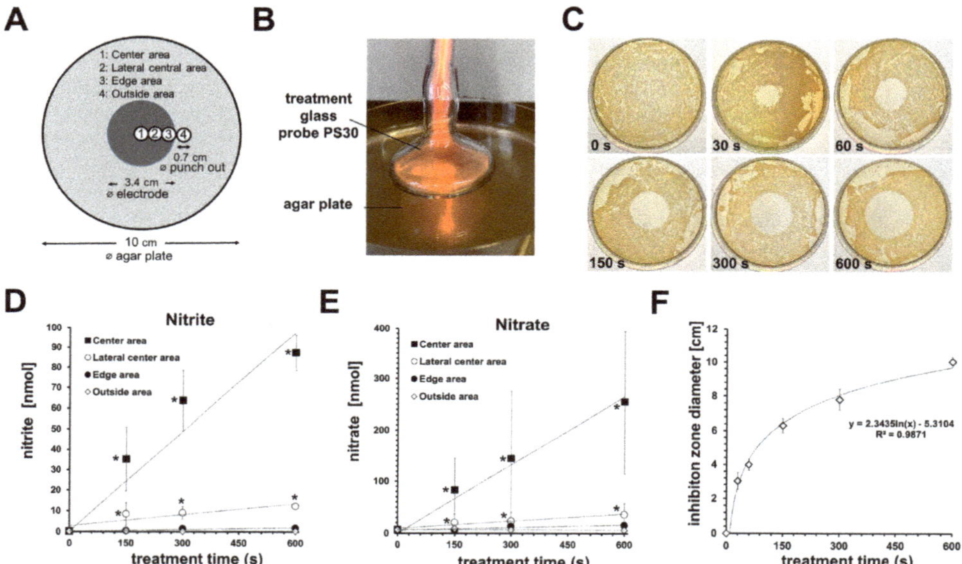

Figure 1. CAP treatment showed antibacterial efficacy against *Pseudomonas aeruginosa* and induced nitrite/nitrate accumulation. (**A**) Overview of the CAP treated area and sample locations within the agar plate. (**B**) Shown here is a representative photograph of a CAP treatment of an agar plate. (**C**) Exemplary photographs of agar plates cultivated with *Pseudomonas aeruginosa*, with inhibition zones induced by different CAP treatments (as indicated). (**D**) shows the amount of nitrite and (**E**) nitrate found in the punch biopsy samples taken from the different areas of the agar plates after CAP treatment as indicated (n = 4; * $p < 0.05$ compared to untreated controls). (**F**) Given are the mean values of the measured diameter of the CAP-induced inhibition zones (n = 3; * $p < 0.05$ compared to untreated controls).

In another experimental setup, 120 µL TSB containing PA was distributed in a circle (diameter: ~2 cm) on a microscope slide (Menzel-Gläser, Thermo Scientific, Waltham, MA, USA). The slide was mounted on a metal laboratory jack and the TSB drop that was spread was connected with the electrode of the CAP device with a thin wire, which was immersed in the liquid at the edge. Using the lifting platform of the laboratory jack, the TSB drop was placed centrally under the tip of the treatment glass probe at a distance of 1–2 mm and CAP was applied for 0, 60, 150, 300, or 600 s. The TSB was aspired and transferred with a pipette into a cooled 1.5 mL centrifugation tube (Eppendorf, Hamburg, Germany), and the amounts of nitrite and nitrated were analyzed.

2.3. Measurement of CAP-Induced Hydrogen Peroxide

The concentration of hydrogen peroxide (H_2O_2) in CAP-treated TSB was determined using the titanium oxide oxalate method, as described elsewhere [46]. In brief, a stock solution of potassium titanium oxide dihydrate (0.1 M) with 2 M sulfuric acid was freshly prepared. The CAP treatment was conducted analogously to the samples for the nitrite/nitrate measurements with 120 µL TSB spread on a microscope slide (see above).

Directly after CAP treatment, 100 µL of the TSB was mixed with 100 µL of the titanium oxide stock solution and transferred to a 96-well plate. The absorbance at 400 nm was measured using a photometer (Epoch II, BioTek, Winooski, VT, USA). The concentrations of H_2O_2 were calculated using calibration curves with known H_2O_2 concentrations (0–500 µM in TSB). In preliminary experiments, the presence of high nitrite/nitrate concentrations (both 2 mM) and low pH values (5.5 and 6.4) had no significant effects on calibration curves.

2.4. Determination of CAP-Induced Changes of pH Value

The pH values of the TSB and skin surface w/o CAP treatment were measured using a pH meter equipped with a flat pH electrode (PH CHECK F, Th. Geyer, Lohmar, Germany).

2.5. Determination of CAP-Induced Antibacterial Effects on P. aeruginosa
2.5.1. Pseudomonas Strain and Culture Conditions

The bacterial culture of *Pseudomonas aeruginosa* (PA) used was provided by the Leibniz Institute DSMZ—German Collection of Microorganisms and Cell Culture (batch No.: 0411). For the experiments, the subculture II of PA was used. Prior to the experiments, a master plate was prepared using the cryopreserved PA sample and cultivated on tryptone soya agar (TSA) plates (Sigma-Aldrich, Munich, Germany) for 24 h and was subsequently kept at 7 °C for a maximum of 2 weeks. A bacterial strain was prepared by picking a colony from the master plate 24 h before CAP treatment and incubated in 25 mL tryptone soya broth (TSB) culture medium at 37 °C. Using a photometer to measure the absorbance at 600 nm (Epoch II, BioTek, Winooski, VT, USA), the PA solution was diluted to a 0.5 McFarland standard, which corresponds to approximately 1.5×10^8 CFU/mL, and then further diluted in TSB to the specific concentration required for the experiments.

2.5.2. Determination of CAP-Induced Inhibitory and Bactericidal Effects on *P. aeruginosa*

Analogous to the determination of nitrite/nitrate, 120 µL TSB containing PA (1.8×10^6 CFU) was distributed in a circle (diameter ~2 cm) on a microscope slide (Menzel-Gläser, Thermo Scientific, Waltham, MA, USA) mounted on a metal laboratory lifting platform, and connected with the electrode of the PlasmaOne CAP device with a thin wire (see Figure 2A). The microscope slide was lifted slowly up to the treatment probe until a distance of 1–2 mm from the probe and CAP was applied for 0, 30, 60, 90, 120, 150, 180, 210, or 240 s. Any contact of the probe with the liquid was avoided. Adapted from common assays to determine minimum inhibitory concentrations, directly after treatment, 100 µL of the CAP-treated TSB drop was transferred into a 96-well plate with a flat bottom and transparent lid (TC-Plate, Sarstedt, Nürmbrecht, Germany). The 96-well microplate was subjected to absorbance readings using a microplate spectrophotometer (EPOCH II, Biotek, Winooski, VT, USA) and read at a wavelength of 595 nm for 20 h, with 20 min intervals and agitation of 10 s. The growth curves obtained from the OD measurements were evaluated to determine the minimum inhibition CAP treatment time. Analogously, 120 µL TSB containing PA (1.8×10^6 CFU) was treated with CAP (0, 60, 150, 300, 600 s). Here, 100 µL of the CAP-treated TSB was serial diluted for the CFU assay; thus, 100 µL of the dilution was plated on TSA plates and incubated for 24 h at 37 °C to determine the bacterial survival rate by counting colonies. Each single experiment was performed independently three times in duplicate.

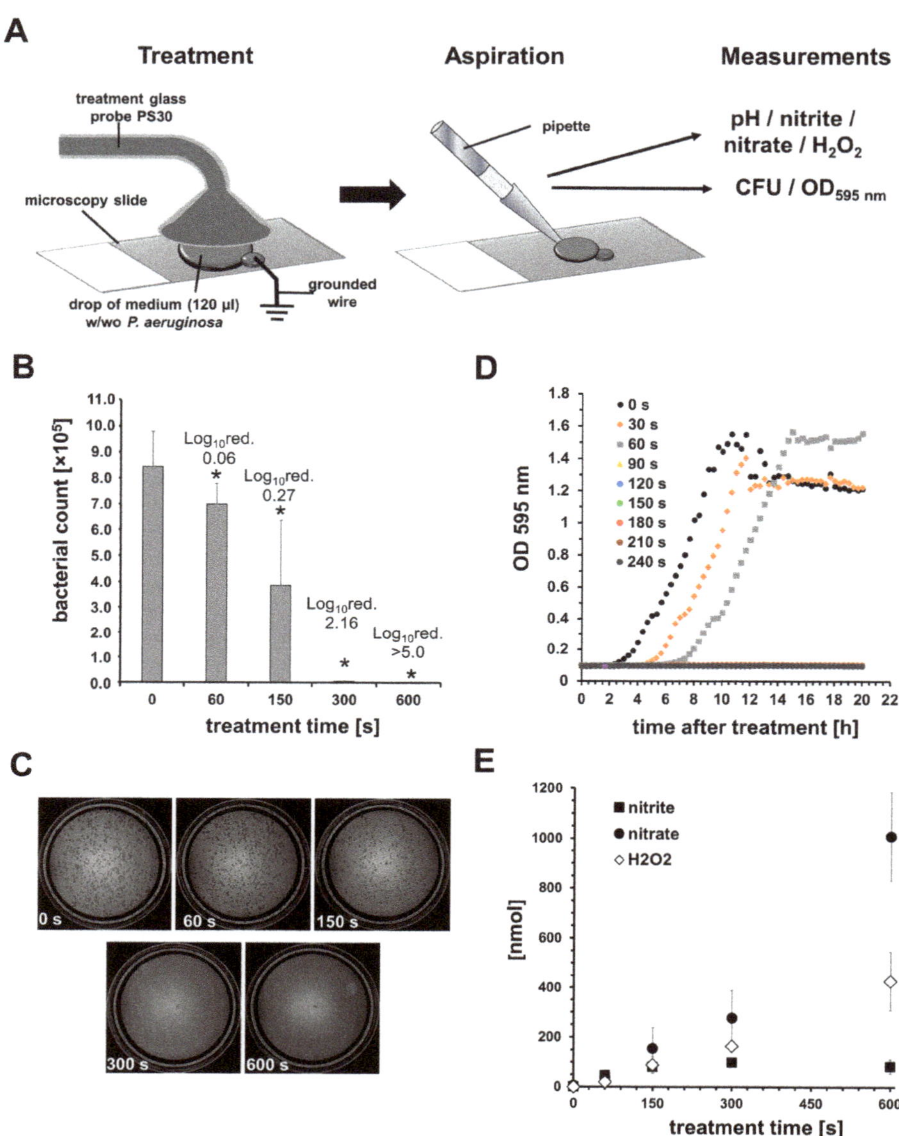

Figure 2. CAP treatment can reduce bacterial load in small volumes of medium. (**A**) Overview of the experimental procedure. (**B**) Given are the mean $\pm$ SD values of the number of *Pseudomanas aeruginosa* (PA) found after CAP treatments of small volumes of PA-containing medium ($1 \times 10^6 / 120$ µL). n = 3; * $p < 0.05$ compared to untreated controls. (**C**) Shown are representative photographs of the respective agar plates after CFU assays (10^4 dilution). (**D**) Exemplary measurements of the optical density at 595 nm of PA-containing media further cultivated after different CAP treatments (as indicated). (**E**) Amounts of nitrite, nitrate, and hydrogen peroxide (H_2O_2) found in small volumes of medium after CAP treatment (n = 3).

2.5.3. Agar Inhibition Zone Assays

On TSA plates (diameter: 10 cm), 100 µL TSB containing 1.5×10^7 CFU was spread and incubated for 20 min at 37 °C prior to CAP treatment. The CAP treatment glass probe was mounted on a stand and the agar plate was placed centrally on a laboratory lifting platform and lifted slowly up to probe tip until a distance of 1–2 mm between probe tip and agar surface. The agar was grounded by a wire and treated with CAP for 0, 30, 60, 150, 300, or 600 s. Thereafter, agar plates were incubated overnight at 37 °C before photos were taken. The diameter of the inhibition zones was determined with the help of the ImageJ® software (v. 1.53k) [47]. These experiments were performed independently three times.

2.5.4. Skin Wound Contamination Model

Human skin specimens from four patients aged between 27 and 49 (mean age 42) were obtained from abdominoplastic surgeries provided by the Clinic for Plastic Surgery, Hand Surgery, and Burn Center at the Cologne-Merheim Hospital. The use of human donor skin was approved by the ethics committee of the University of Witten/Ethics Herdecke's Committee (Votum No. 15/2018), and all experiments were conducted in compliance with the Declaration of Helsinki Principles. Patient consent was obtained for research purposes. The donor skin specimens were transported postoperatively in a sterile container on ice to the laboratory. Here, the skin was washed with sterile NaCl and incubated for one minute with 70% ethanol (Carl Roth, Karlsruhe, Germany) before further processing. In a previous study, we established a wound skin model [44], which was modified for this study.

Standardized skin samples were prepared, as shown in Figure 3A. First, wound areas 5 mm and 1 mm in depth were prepared using a biopsy punch (5 mm, Acuderm Inc., Fort Lauderdale, FL, USA) for a superficial cut of the epidermis, which was then carefully removed with a scalpel and forceps. Second, using a larger biopsy punch (12 mm; Acuderm Inc., Fort Lauderdale, FL, USA), round skin samples were punched around the 5 mm wounds. These round skin samples were placed onto sterile gauze pads (1.5×1.5 cm) in single culture plates (35 mm, Greiner, Frickenhausen, Germany) and 5.5 mL of cell culture medium (DMEM w/o phenol red; 1.0 g/L glucose, with 10% fetal calf serum, PAN Biotech, Aidenbach, Germany) was added to each plate. Finally, defined volumes of 3.3 µL TSB containing PA (50 CFU) were applied onto each wound area and the skin/wound samples were incubated for 6 h (37 °C, 5% CO_2) prior to CAP treatment, which were performed in duplicates. The CAP treatment glass probe was mounted on a stand and the culture plate with the skin sample was placed on a laboratory lifting platform with a distance of 1–2 mm between probe tip and skin surface, whereas the wound was in the central area of the probe tip. The skin was grounded by a wire, which was immersed in the medium, and treated with CAP for 0, 30, 60, 150, 300, or 600 s.

Subsequently, the skin samples were incubated for 16 h at 37 °C. After incubation, the duplicates were pooled and cut into small pieces, transferred into centrifugation tubes (50 mL, Greiner, Frickenhausen, Germany), and enzymatically digested in 5 mL of 0.25% trypsin/HBSS solution (PAN Biotech, Aidenbach, Germany) in an incubator on a shaker (CO_2-resistant, 3 mm Orbit, Thermo Fisher Scientific, Waltham, MA, USA) for 30 min at 150 rpm. After a centrifugation of 3 s, 100 µL of the solution was plated on agar plates to determine the bacterial survival rate (CFU/mL). These experiments were performed four times independently with different skin specimens. Before cutting and enzymatic digestion, we used a commercially available LED black light torch (395 nm, Bestsun, Jiaxing, Zhejiang, China) to capture exemplary photos of the bacterial biofilm fluorescence on the skin/wound samples under black light illumination in a dark room. Furthermore, we measured the biofilm fluorescence using a multiplate reader (VictorNivo, PerkinElmer, Waltham, MA, USA). Thereafter, the skin/wound samples with or without PA contamination were transferred to a 12-well cell culture plate (Greiner, Frickenhausen, Germany) without medium for fluorescence intensity scans by top measurement (ext. 405 nm; em. 530 nm; rows and columns 24×24 bi-directional; row/column range 12 mm; measurement time 100 ms; flash energy 10 mJ, PMT HV 500 V; excitation spot size 0.5 mm; emission spot size 1.0 mm).

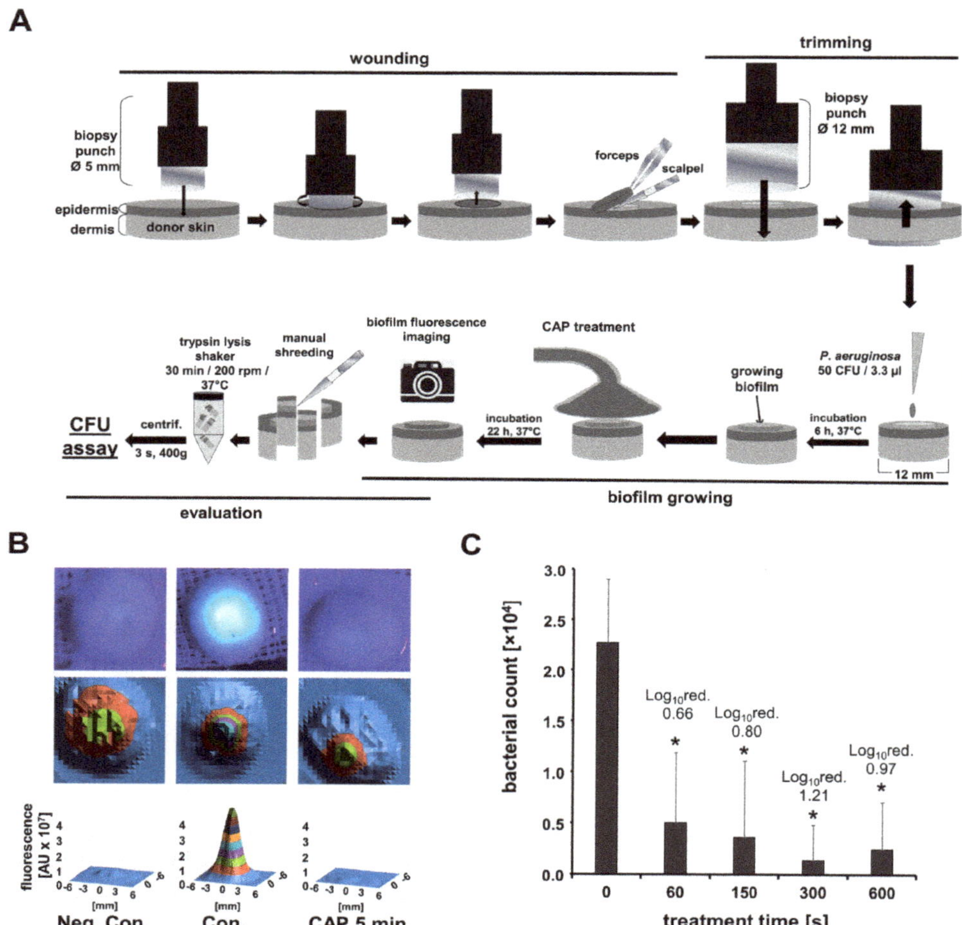

Figure 3. CAP treatment reduced bacterial burden in a skin/wound model. (**A**) The experimental procedure of the skin/wound model, CAP treatment, and evaluation of the bacterial burden is illustrated. (**B**) Exemplary photos of black light-induced fluorescence of contaminated (Neg. Con = uncontaminated) wounds in the skin/wound model after different treatments, as indicated. The fluorescence signal of the bacterial biofilm was evaluated further using a multiplate photometer. (**C**) Given are the means of the relative bacterial counts (to the respective untreated control) and the calculated Log_{10} reduction after different CAP treatments (0–10 min), assessed by CFU assay after trypsin lysis of the skin samples (n = 4; * $p < 0.05$).

2.6. Statistic

The software GraphPad Prism Version 8.4.3 (San Diego, CA, USA) was used for statistical analyses; significant differences with a *p*-value of <0.05 were evaluated using one-way ANOVA. Each experiment was independently performed 3–4 times.

3. Results

3.1. CAP Treatment Induced Physical/Chemical Changes

3.1.1. CAP Treatment Locally Increased the Nitrite and Nitrate Content of Agar

Our experiments showed that plasma treatment led to a local accumulation of nitrite and nitrate in the agar. The degree of enrichment depended on the localization under the treatment probe during CAP treatment. The accumulation in the area below the center of the probe was significantly higher ($\sim 10\times$) than in the lateral areas (see Figure 1). In the center, ~ 250 nmol of nitrate and ~ 85 nmol of nitrite were found in an area of 0.385 cm^2 after a 10 min CAP treatment. With respect to the volume of the agar, we estimate that the concentrations of nitrite obtained in this spot were around 1.75 mM, and of nitrate, around 5 mM.

3.1.2. CAP Treatment Increased the Nitrite, Nitrate, and H$_2$O$_2$ Contents in Medium

The exposure of liquid—in our case, TSB—to CAP led a linear increase in nitrate and H$_2$O$_2$ concentrations, as shown in Figure 2E. In total, around 1000 nmol of nitrate was found in 120 µL of TSB after a 10 min CAP treatment, which equals a concentration of ~ 8.3 mM. The amount of accumulated H$_2$O$_2$ was around 450 nmol, which is equal to a concentration of ~ 3.5 mM. The amount of nitrite did not increase linearly. Here, after CAP treatment for 10 min, the nitrite amount/concentration of ~ 83 nmol/690 µM was lower than after a 5 min CAP treatment, which achieved ~ 98 nmol/816 µM.

3.1.3. CAP Treatment Induced Acidification

The surface pH value 7.0 in the middle of a TSA plate was decreased by a 5 min CAP treatment down to 4.9 and after 10 min to 4.4. When dealing with small volumes of TSB (120 µL) spread on a microscopy slide, a pH value of 5.6 and 4.7 could be found after CAP treatment of 5 min and 10 min, respectively. The surface pH value of intact human skin was, in our experiments, around 5. After a single CAP treatment longer than 150 s, the skin surface was significantly acidified, achieving pH values between 1 and 2 (Figure S1).

3.2. CAP Induced Antibacterial Effects

With regard to a possible antibacterial effect of the CAP treatment, our experiments showed that this depended on the treatment time and on the type of test system. When a PA-contaminated agar plate was CAP-treated, a clear inhibition zone could already be seen after 30 s of CAP exposure, which became larger with increasing treatment time. After 10 min of CAP treatment, the inhibition zone extended across the entire area covered by the treatment probe (Figure 1C,F).

When treating small volumes in a flat volume of bacterial suspension with CAP, a relevant bactericidal effect of CAP was only detected after 300 s of treatment time. An almost complete elimination of the bacterial load in a small and flat volume of liquid could be observed in our experiments after a 10 min CAP treatment (Figure 2B,C). In addition, an inhibition of bacterial growth was observed after a treatment time of 90–120 s, when PA had been further cultivated in the treatment medium after CAP treatment (Figure 2D). In preliminary tests with larger and higher volumes, we could not detect any significant antibacterial efficacy even after longer treatment times.

A reduction in the number of bacteria by the CAP treatment was observed in the more realistic skin wound contamination model as well. Here, a 1 min treatment resulted in about a 65%, and a 2.5 min treatment resulted in an 80% lower bacterial load compared to the untreated control. Only after 5 and 10 min of treatment, a reduction in the bacterial count of over 90% could be achieved. Here, the 10 min treatment was not more effective than the 5 min treatment. The reduction in the bacterial load was also associated with a lower bacteria/biofilm fluorescence signal in the wound area one day after CAP treatment.

4. Discussion

The Gram-negative bacterium *Pseudomonas aeruginosa* is one of the most commonly isolated organisms from infected burn wounds [14,15]. The ability of PA to maintain persistent infections through biofilm formation can result in significant delays in the healing of burn wounds. The main goal of treating infected wounds in burn patients is to reduce the bacterial load and prevent complications like sepsis and death and allow the skin to heal without further complications [16,18]. Some CAP systems showed antibacterial activity, albeit moderate, against a wide range of pathogens [48–50].

Therefore, in the present study, we investigated the antimicrobial efficacy of CAP generated by the PlasmaOne device, which showed promising antibacterial effects on *Staphylococcus aureus* and also PA in in vitro studies, and could be an alternative option for the treatment of (burn) wound infections caused by PA [51,52]. However, it was observed that PA was more resistant against CAP than *Staphylococcus aureus*, when contaminated solid surfaces (agar plates) were treated [52].

In various test systems, we were also able to demonstrate an antibacterial effect of plasma treatment with this CAP source. Depending on the test system, the antibacterial efficacy varied significantly. With a low volume of liquid and a flat surface, such as a thin film of liquid on an agar plate, even short treatment times were sufficient to achieve a bacteria-free zone. If the liquid volume was somewhat larger and higher, as in the suspension experiment with a spread droplet containing PA, longer treatment times were necessary to achieve a significant reduction in the bacterial load. With the larger liquid volumes (1 mL/6-well cell culture dish) used in preliminary experiments, we could not detect a significant reduction in the number of bacteria after CAP treatment. In the skin wound model, the liquid volume was low and flat, but the wound surface was uneven. In addition, there was biofilm formation, which can protect the bacteria against CAP. Additionally, under these conditions, longer CAP treatments up to 10 min achieved only a 90% reduction. In comparison, the 15 min application of antibacterial Polyhexanide-containing wound irrigation solutions (Pronotosan) achieved a bacterial reduction of over 98% in a similar assay (data not published).

Hence, the use of compresses fully saturated with antibacterial wound irrigation solutions appears to be a much easier, faster, and also more effective way for the prophylaxis and treatment of wound infection in everyday clinical practice than plasma treatment. In particular, for more extensive burn wounds, CAP treatment would be very time-consuming. Another shortcoming of plasma treatment is the lack of homogeneity of the plasma distribution over a wide area. Our experiments showed that on smooth surfaces, the bioactive plasma was more effective in the center under the treatment electrode, where it causes an accumulation of nitrite and nitrate and reduced bacterial growth after a short treatment, while under the margin area of the electrode a longer CAP treatment (5–10 min) was required, probably mediated indirectly by diffusion of CAP-treated medium. On the one hand, in the case of uneven surfaces—such as wound edges—it can be observed that direct plasma preferentially ignites at the higher points, thereby reaching the intact skin at the wound edge rather than the wound surface itself. On the other hand, treatment of the wound margin could ensure that skin bacteria do not migrate from the skin area into the wound. Here, the accumulation of nitrite and nitrate as well as a strong decrease in the skin pH value at the wound edge may lead to a chemical barrier for many pathogenic bacteria.

In addition to the antibacterial effects, other possible wound healing effects of plasma treatment should not be underestimated. Via acidification of the wound area and enrichment with nitrite/nitrate as described above, it is known that some types of CAPs produced by dermal barrier discharge (DBD) can induce a local increase in microcirculation [53,54].

Furthermore, it is conceivable that many radicals produced by CAP may positively influence the cell physiology of fibroblasts and keratinocytes at lower doses, as observed in some studies using other types of CAPs [46,55].

However, the dose makes the poison, and, in many studies, toxic effects on cells induced by CAP, particularly at higher doses, could be observed [42,56,57].

5. Conclusions

Our results indicate that CAP therapy alone, using the PlasmaOne device, is probably not sufficient for the treatment of PA-contaminated burn wounds. In comparison with antibacterial wound irrigation solutions, the antibacterial efficacy that was observed was significantly lower; hence, CAP treatment of larger PA-infected burns would be very time-consuming. Nevertheless, additional CAP therapy with PlasmaOne could potentially provide benefits. In particular, apart from the better antibacterial efficacy against other common wound pathogens compared to PA [52], CAP-induced effects, such as the observed acidification of the wound area, could have a positive impact on wound healing [58]. Furthermore, possible improvement of local microcirculation and stimulation of skin cells—as observed with other CAP devices [53,54,59]—may have beneficial effects on the healing process. However, further clinical and experimental studies are necessary to verify this assumption.

Supplementary Materials: The following supporting information can be downloaded at: https://www.mdpi.com/article/10.3390/biomedicines11051239/s1, Figure S1: CAP-induced acidification

Author Contributions: Conceptualization, C.O., A.Z. and M.B.; methodology, M.B., A.Z. and C.O.; validation, C.O., P.C.F. and J.L.S.; formal analysis, A.Z., M.B. and C.O.; investigation, M.B., M.v.K., A.Z. and C.O.; resources, C.O., P.C.F. and J.L.S.; data curation, M.B. and C.O.; writing—original draft preparation, M.B. and C.O.; writing—review and editing, J.L.S. and P.C.F.; visualization, M.B., M.v.K., A.Z. and C.O.; supervision, C.O. and J.L.S.; project administration, C.O. and M.B.; funding acquisition, C.O. and M.B. All authors have read and agreed to the published version of the manuscript.

Funding: This research received funding from the internal grant program (project IFF 2022-12) of the Faculty of Health at Witten/Herdecke University, Germany, and from the German Research Foundation DFG (OP 207/11-1).

Institutional Review Board Statement: The study was conducted in accordance with the Declaration of Helsinki and approved by the Ethics Committee of Witten/Herdecke University (ID 15/2018).

Informed Consent Statement: Informed consent was obtained from all subjects involved in the study.

Data Availability Statement: The data that support the findings of this study are available from the corresponding author upon reasonable request.

Acknowledgments: We thank Isabell Blassnig and Ayperi Eyüboglu for technical assistance.

Conflicts of Interest: The authors declare no conflict of interest.

References

1. Hänel, K.H.; Cornelissen, C.; Lüscher, B.; Baron, J.M. Cytokines and the skin barrier. *Int. J. Mol. Sci.* **2013**, *14*, 6720–6745. [CrossRef]
2. Grice, E.A.; Segre, J.A. The skin microbiome. *Nat. Rev. Microbiol.* **2011**, *9*, 244–253. [CrossRef]
3. Evers, L.H.; Bhavsar, D.; Mailänder, P. The biology of burn injury. *Exp. Dermatol.* **2010**, *19*, 777–783. [CrossRef]
4. Gacto-Sanchez, P. Surgical treatment and management of the severely burn patient: Review and update. *Med. Intensiv.* **2017**, *41*, 356–364. [CrossRef] [PubMed]
5. Appelgren, P.; Björnhagen, V.; Bragderyd, K.; Jonsson, C.E.; Ransjö, U. A Prospective Study of Infections in Burn Patients. *Burn. J. Int. Soc. Burn. Inj.* **2002**, *28*, 39–46. [CrossRef] [PubMed]
6. Dodd, D.; Stutman, H.R. Current Issues in Burn Wound Infections. *Adv. Pediatr. Infect. Dis.* **1991**, *6*, 137–162.
7. Griswold, J.A. White blood cell response to burn injury. *Semin. Nephrol.* **1993**, *13*, 409–415.
8. Hansbrough, J.F.; Field, T.O., Jr.; Gadd, M.A.; Soderberg, C. Immune response modulation after burn injury: T cells and antibodies. *J. Burn. Care Rehabil.* **1987**, *8*, 509–512. [CrossRef] [PubMed]
9. Heideman, M.; Bengtsson, A. The immunologic response to thermal injury. *World J. Surg.* **1992**, *16*, 53–56. [CrossRef] [PubMed]
10. Erol, S.; Altoparlak, U.; Akcay, M.N.; Celebi, F.; Parlak, M. Changes of microbial flora and wound colonization in burned patients. *Burn. J. Int. Soc. Burn. Inj.* **2004**, *30*, 357–361. [CrossRef]
11. Tiwari, V.K. Burn wound: How it differs from other wounds? *Indian J. Plast. Surg.* **2012**, *45*, 364–373. [CrossRef]
12. Barret, J.P.; Herndon, D.N. Effects of burn wound excision on bacterial colonization and invasion. *Plast. Reconstr. Surg.* **2003**, *111*, 744–750; discussion 742–751. [CrossRef]
13. Geyik, M.F.; Aldemir, M.; Hosoglu, S.; Tacyildiz, H.I. Epidemiology of burn unit infections in children. *Am. J. Infect. Control* **2003**, *31*, 342–346. [CrossRef] [PubMed]

14. Rennie, R.P.; Jones, R.N.; Mutnick, A.H. Occurrence and antimicrobial susceptibility patterns of pathogens isolated from skin and soft tissue infections: Report from the SENTRY Antimicrobial Surveillance Program (United States and Canada, 2000). *Diagn. Microbiol. Infect. Dis.* **2003**, *45*, 287–293. [CrossRef]

15. Revathi, G.; Puri, J.; Jain, B.K. Bacteriology of burns. *Burn. J. Int. Soc. Burn. Inj.* **1998**, *24*, 347–349. [CrossRef] [PubMed]

16. Neely, C.J.; Kartchner, L.B.; Mendoza, A.E.; Linz, B.M.; Frelinger, J.A.; Wolfgang, M.C.; Maile, R.; Cairns, B.A. Flagellin treatment prevents increased susceptibility to systemic bacterial infection after injury by inhibiting anti-inflammatory IL-10+ IL-12- neutrophil polarization. *PLoS ONE* **2014**, *9*, e85623. [CrossRef]

17. Yildirim, S.; Nursal, T.Z.; Tarim, A.; Torer, N.; Noyan, T.; Demiroglu, Y.Z.; Moray, G.; Haberal, M. Bacteriological profile and antibiotic resistance: Comparison of findings in a burn intensive care unit, other intensive care units, and the hospital services unit of a single center. *J. Burn Care Rehabil.* **2005**, *26*, 488–492. [CrossRef] [PubMed]

18. Ugburo, A.O.; Atoyebi, O.A.; Oyeneyin, J.O.; Sowemimo, G.O. An evaluation of the role of systemic antibiotic prophylaxis in the control of burn wound infection at the Lagos University Teaching Hospital. *Burns* **2004**, *30*, 43–48. [CrossRef]

19. Tredget, E.E.; Shankowsky, H.A.; Rennie, R.; Burrell, R.E.; Logsetty, S. Pseudomonas infections in the thermally injured patient. *Burn. J. Int. Soc. Burn. Inj.* **2004**, *30*, 3–26. [CrossRef]

20. Hof, H.; Dörries, R. *Medizinische Mikrobiologie*, 6th ed.; Thieme: Stuttgart, Germany, 2017.

21. Van Delden, C.; Iglewski, B.H. Cell-to-cell signaling and Pseudomonas aeruginosa infections. *Emerg. Infect. Dis.* **1998**, *4*, 551–560. [CrossRef]

22. Fuchs, G. *Allgemeine Mikrobiologie*, 10th ed.; Thieme: Stuttgart, Germany, 2017.

23. Pang, Z.; Raudonis, R.; Glick, B.R.; Lin, T.-J.; Cheng, Z. Antibiotic resistance in Pseudomonas aeruginosa: Mechanisms and alternative therapeutic strategies. *Biotechnol. Adv.* **2019**, *37*, 177–192. [CrossRef]

24. Breidenstein, E.B.M.; de la Fuente-Núñez, C.; Hancock, R.E.W. Pseudomonas aeruginosa: All roads lead to resistance. *Trends Microbiol.* **2011**, *19*, 419–426.

25. Ruppé, É.; Woerther, P.-L.; Barbier, F. Mechanisms of antimicrobial resistance in Gram-negative bacilli. *Ann. Intensive Care* **2015**, *5*, 61. [CrossRef] [PubMed]

26. Gortner, L.; Meyer, S. *Duale Reihe: Pädiatrie*, 5th ed.; Thieme: Stuttgart, Germany, 2018.

27. Afshari, A.; Nguyen, L.; Kahn, S.A.; Summitt, B. 2.5% Mafenide Acetate: A Cost-Effective Alternative to the 5% Solution for Burn Wounds. *J. Burn. Care Res.* **2017**, *38*, e42–e47. [CrossRef] [PubMed]

28. Nagoba, B.S.; Gandhi, R.C.; Wadher, B.J.; Deshmukh, S.R.; Gandhi, S.P. Citric acid treatment of severe electric burns complicated by multiple antibiotic resistant Pseudomonas aeruginosa. *Burns* **1998**, *24*, 481–483. [CrossRef]

29. Abbaspour, M.; Sharif Makhmalzadeh, B.; Rezaee, B.; Shoja, S.; Ahangari, Z. Evaluation of the Antimicrobial Effect of Chitosan/Polyvinyl Alcohol Electrospun Nanofibers Containing Mafenide Acetate. *Jundishapur J. Microbiol.* **2015**, *8*, e24239. [CrossRef] [PubMed]

30. Church, D.; Elsayed, S.; Reid, O.; Winston, B.; Lindsay, R.o. Burn wound infections. *Clin. Microbiol. Rev.* **2006**, *19*, 403–434. [CrossRef]

31. Zahmatkesh, M.; Manesh, M.J.; Babashahabi, R. Effect of Olea ointment and Acetate Mafenide on burn wounds—A randomized clinical trial. *Iran. J. Nurs. Midwifery Res.* **2015**, *20*, 599–603. [CrossRef] [PubMed]

32. Nagoba, B.S.; Gandhi, R.C.; Hartalkar, A.R.; Wadher, B.J.; Selkar, S.P. Simple, effective and affordable approach for the treatment of burns infections. *Burn. J. Int. Soc. Burn. Inj.* **2010**, *36*, 1242–1247. [CrossRef]

33. Lachiewicz, A.M.; Hauck, C.G.; Weber, D.J.; Cairns, B.A.; van Duin, D. Bacterial Infections After Burn Injuries: Impact of Multidrug Resistance. *Clin. Infect. Dis.* **2017**, *65*, 2130–2136. [CrossRef]

34. Gonzalez, M.R.; Fleuchot, B.; Lauciello, L.; Jafari, P.; Applegate, L.A.; Raffoul, W.; Que, Y.A.; Perron, K. Effect of Human Burn Wound Exudate on Pseudomonas aeruginosa Virulence. *mSphere* **2016**, *1*, e00111-15. [CrossRef] [PubMed]

35. Heinlin, J.; Morfill, G.; Landthaler, M.; Stolz, W.; Isbary, G.; Zimmermann, J.L.; Shimizu, T.; Karrer, S. Plasma medicine: Possible applications in dermatology. *J. Dtsch. Dermatol. Ges.* **2010**, *8*, 968–976. [CrossRef] [PubMed]

36. Helmke, A. Niedertemperaturplasma: Eigenschaften, Wirkungen und Gerätetechnik. In *Plasmamedizin*; von Woedtke, T., Metelmann, H.-R., Weltmann, K.-D., Eds.; Springer: Berlin/Heidelberg, Germany, 2016; pp. 33–43. [CrossRef]

37. Isbary, G.; Shimizu, T.; Li, Y.F.; Stolz, W.; Thomas, H.M.; Morfill, G.E.; Zimmermann, J.L. Cold atmospheric plasma devices for medical issues. *Expert Rev. Med. Devices* **2013**, *10*, 367–377. [CrossRef]

38. Heinlin, J.; Isbary, G.; Stolz, W.; Morfill, G.; Landthaler, M.; Shimizu, T.; Steffes, B.; Nosenko, T.; Zimmermann, J.; Karrer, S. Plasma applications in medicine with a special focus on dermatology. *J. Eur. Acad. Dermatol. Venereol.* **2011**, *25*, 1–11. [CrossRef]

39. Eggers, B.; Stope, M.B.; Mustea, A.; Nokhbehsaim, M.; Heim, N.; Kramer, F.-J. Non-Invasive Physical Plasma Treatment after Tooth Extraction in a Patient on Antiresorptive Medication Promotes Tissue Regeneration. *Appl. Sci.* **2022**, *12*, 3490. [CrossRef]

40. Werra, U.E.M.; Dorweiler, B. Kaltplasmatherapie in der Wundbehandlung—Was wissen wir? *Gefässchirurgie* **2022**, *28*, 7–14. [CrossRef]

41. Gupta, T.T.; Ayan, H. Application of Non-Thermal Plasma on Biofilm: A Review. *Appl. Sci.* **2019**, *9*, 3548. [CrossRef]

42. Plattfaut, I.; Besser, M.; Severing, A.L.; Sturmer, E.K.; Oplander, C. Plasma medicine and wound management: Evaluation of the antibacterial efficacy of a medically certified cold atmospheric argon plasma jet. *Int. J. Antimicrob. Agents* **2021**, *57*, 106319. [CrossRef]

43. Liu, D.; Zhang, Y.; Xu, M.; Chen, H.; Lu, X.; Ostrikov, K. Cold atmospheric pressure plasmas in dermatology: Sources, reactive agents, and therapeutic effects. *Plasma Process. Polym.* **2020**, *17*, 1900218. [CrossRef]
44. Leder, M.D.; Bagheri, M.; Plattfaut, I.; Fuchs, P.C.; Bruning, A.K.E.; Schiefer, J.L.; Oplander, C. Phototherapy of Pseudomonas aeruginosa-Infected Wounds: Preclinical Evaluation of Antimicrobial Blue Light (450–460 nm) Using In Vitro Assays and a Human Wound Skin Model. *Photobiomodul. Photomed. Laser Surg.* **2022**, *40*, 800–809. [CrossRef]
45. Küçük, D.; Savran, L.; Ercan, U.K.; Yarali, Z.B.; Karaman, O.; Kantarci, A.; Sağlam, M.; Köseoğlu, S. Evaluation of efficacy of non-thermal atmospheric pressure plasma in treatment of periodontitis: A randomized controlled clinical trial. *Clin. Oral. Investig.* **2020**, *24*, 3133–3145. [CrossRef]
46. Balzer, J.; Heuer, K.; Demir, E.; Hoffmanns, M.A.; Baldus, S.; Fuchs, P.C.; Awakowicz, P.; Suschek, C.V.; Oplander, C. Non-Thermal Dielectric Barrier Discharge (DBD) Effects on Proliferation and Differentiation of Human Fibroblasts Are Primary Mediated by Hydrogen Peroxide. *PLoS ONE* **2015**, *10*, e0144968. [CrossRef]
47. Schneider, C.A.; Rasband, W.S.; Eliceiri, K.W. NIH Image to ImageJ: 25 years of image analysis. *Nat. Methods* **2012**, *9*, 671–675. [CrossRef]
48. Brehmer, F.; Haenssle, H.A.; Daeschlein, G.; Ahmed, R.; Pfeiffer, S.; Gorlitz, A.; Simon, D.; Schon, M.P.; Wandke, D.; Emmert, S. Alleviation of chronic venous leg ulcers with a hand-held dielectric barrier discharge plasma generator (PlasmaDerm((R)) VU-2010): Results of a monocentric, two-armed, open, prospective, randomized and controlled trial (NCT01415622). *J. Eur. Acad. Dermatol. Venereol.* **2015**, *29*, 148–155. [CrossRef]
49. Isbary, G.; Heinlin, J.; Shimizu, T.; Zimmermann, J.L.; Morfill, G.; Schmidt, H.U.; Monetti, R.; Steffes, B.; Bunk, W.; Li, Y.; et al. Successful and safe use of 2 min cold atmospheric argon plasma in chronic wounds: Results of a randomized controlled trial. *Br. J. Dermatol.* **2012**, *167*, 404–410. [CrossRef] [PubMed]
50. Stratmann, B.; Costea, T.C.; Nolte, C.; Hiller, J.; Schmidt, J.; Reindel, J.; Masur, K.; Motz, W.; Timm, J.; Kerner, W.; et al. Effect of Cold Atmospheric Plasma Therapy vs. Standard Therapy Placebo on Wound Healing in Patients With Diabetic Foot Ulcers: A Randomized Clinical Trial. *JAMA Netw. Open* **2020**, *3*, e2010411. [CrossRef]
51. Ulu, M.; Pekbagriyanik, T.; Ibis, F.; Enhos, S.; Ercan, U.K. Antibiofilm efficacies of cold plasma and er: YAG laser on Staphylococcus aureus biofilm on titanium for nonsurgical treatment of peri-implantitis. *Niger. J. Clin. Pract.* **2018**, *21*, 758–765. [CrossRef] [PubMed]
52. Zashev, M.; Donchev, D.; Ivanov, I.; Gornev, R. Efficacy of Plasma ONE apparatus for disinfection of *S. aureus*, *P. aeruginosa* and *E. coli* bacteria from the solid surface. *J. Theor. Appl. Phys.* **2020**, *14*, 41–49. [CrossRef]
53. Heuer, K.; Hoffmanns, M.A.; Demir, E.; Baldus, S.; Volkmar, C.M.; Rohle, M.; Fuchs, P.C.; Awakowicz, P.; Suschek, C.V.; Oplander, C. The topical use of non-thermal dielectric barrier discharge (DBD): Nitric oxide related effects on human skin. *Nitric Oxide* **2015**, *44*, 52–60. [CrossRef]
54. Kisch, T.; Helmke, A.; Schleusser, S.; Song, J.; Liodaki, E.; Stang, F.H.; Mailaender, P.; Kraemer, R. Improvement of cutaneous microcirculation by cold atmospheric plasma (CAP): Results of a controlled, prospective cohort study. *Microvasc. Res.* **2016**, *104*, 55–62. [CrossRef]
55. Arndt, S.; Unger, P.; Berneburg, M.; Bosserhoff, A.K.; Karrer, S. Cold atmospheric plasma (CAP) activates angiogenesis-related molecules in skin keratinocytes, fibroblasts and endothelial cells and improves wound angiogenesis in an autocrine and paracrine mode. *J. Dermatol. Sci.* **2018**, *89*, 181–190. [CrossRef] [PubMed]
56. Dezest, M.; Chavatte, L.; Bourdens, M.; Quinton, D.; Camus, M.; Garrigues, L.; Descargues, P.; Arbault, S.; Burlet-Schiltz, O.; Casteilla, L.; et al. Mechanistic insights into the impact of Cold Atmospheric Pressure Plasma on human epithelial cell lines. *Sci. Rep.* **2017**, *7*, 41163. [CrossRef]
57. Duval, A.; Marinov, I.; Bousquet, G.; Gapihan, G.; Starikovskaia, S.M.; Rousseau, A.; Janin, A. Cell death induced on cell cultures and nude mouse skin by non-thermal, nanosecond-pulsed generated plasma. *PLoS ONE* **2013**, *8*, e83001. [CrossRef] [PubMed]
58. Sim, P.; Strudwick, X.L.; Song, Y.; Cowin, A.J.; Garg, S. Influence of Acidic pH on Wound Healing In Vivo: A Novel Perspective for Wound Treatment. *Int. J. Mol. Sci.* **2022**, *23*, 3655. [CrossRef] [PubMed]
59. Marches, A.; Clement, E.; Alberola, G.; Rols, M.P.; Cousty, S.; Simon, M.; Merbahi, N. Cold Atmospheric Plasma Jet Treatment Improves Human Keratinocyte Migration and Wound Closure Capacity without Causing Cellular Oxidative Stress. *Int. J. Mol. Sci.* **2022**, *23*, 650. [CrossRef] [PubMed]

biomedicines

MDPI

Article

Effects of Cold Atmospheric Plasma Pre-Treatment of Titanium on the Biological Activity of Primary Human Gingival Fibroblasts

Madline P. Gund [1,*], Jusef Naim [1], Antje Lehmann [2,3], Matthias Hannig [1], Constanze Linsenmann [1], Axel Schindler [2,4] and Stefan Rupf [1,5]

[1] Clinic of Operative Dentistry, Periodontology and Preventive Dentistry, Saarland University, 66421 Homburg, Germany
[2] Leibniz Institute of Surface Modification (IOM), 04318 Leipzig, Germany
[3] ADMEDES GmbH, 75179 Pforzheim, Germany
[4] Piloto Consulting Ion Beam and Plasma Technologies, 04668 Grimma, Germany
[5] Synoptic Dentistry, Saarland University, 66421 Homburg, Germany
* Correspondence: madline.gund@uks.eu; Tel.: +49-6841-1624998

Abstract: Cold atmospheric plasma treatment (CAP) enables the contactless modification of titanium. This study aimed to investigate the attachment of primary human gingival fibroblasts on titanium. Machined and microstructured titanium discs were exposed to cold atmospheric plasma, followed by the application of primary human gingival fibroblasts onto the disc. The fibroblast cultures were analyzed by fluorescence, scanning electron microscopy and cell-biological tests. The treated titanium displayed a more homogeneous and denser fibroblast coverage, while its biological behavior was not altered. This study demonstrated for the first time the beneficial effect of CAP treatment on the initial attachment of primary human gingival fibroblasts on titanium. The results support the application of CAP in the context of pre-implantation conditioning, as well as of peri-implant disease treatment.

Keywords: biological cell activity; cell attachment; cold atmospheric plasma; primary human gingival fibroblasts

Citation: Gund, M.P.; Naim, J.; Lehmann, A.; Hannig, M.; Linsenmann, C.; Schindler, A.; Rupf, S. Effects of Cold Atmospheric Plasma Pre-Treatment of Titanium on the Biological Activity of Primary Human Gingival Fibroblasts. *Biomedicines* **2023**, *11*, 1185. https://doi.org/10.3390/biomedicines11041185

Academic Editor: Christoph Viktor Suschek

Received: 1 March 2023
Revised: 6 April 2023
Accepted: 7 April 2023
Published: 16 April 2023

1. Introduction

Titanium provides excellent biocompatibility, a high flexural strength, and corrosion resistance. Therefore, it is widely used in medicine, particularly in the field of implantology, as well as for osteosynthesis plates, or bone screws in all fields of surgery. In dentistry, titanium is the most used implant material of choice for replacement of missing teeth [1]. Various factors, such as wettability or hydrophilicity, influence cell behavior and attachment [2,3]. Technical methods such as sandblasting, etching and coating are used to modify surface properties, aiming to improve the adhesion of the surrounding tissue to the titanium implant and promoting an early adhesion of osteoblasts and osteosynthesis [4–6]. Furthermore, a dense attachment of gingival fibroblasts and epithelia cells is essential for successful implant therapy [7]. The peri-implant tissue establishes a barrier and protects against colonization by bacteria of the oral environment [8]. Thus, colonization of the implant surface with microorganisms and inflammation of peri-implant bone can be prevented [9]. Nevertheless, the excellent biocompatibility of titanium also allows the attachment of microorganisms and therefore the adsorption of biofilms on the implant surface [10]. Microbial biofilms are playing the decisive role in the development of peri-implant diseases such as perimucositis and peri-implantitis [11]. Again, fostering human cell adhesion on the implant surface is essential. However, different cell species have a variable interaction with the titanium surface. Fibroblasts and epithelia cells show the best adhesion on machined surfaces, while osteoblasts favor microstructured surfaces [12,13].

Cold atmospheric plasma (CAP) can be used for the removal of biofilms [14–16]. CAP also enables a contactless surface modification, in particular, fissures and cavities, leading to an improved wettability, also on dental implant materials [17,18]. There are already studies available that investigated cell behavior and adhesion patterns of human gingival fibroblasts (HGF-1) and osteoblast-like cells (MG-63) on coated and uncoated titanium and on zirconia discs after cold plasma treatment, pointing out the faster proliferation and adhesion of cells from surrounding tissue [19].

The aim of this study was to investigate the attachment and the biological activity of primary human gingival fibroblasts on machined and microstructured titanium after CAP modification.

2. Materials and Methods

Titanium discs: A total of 120 titanium discs (titanium grade 2, Friadent, Mannheim, Germany, diameter 5 mm or 10 mm, height 1 or 2 mm) with machined and microstructured (sand-blasted, air-abraded, roughness of 2 μm, mean maximum value of the tread depth of 21.35 μm) surfaces were used for test specimen. A total of 84 titanium discs were treated, with 36 samples being treated with the gas flow without ignition of the plasma. Titanium discs with a diameter of 5 mm were utilized for the fluorescence microscopic evaluation and with a diameter of 10 mm for scanning electron microscopic examination.

Plasma Source: A plasma source of the Leibniz Institute for surface modification was used for the plasma treatment, with the plasma jet being generated by a pulsed microwave (2.45 GHz). The plasma treatment parameters were adjustable in terms of pulse power, pulse width, and mean microwave power. The plasma source was mounted on a computer-controlled 3-axis movement system (Steinmeyer MC-G047, Feinmess Dresden GmbH, Dresden, Germany). The chosen processing parameters—5 μs pulse width and 300 W peak power—resulted in a mean power of 5 W and a scan velocity of 8 mm/s. The surface was processed by means of meander-like line scans through the plasma jet. The working distance was 2 mm, and the gas flow was 2 L/min, made up of two different gas mixtures. The noble gas Helium was chosen as carrier gas. The gas mixture contained 2000 sccm helium and 5 sccm oxygen.

Primary human gingival fibroblasts: Primary human gingival fibroblasts were obtained during osteotomy of third molars in the Department of Oral and Maxillofacial Surgery, Saarland University Hospital, Homburg. The subsequent cultivation took place under constant conditions (37 °C, 5% CO_2, 2 times per week change of medium), cells of the sixth passage were selected for the series of experiments. The use of the cells was approved by the Ethics Committee of the Saarland Medical Association (E168/09), and the donors gave their written consent.

Sample treatment: The samples were treated under stable conditions in an S2-laboratory (air temperature: 22 ± 2 °C, eightfold air change per hour, light overpressure). The application of gingival fibroblasts was performed immediately after plasma treatment in a well plate under an aseptic bench. The 10^4 cells were applied on treated samples and the non-treated controls. The titanium samples were constantly kept moist over a period of 30 min with DMEM liquid medium before the cavities were filled with 200 μL DMEM liquid medium. Afterwards, cultivation was performed in an incubator (Hera Cell 240, Thermo Scientific, Waltham, MA, USA) at 37 °C and 5% CO_2 for 4, 12 and 24 h.

Sample analysis: The analysis of the attachment of gingival fibroblasts was performed by fluorescence microscopy. Vinculin, Phalloidin and DAPI staining were used. The fixation of cells on specimens was performed using 2.5% glutaraldehyde (Carl Roth GmbH, Karlsruhe, Germany) for 30 min. To remove the glutaraldehyde, the specimens were rinsed five times with PBS, followed by permeabilization of the cell wall with 0.1% Triton-X (Roche, Grenzach-Wyhlen, Germany) in PBS. Thereafter, the samples were washed two times by PBS, the nonspecific background being reduced using 1 % BSA (bovine serum albumin in PBS, Gibco, Invitrogen, Carlsbad, CA, USA). Afterwards, the specimens were incubated in a humid chamber with a primary antibody against vinculin (Monoclonaler Anti-Vinculin, antibodies of mice; Sigma-Aldrich Chemie GmbH, Steinheim, Germany) being dissolved in 5% BSA, followed by a double lavation with 0.1% Tween 20 (Roche,

Grenzach-Wyhlen, Germany) in PBS and removal of the Tween 20 by PBS. Subsequently, the specimens were washed twice with 0.1% Tween 20 (Roche, Grenzach-Wyhlen, Germany) in PBS, followed by the removal of Tween 20 in PBS. After incubation, the specimens graduated with a secondary antibody (anti-mouse IgG, RD Systems, Minneapolis, MN, USA) for one hour in a humidified chamber; re-washing in 0.1% Tween 20 and a subsequent washing by PBS followed. Next, the incubation concluded with a phalloidin stock solution (1 mg/mL MetOH) for 25 min, in the dark, at room temperature. This was followed by the application of DAPI solution (Roche, Grenzach-Wyhlen, Germany) for ten minutes, at room temperature, in the dark.

The titanium disks were mounted on slides (Superfrost; Menzel-glasses, Braunschweig, Germany) and evaluated under a reflected-light microscope (Axio Scope A1, Carl Zeiss, Göttingen, Germany). Recordings were made with lenses of 2.5× magnification, 10× and 20×. The recordings at a magnification of 25 (2.5 (lens) × 10 (ocular)) were used for each sample to create a general overview, as well as to evaluate the percentage of cell cover. Five shots were customized at 100× magnification, each from the center of the sample and from the border area. These images were used for analyzing the number of cells, the attachment behavior and the average nucleus size. AxioVision 4.8. software (Carl Zeiss Microlmaging, Göttingen, Germany) was used to evaluate attachment, average nucleus size and number of cells. The scanning electron microscopy examination was performed with a FEI XL 30 ESEM FEG (Fei, Eindhoven, the Netherlands). A total of 36 treated titanium discs and 16 untreated controls were analyzed. These were fixed in 2.5% glutaraldehyde for 30 min, rinsed five times for ten minutes with PBS and followed by 4% osmic acid (osmium tetroxide, Carl Roth GmbH, Karlsruhe, Germany), and then they were fixed the second time for 20 min. This was followed by five additional washes of ten minutes each with PBS, dehydration with an ascending alcohol series and 1,1,1,3,3,3-hexamethyldisilazane (HMDS, Merck AG, Darmstadt, Germany). The samples were applied on SEM stubs (Plano, Wetzlar, Germany) and coated with a 2–3 nm thick platinum layer (Sputter Coater SC 7640, Quorum Technologies Ltd., Lewes, UK). Each sample was then investigated under the scanning electron microscope, under three levels of magnification (radiograph 25-, 500- and 1000-fold). The overview screen was used for the evaluation of cell coverage on the test specimens. In some cases, higher or lower magnifications were used (200- to 8000-fold) to obtain images of the cells. The scanning electron microscopic examination was used to verify the fluorescence microscopic analysis, allowing statements on the phenotypic formation of the fibroblast cell body.

The colorimetric determination of alkaline phosphatase activity was performed by using a kit (Biovison, Mountain View, CA, USA). A volume of 80 μL medium was taken from each well of the culture plates containing titanium discs and transferred into a 96-well plate (Greiner bio-one, Frickenhausen, Germany). Subsequently, 50 μL of pNPP solution was added, and the plate was then left for 60 min, in darkness, at room temperature. Simultaneously, a color sample (blank) was prepared. As a standard, 40 μL of 5 mM pNPP solution was used in 160 μL of assay buffer. A series of ascending concentrations was prepared for the standard curve. All experiments were performed twice. After one hour, the reaction was interrupted using 20 μL stop solution to measure the absorbance in the ELISA reader (Tecan Infinite 200, Magellan V6.6, Tecan, Groedig, Austria) at a wavelength of 405 nm.

After cultivation for 4, 12 and 24 h, 20 μL of WST-1 reagent (Cell proliferation kit WST-1, Water soluble tetrazolium) was added to every cavity of the culture plates. In addition, a control well containing pure medium without WST-1 reagent was prepared, and a blank with WST-1 reagent carried. After two hours of incubation at above culture conditions, the well plate was shaken for a minute, and light absorption was measured in the ELISA reader at a wavelength of 420 nm. The reference was determined at a wavelength of 600 nm. The use of cells was approved by the Ethics Committee of the Saarland Medical Association (E168/09).

3. Results

3.1. Fluorescence Microscopy

Overall, attachment areas of fibroblasts on machined titanium were significantly larger after 4 h, 12 h and 24 h of cultivation (Figures 1 and 2). During the initial colonization (4 h), this was significant for both machined and microstructured (Figures 3 and 4) surfaces. The cell count was higher in treated samples for both the machined and microstructured titanium during the entire period (Figures 5 and 6).

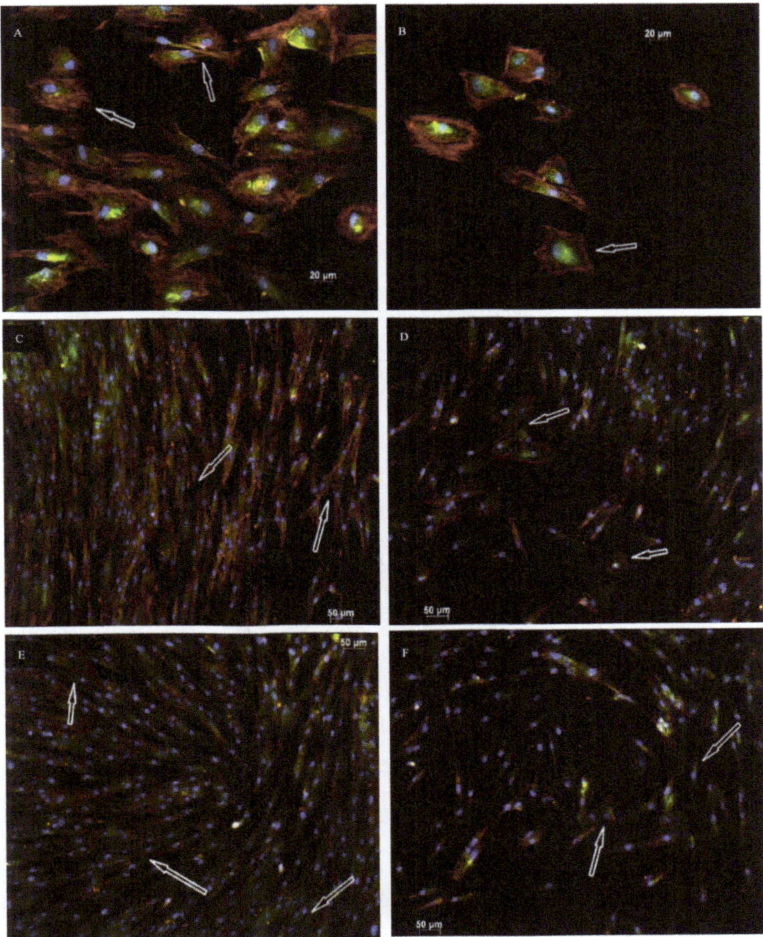

Figure 1. Representative fluorescence microscopic images of fibroblasts on machined titanium discs after 4, 12 and 24 h of cultivation (Vinculin, Phalloidin, DAPI stains). (**A**) Plasma-treated sample after 4 h of culture. Cells show planar attachment of the individual cells (arrow: green staining, vinculin). (**B**) Control sample, after 4 h of culture with rounded cell bodies (arrow) with small areas of adhesion (green). (**C**) Treated sample after 12 h of culture. Arrows: elongated cells with intensive attachment and cell interaction (green). (**D**) Control sample, after 12 h. Arrows: more round cells are present and less attachment (green). (**E**) Treated sample after 24 h. Arrow: intensive attachment of cells, elongated configuration. (**F**) Control sample after 24 h of culture. Arrow: less cells and lower focal attachment (green) of the individual fibroblasts in comparison to the treated samples (**E**).

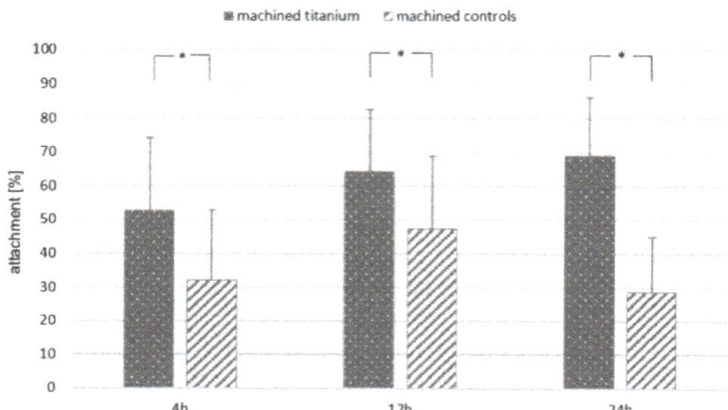

Figure 2. Comparison of relative proportion of focal adhesion (Vinculin) after 4, 12 and 24 h of fibroblast cultivation on machined titanium probes with controls. The bars correspond to the mean value, and the line on top corresponds to the ±standard deviation. Statistically significant differences ($p < 0.05$) are marked with an asterisk.

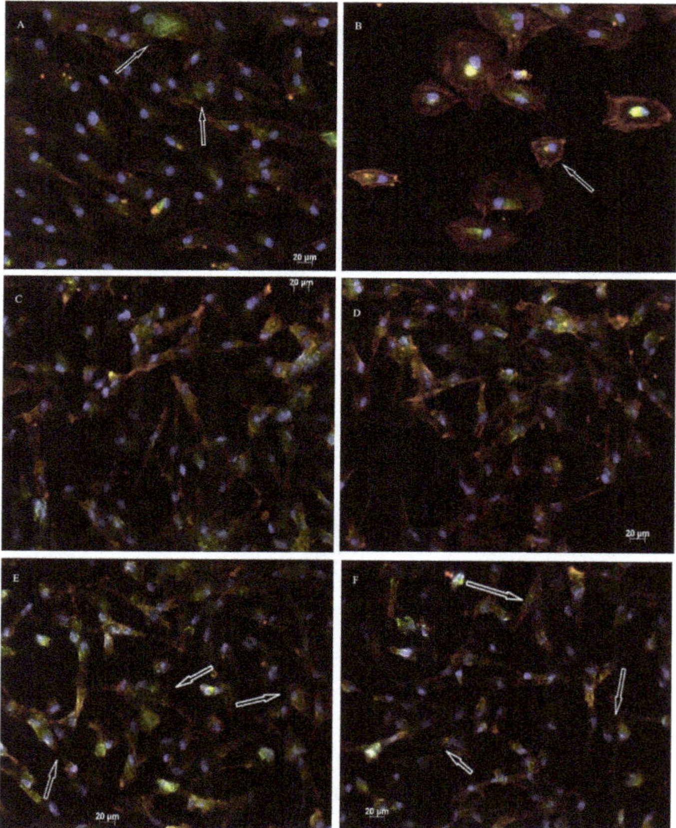

Figure 3. Representative fluorescence microscopic images of fibroblasts on microstructured titanium discs after 4, 12 and 24 h of cultivation (Vinculin, Phalloidin and DAPI stains). (**A**) Plasma-treated titanium

after 4 h of culture. Cells show larger adhesion areas (green, arrows) in comparison to (**B**) control sample, gas stream without plasma ignition (red, arrow). (**C**) Treated and (**D**) untreated titanium discs after 12 h of culture. No differences are visible for the attachment of the fibroblasts. (**E**) Treated sample after 24 h of culture. Arrows: minimally more cells and stronger attachment of the individual cells and larger cell bodies in comparison to (**F**) untreated titanium discs.

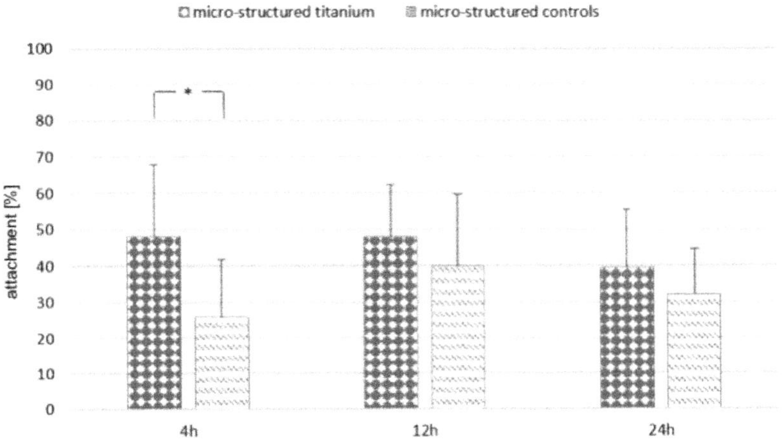

Figure 4. Comparison of relative proportion of focal adhesion (Vinculin) after 4, 12 and 24 h of fibroblast cultivation on microstructured titanium probes with controls. The bars correspond to the mean value, and the line on top corresponds to the ±standard deviation. Statistically significant differences ($p < 0.05$) are marked with an asterisk.

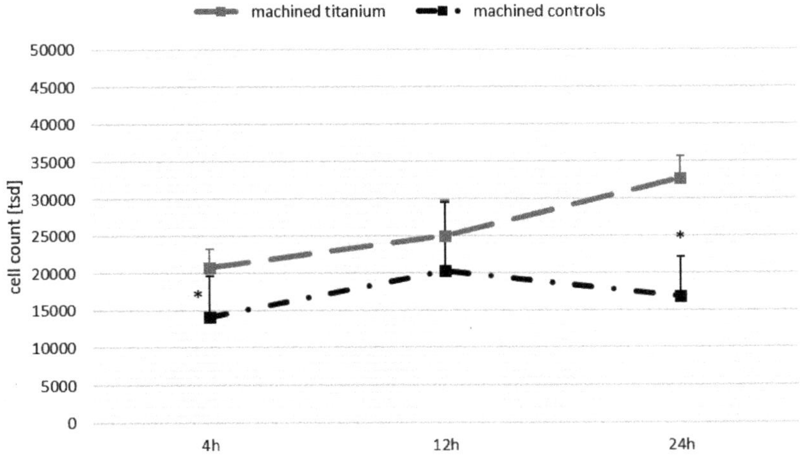

Figure 5. Comparison of cell count after 4, 12 and 24 h between machined titanium samples and controls. The squares correspond to the mean value, and the line on top corresponds to the ±standard deviation. Statistically significant differences ($p < 0.05$) are marked with an asterisk.

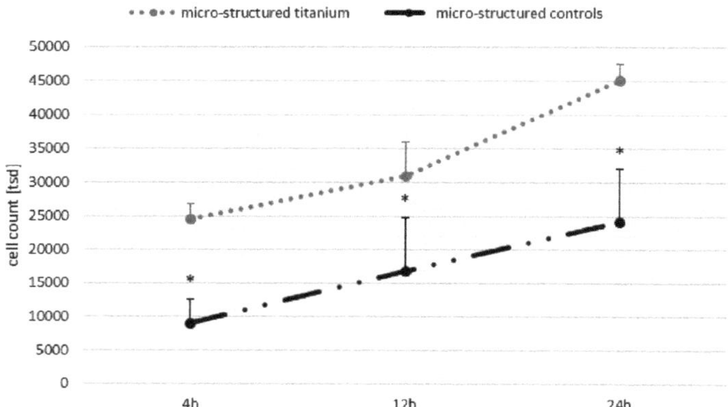

Figure 6. Comparison of cell count after 4, 12 and 24 h between microstructured titanium samples and controls. The circles correspond to the mean value, and the line on top corresponds to the ±standard deviation. Statistically significant differences ($p < 0.05$) are marked with an asterisk.

3.2. Scanning Electron Microscopy

After 4 h of culture, the fibroblasts on the machined and microstructured titanium discs showed a dense and homogeneous colonization pattern. The cell bodies of the fibroblasts on the plasma-treated machined surfaces showed a more elongated configuration with a greater and stronger attachment to the surface and stronger intercellular contacts compared to the controls without plasma treatment (Figure 7). On the microstructured titanium, the cells displayed no clear differences in their phenotypes on plasma-treated and untreated titanium. However, the attachments of the fibroblasts to plasma-treated titanium appeared stronger (Figures 7 and 8). After 12 h, the results were comparable to those observed after 4 h of culture, and after 24 h, the results were inconclusive because the fibroblasts detached from some titanium surfaces during SEM preparation.

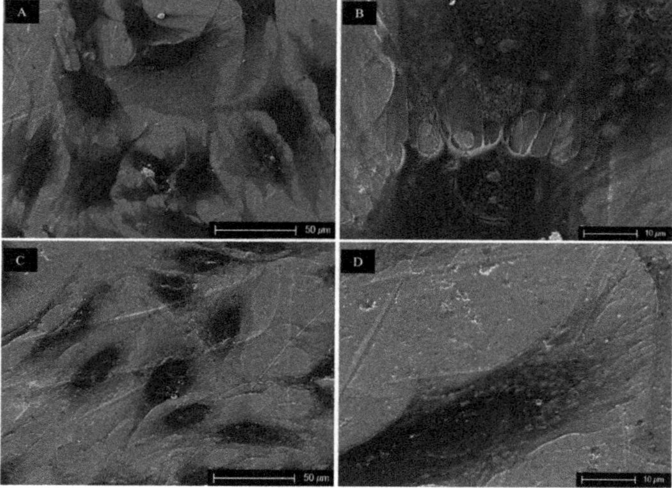

Figure 7. Representative scanning electron microscopy images of fibroblasts on machined titanium discs after 4 h of cultivation. (**A,B**) Plasma-treated samples. Fibroblasts are larger and show more intensive attachment and intercellular contacts in comparison to (**C,D**) controls without plasma treatment.

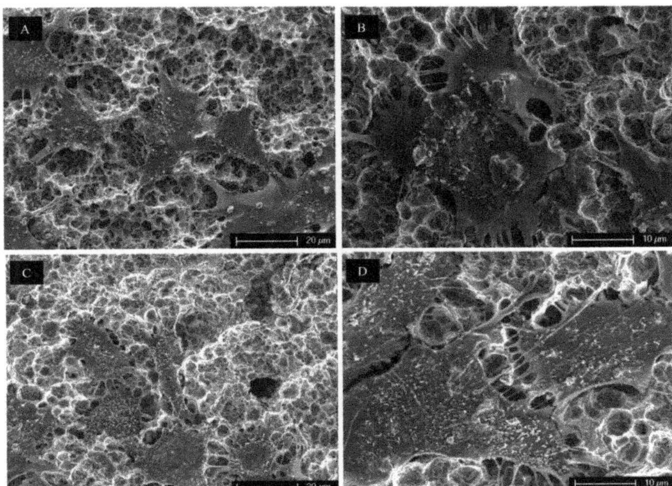

Figure 8. Representative scanning electron microscopy images of fibroblasts on microstructured titanium discs after 4 h of cultivation. (**A,B**) Plasma-treated samples. (**C,D**) Controls without plasma treatment. Fibroblasts show intensive intercellular contacts and attachment to the titanium, suggesting favorable conditions for cell–surface interaction.

3.3. Biological Activity

The analysis of the biological activity showed no negative effect of the plasma treatment on cell proliferation over the cultivation periods of 4 h to 24 h. However, no clear beneficial effect was observed by plasma treatment either. The results for alkaline phosphatase activity, cell proliferation assay and the measurement of nucleus size are presented in the Appendix Materials (Figures A1–A6 → Appendices A–F).

4. Discussion

To the best of the authors' knowledge, this is the first study demonstrating the beneficial effects of the conditioning of microstructured or machined titanium with cold atmospheric plasma on the initial attachment of primary human gingival fibroblasts.

It has already been demonstrated in several studies that osteoblasts respond positively to the treatment of titanium surfaces with cold atmospheric plasma [14].

In this in vitro study, primary human gingival fibroblasts were investigated. They are the main cells of peri-implant mucosa surrounding the implant. Tight and inflammation-free enclosure of the implant is essential for long-term implant success [20].

It has been demonstrated that the treatment of machined and microstructured titanium surfaces with cold atmospheric plasma results in more homogeneous and more intense initial adhesion of gingival fibroblasts. This is discussed as an advantage for the re-osseointegration of titanium implants [21,22].

To analyze the initial colonization of plasma-treated surfaces, primary human gingival fibroblasts of the sixth passage were used. These cells allow a closer approach to the real situation; nevertheless, they are more difficult to culture than immortalized cells. The titanium test specimens used were made of medical-grade pure titanium (titanium grade 2) with different surface configurations. Both microstructured surfaces, which were sandblasted and etched, and machined titanium specimens were used to match the implant structure of common implants, which have a machined and a microstructured part [23]. The treatment was performed using a miniaturized plasma source, which has been used and characterized several times [24,25]. The experiments were performed under ambient conditions. Care was taken to select parameters guaranteeing biologically acceptable temperatures in the treatment area. Thus, the closest possible approximation to realistic conditions in practice was simulated. The cultivation period was limited to a maximum of 24 h, so that the prophylactic addition of antibiotics to the medium could be avoided. The analytical methods used were aimed at visualizing the attachment, morphology and biological activity of fibroblasts.

Both plasma-treated and non-treated control samples showed cell proliferation up to 24 h. The scanning electron microscopy analysis of the microstructured samples showed that the fibroblasts grew into the structure of the titanium sample surface. This confirms the results of older studies underlining the excellent biocompatibility of titanium [26,27].

Overall, the plasma-treated titanium surfaces showed benefits especially for initial colonization. After 4 h of cultivation, slight advantages were shown for all investigated parameters, such as cell attachment, cell number, metabolic activity and nucleus size. After a longer culture period, these advantages were not as distinct. The results of the analysis of cell proliferation (WST-1) showed no clear advantage for plasma treatment. For the intensity of initial attachment as the most important parameter, treatment with cold atmospheric plasma showed a clear improvement over the entire analysis period [28,29].

Adhesion is a multistep process [30]. Firstly, the titanium surface encounters an aqueous medium in which an electrochemical bilayer forms [31], possibly modified by plasma treatment. The surrounding medium also contains proteins such as fibronectin or growth factors helping the cell adhere to the surface. The proteins absorb at the surface [32]. After the first cells have adhered to the surface, an extracellular matrix (ECM) is established. The ECM is the basis for a firm adhesion of cells on a surface, and its structural basis is formed by integrins. For dental practice, the results of this study may be of importance. Treatment of titanium with cold atmospheric plasma improves the wettability and thus the distribution of applied biological substances and the cells on surfaces. This is in line with other findings [22]. The improvement in initial attachment demonstrated in this study provides the opportunity for use of cold atmospheric plasmas in implantology, both in primary implant placement and in the regeneration of lost attachment due to peri-implantitis [33,34]. Additionally, cold atmospheric plasmas can be used for the disintegration of biofilms on implant surfaces [35].

Further studies are necessary to assess the potential of cold atmospheric plasma in dentistry. However, there is no question that its application offers interesting possibilities, especially for the re-osseointegration of titanium implants.

5. Conclusions

The present in vitro study demonstrated the beneficial effect of plasma treatment on the initial attachment of primary human gingival fibroblasts to titanium. This was observed for both the machined and the microstructured titanium surfaces. The results support considerations for the clinical application of cold atmospheric plasma in the context of pre-implantation/peri-implantitis treatment.

Author Contributions: The study was planned by S.R., M.H. and A.S. The experimental work was conducted by S.R. and C.L. The data analysis and interpretation were performed by M.P.G., J.N., S.R., A.L., M.H., A.S. and C.L. The manuscript draft was written by M.P.G., J.N. and S.R. The manuscript revision was conducted by S.R., M.P.G., J.N., M.H., A.L., C.L. and A.S. All authors have read and agreed to the published version of the manuscript.

Funding: This study was funded partly by the German Federal Ministry of Education and Research (BMBF FKZ 01 EZ 0730/0731).

Institutional Review Board Statement: Not applicable.

Informed Consent Statement: Not applicable.

Data Availability Statement: The data supporting the findings of this study are available from the corresponding author upon reasonable request.

Acknowledgments: The authors thank Andreas Schubert, Markus Lange and Heiko Landau for their technical support.

Conflicts of Interest: The authors declare no conflict of interest.

Abbreviations

CAP 1: cold atmospheric plasma 1; ma-Ti 2, machined Titanium 2; mi-Ti 3, microstructured Titanium 3; WST-1 4, water soluble tetrazolium 4.

Appendix A

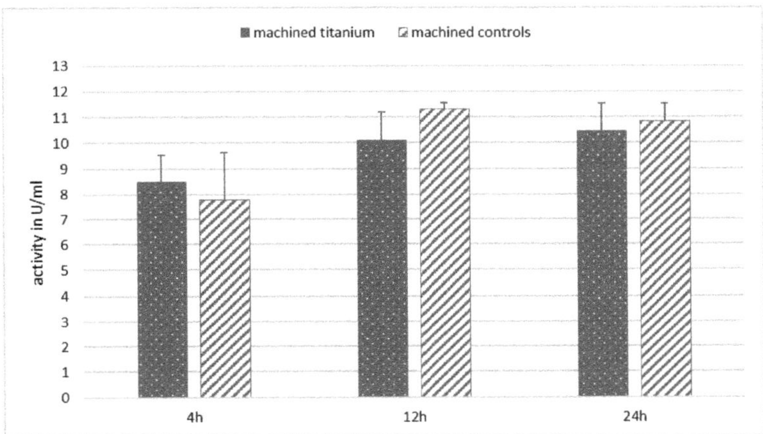

Figure A1. Comparison of alkaline phosphatase activity after 4, 12, 24 h between machined titanium samples and controls. The bars correspond to the mean value, the line on top to the ±standard deviation.

Appendix B

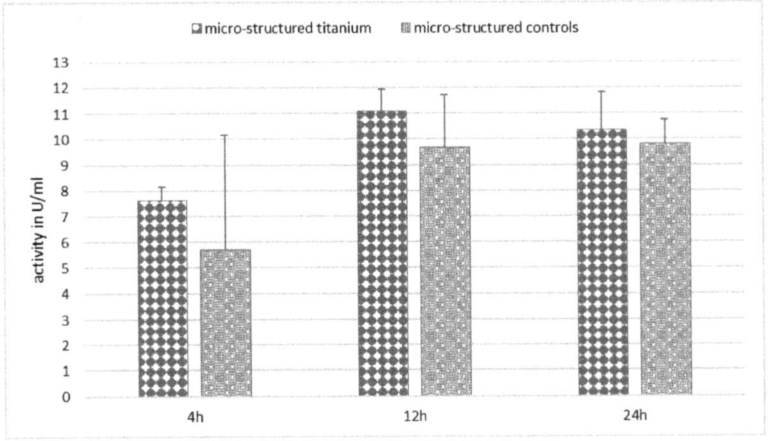

Figure A2. Comparison of alkaline phosphatase activity after 4, 12, 24 h between microstructured titanium samples and controls. The bars correspond to the mean value, the line on top to the ±standard deviation.

Appendix C

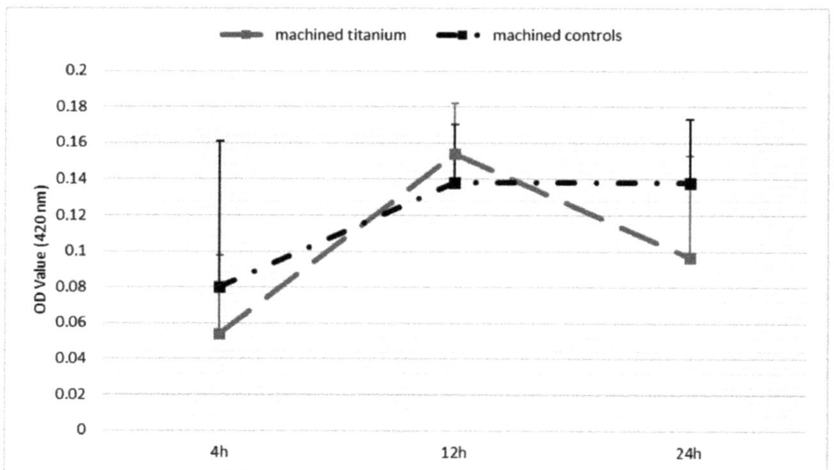

Figure A3. Comparison of cell proliferation using the WST-1 assay after 4, 12, 24 h between machined titanium samples and controls. The squares correspond to the mean value, the line on top to the ±standard deviation.

Appendix D

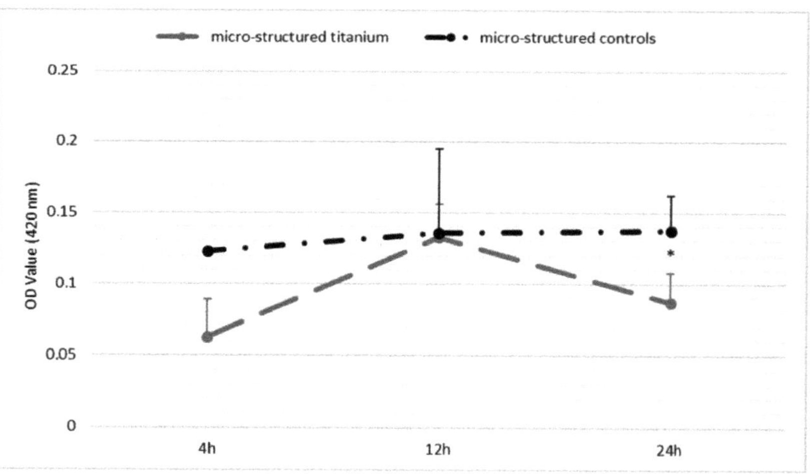

Figure A4. Comparison of cell proliferation in terms of WST-1 assay after 4, 12, 24 h between microstructured titanium samples and controls. The circles correspond to the mean value, the line on top to the ±standard deviation. Statistically significant differences ($p < 0.05$) are marked with an asterisk.

Appendix E

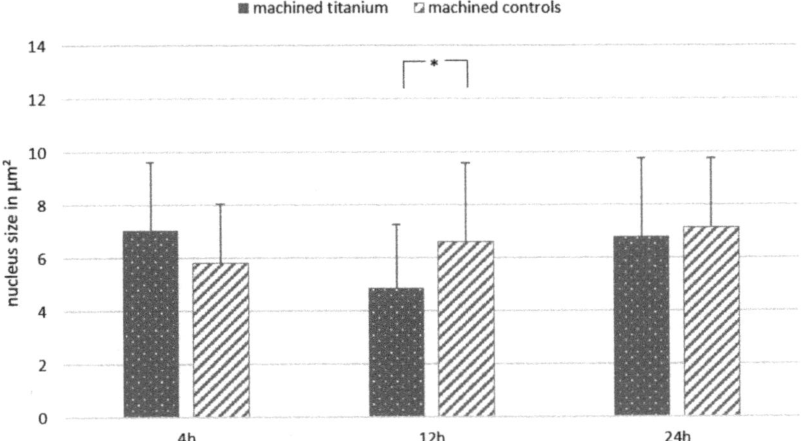

Figure A5. Comparison of the nucleus size after 4, 12 and 24 h of cultivation of machined titanium probes with controls. The bars correspond to the mean value, the line on top to the ±standard deviation. Statistically significant differences ($p < 0.05$) are marked with an asterisk.

Appendix F

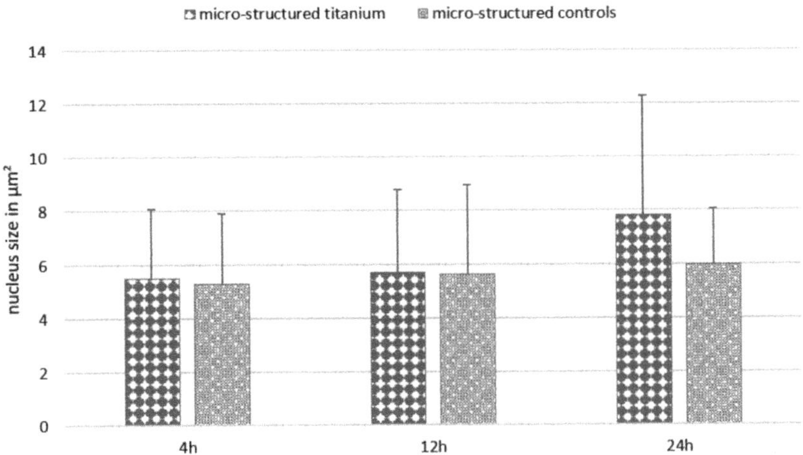

Figure A6. Comparison of the nucleus size after 4, 12 and 24 h of cultivation of microstructured titanium probes with controls. The bars correspond to the mean value, the line on top to the ±standard deviation.

References

1. Subramani, K.; Jung, R.E.; Molenberg, A.; Hammerle, C.H. Biofilm on dental implants: A review of the literature. *Int. J. Oral. Maxillofac. Implant.* **2009**, *24*, 616–626.
2. Sartoretto, S.C.; Alves, A.T.N.N.; Resende, R.F.B.; Calasans-Maia, J.; Granjeiro, J.M.; Calasans-Maia, M.D. Early osseointegration driven by the surface chemistry and wettability of dental implants. *J. Appl. Oral Sci.* **2015**, *23*, 279–287. [CrossRef] [PubMed]
3. Toffoli, A.; Parisi, L.; Tatti, R.; Lorenzi, A.; Verucchi, R.; Manfredi, E.; Lumetti, S.; Macaluso, G.M. Thermal-induced hydrophilicity enhancement of titanium dental implant surfaces. *J. Oral Sci.* **2020**, *62*, 217–221. [CrossRef]

4. Smeets, R.; Stadlinger, B.; Schwarz, F.; Beck-Broichsitter, B.; Jung, O.; Precht, C.; Kloss, F.; Gröbe, A.; Heiland, M.; Ebker, T. Impact of Dental Implant Surface Modifications on Osseointegration. *BioMed Res. Int.* **2016**, *2016*, 217–221. [CrossRef] [PubMed]
5. Jiang, X.; Yao, Y.; Tang, W.; Han, D.; Zhang, L.; Zhao, K.; Wang, S.; Meng, Y. Design of dental implants at materials level: An overview. *J. Biomed. Mater. Res. Part A* **2020**, *108*, 1634–1661. [CrossRef] [PubMed]
6. Jemat, A.; Ghazali, M.J.; Razali, M.; Otsuka, Y. Surface Modifications and Their Effects on Titanium Dental Implants. *BioMed Res. Int.* **2015**, *2015*, 1634–1661. [CrossRef]
7. Kawase, T.; Tanaka, T.; Minbu, H.; Kamiya, M.; Oda, M.; Hara, T. An atmospheric-pressure plasma-treated titanium surface potentially supports initial cell adhesion, growth, and differentiation of cultured human prenatal-derived osteoblastic cells. *J. Biomed. Mater. Res. Part B Appl. Biomater.* **2014**, *102*, 1289–1296. [CrossRef] [PubMed]
8. Lademann, J.; Richter, H.; Alborova, A.; Humme, D.; Patzelt, A.; Kramer, A.; Weltmann, K.-D.; Hartmann, B.; Ottomann, C.; Fluhr, J.W.; et al. Risk assessment of the application of a plasma jet in dermatology. *J. Biomed. Opt.* **2009**, *14*, 1289–1296. [CrossRef]
9. Liebmann, J.; Scherer, J.; Bibinov, N.; Rajasekaran, P.; Kovacs, R.; Gesche, R.; Awakowicz, P.; Kolb-Bachofen, V. Biological effects of nitric oxide generated by an atmospheric pressure gas-plasma on human skin cells. *Nitric Oxide* **2011**, *24*, 8–16. [CrossRef]
10. Matthes, R.; Bekeschus, S.; Bender, C.; Koban, I.; Hübner, N.O.; Kramer, A. Pilot-study on the influence of carrier gas and plasma application (open resp. delimited) modifications on physical plasma and its antimicrobial effect against Pseudomonas aeruginosa and Staphylococcus aureus. *GMS Krankenhhyg Interdiszip* **2012**, *7*, Doc02. [CrossRef] [PubMed]
11. Daubert, D.M.; Weinstein, B.F. Biofilm as a risk factor in implant treatment. *Periodontol. 2000* **2019**, *81*, 29–40. [CrossRef]
12. Pohler, O.E. Unalloyed titanium for implants in bone surgery. *Injury* **2000**, *31*, 7–13. [CrossRef] [PubMed]
13. Ptasińska, S.; Bahnev, B.; Stypczyńska, A.; Bowden, M.; Mason, N.J.; Braithwaite, N.S.J. DNA strand scission induced by a non-thermal atmospheric pressure plasma jet. *Phys. Chem. Chem. Phys.* **2010**, *12*, 7779–7781. [CrossRef]
14. Jiao, Y.; Tay, F.R.; Niu, L.N.; Chen, J.H. Advancing antimicrobial strategies for managing oral biofilm infections. *Int. J. Oral Sci.* **2019**, *11*, 28. [CrossRef] [PubMed]
15. Kamionka, J.; Matthes, R.; Holtfreter, B.; Pink, C.; Schlüter, R.; Von Woedtke, T.; Kocher, T.; Jablonowski, L. Efficiency of cold atmospheric plasma, cleaning powders and their combination for biofilm removal on two different titanium implant surfaces. *Clin. Oral Investig.* **2022**, *26*, 3179–3187. [CrossRef] [PubMed]
16. Flörke, C.; Janning, J.; Hinrichs, C.; Behrens, E.; Liedtke, K.R.; Sen, S.; Christofzik, D.; Wiltfang, J.; Gülses, A. In-vitro assessment of the efficiency of cold atmospheric plasma on decontamination of titanium dental implants. *Int. J. Implant. Dent.* **2022**, *8*, 12. [CrossRef]
17. Duske, K.; Koban, I.; Kindel, E.; Schröder, K.; Nebe, B.; Holtfreter, B.; Jablonowski, L.; Weltmann, K.D.; Kocher, T. Atmospheric plasma enhances wettability and cell spreading on dental implant metals. *J. Clin. Periodontol.* **2012**, *39*, 400–407. [CrossRef]
18. Hui, W.L.; Perrotti, V.; Iaculli, F.; Piattelli, A.; Quaranta, A. The Emerging Role of Cold Atmospheric Plasma in Implantology: A Review of the Literature. *Nanomaterials* **2020**, *10*, 1505. [CrossRef]
19. Wagner, G.; Eggers, B.; Duddeck, D.; Kramer, F.J.; Bourauel, C.; Jepsen, S.; Deschner, J.; Nokhbehsaim, M. Influence of cold atmospheric plasma on dental implant materials—An in vitro analysis. *Clin. Oral Investig.* **2022**, *26*, 2949–2963. [CrossRef] [PubMed]
20. Chung, D.M.; Oh, T.J.; Shotwell, J.L.; Misch, C.E.; Wang, H.-L. Significance of Keratinized Mucosa in Maintenance of Dental Implants With Different Surfaces. *J. Periodontol.* **2006**, *77*, 1410–1420. [CrossRef]
21. Lopez-Heredia, M.A.; Legeay, G.; Gaillard, C.; Layrolle, P. Radio frequency plasma treatments on titanium for enhancement of bioactivity. *Acta Biomater.* **2008**, *4*, 1953–1962. [CrossRef]
22. Yoshinari, M.; Wei, J.; Matsuzaka, K.; Inoue, T. Effect of Cold Plasma-Surface Modification on Surface Wettability and Initial Cell Attachment. *World Acad. Sci. Eng. Technol.* **2009**, *58*, 171–175. [CrossRef]
23. Bürgers, R.; Gerlach, T.; Hahnel, S.; Schwarz, F.; Handel, G.; Gosau, M. In vivo and in vitro biofilm formation on two different titanium implant surfaces. *Clin. Oral Implant. Res.* **2010**, *21*, 156–164. [CrossRef]
24. Evert, K.; Kocher, T.; Schindler, A.; Müller, M.; Müller, K.; Pink, C.; Holtfreter, B.; Schmidt, A.; Dombrowski, F.; Schubert, A.; et al. Repeated exposure of the oral mucosa over 12 months with cold plasma is not carcinogenic in mice. *Sci. Rep.* **2021**, *11*, 20672. [CrossRef]
25. Lehmann, A.; Pietag, F.; Arnold, T. Human health risk evaluation of a microwave-driven atmospheric plasma jet as medical device. *Clin. Plasma Med.* **2017**, *7*, 16–23. [CrossRef]
26. Aita, H.; Hori, N.; Takeuchi, M.; Suzuki, T.; Yamada, M.; Anpo, M.; Ogawa, T. The effect of ultraviolet functionalization of titanium on integration with bone. *Biomaterials* **2009**, *30*, 1015–1025. [CrossRef] [PubMed]
27. Brunette, D.M.; Tengvall, P.; Textor, M.; Thomsen, P.; Vörös, J.; Wieland, M.; Ruiz-Taylor, L.; Textor, M.; Brunette, D.M. Characterization of titanium surfaces. In *Titanium in Medicine: Material Science, Surface Science, Engineering, Biological Responses and Medical Applications*; Springer: Berlin/Heidelberg, Germany, 2001; pp. 87–144.
28. Jeong, W.S.; Kwon, J.S.; Choi, E.H.; Kim, K.M. The Effects of Non-Thermal Atmospheric Pressure Plasma treated Titanium Surface on Behaviors of Oral Soft Tissue Cells. *Sci. Rep.* **2018**, *8*, 15963. [CrossRef] [PubMed]
29. González-Blanco, C.; Rizo-Gorrita, M.; Luna-Oliva, I.; Serrera-Figallo, M.; Torres-Lagares, D.; Gutiérrez-Pérez, J.L. Human Osteoblast Cell Behaviour on Titanium Discs Treated with Argon Plasma. *Materials* **2019**, *12*, 1735. [CrossRef] [PubMed]
30. Gongadze, E.; Kabaso, D.; Bauer, S.; Slivnik, T.; Schmuki, P.; van Rienen, U.; Iglič, A. Adhesion of osteoblasts to a nanorough titanium implant surface. *Int. J. Nanomed.* **2011**, *6*, 1801–1816. [CrossRef]

31. Gongadze, E.; Van Rienen, U.; Iglič, A. Generalized stern models of the electric double layer considering the spatial variation of permittvity and finite size of ions in saturation regime. *Cell. Mol. Biol. Lett.* **2011**, *16*, 576–594. [CrossRef]
32. Calazans Neto, J.V.; Kreve, S.; Valente, M.; Reis, A.C.D. Protein absorption on titanium surfaces treated with a high-power laser: A systematic review. *J. Prosthet. Dent.* **2022**. *online ahead of print*. [CrossRef] [PubMed]
33. Rupf, S.; Idlibi, A.; Marrawi, F.; Hannig, M.; Schubert, A.; von Müller, L.; Spitzer, W.; Holtmann, H.; Lehmann, A.; Rueppell, A.; et al. Removing Biofilms from Microstructured Titanium Ex Vivo: A Novel Approach Using Atmospheric Plasma Technology. *PLoS ONE* **2011**, *6*, e25893. [CrossRef] [PubMed]
34. Matthes, R.; Jablonowski, L.; Pitchika, V.; Holtfreter, B.; Eberhard, C.; Seifert, L.; Gerling, T.; Vilardell Scholten, L.; Schlüter, R.; Kocher, T. Efficiency of biofilm removal by combination of water jet and cold plasma: An in-vitro study. *BMC Oral Health* **2022**, *22*, 157. [CrossRef]
35. Idlibi, A.N.; Al-Marrawi, F.; Hannig, M.; Lehmann, A.; Rueppell, A.; Schindler, A.; Jentsch, H.; Rupf, S. Destruction of oral biofilms formed in situ on machined titanium (Ti) surfaces by cold atmospheric plasma. *Biofouling* **2013**, *29*, 369–379. [CrossRef] [PubMed]

Article

Cold Atmospheric Plasma Improves the Colonization of Titanium with Primary Human Osteoblasts: An In Vitro Study

Madline P. Gund [1,*], Jusef Naim [1], Antje Lehmann [2,3], Matthias Hannig [1], Markus Lange [4], Axel Schindler [2,5] and Stefan Rupf [4]

1 Clinic of Operative Dentistry, Periodontology and Preventive Dentistry, Saarland University, 66421 Homburg, Germany; matthias.hannig@uks.eu (M.H.)
2 Leibniz Institute of Surface Modification (IOM), 04318 Leipzig, Germany; aschindler@t-online.de (A.S.)
3 ADMEDES GmbH, 75179 Pforzheim, Germany
4 Synoptic Dentistry, Saarland University, 66421 Homburg, Germany; stefan.rupf@uks.eu (S.R.)
5 Piloto Consulting Ion Beam and Plasma Technologies, 04668 Grimma, Germany
* Correspondence: madline.gund@uks.eu; Tel.: +49-6841-1624478

Abstract: Several studies have shown that cold atmospheric plasma (CAP) treatment can favourably modify titanium surfaces to promote osteoblast colonization. The aim of this study was to investigate the initial attachment of primary human osteoblasts to plasma-treated titanium. Micro-structured titanium discs were treated with cold atmospheric plasma followed by the application of primary human osteoblasts. The microwave plasma source used in this study uses helium as a carrier gas and was developed at the Leibniz Institute for Surface Modification in Leipzig, Germany. Primary human osteoblasts were analyzed by fluorescence and cell biological tests (alkaline phosphatase activity and cell proliferation using WST-1 assay). The tests were performed after 4, 12, and 24 h and showed statistically significant increased levels of cell activity after plasma treatment. The results of this study indicate that plasma treatment improves the initial attachment of primary human osteoblasts to titanium. For the first time, the positive effect of cold atmospheric plasma treatment of micro-structured titanium on the initial colonization with primary human osteoblasts has been demonstrated. Overall, this study demonstrates the excellent biocompatibility of micro-structured titanium. The results of this study support efforts to use cold atmospheric plasmas in implantology, both for preimplantation conditioning and for regeneration of lost attachment due to peri-implantitis.

Keywords: biological cell activity; cell attachment; cold atmospheric plasma; primary human osteoblasts

Citation: Gund, M.P.; Naim, J.; Lehmann, A.; Hannig, M.; Lange, M.; Schindler, A.; Rupf, S. Cold Atmospheric Plasma Improves the Colonization of Titanium with Primary Human Osteoblasts: An In Vitro Study. *Biomedicines* **2024**, *12*, 673. https://doi.org/10.3390/biomedicines12030673

Academic Editor: Christoph Viktor Suschek

Received: 11 December 2023
Revised: 12 February 2024
Accepted: 21 February 2024
Published: 18 March 2024

1. Introduction

Cold atmospheric plasma (CAP) has a broad spectrum of medical applications. In dentistry, it can be used to eliminate biofilms [1,2] or for non-contact surface modification to improve the wettability of dental implants [3].

Successful implantation requires osseointegration of the implant. This requires the differentiation of progenitor cells into osteoblasts which later become osteocytes embedded in the mineralized matrix. During osseointegration, osteoprogenitor cells and vessels grow and osteoblasts adhere to the implant surface [4]. Therefore, the ability to promote the rapid and effective attachment of the osteoblasts is crucial.

The long-term clinical success rate of titanium dental implants is reported to be 87.8% with follow-up periods of up to 36 years. Successful osseointegration and long-term stability of the implant–bone interface enable such results to be achieved in clinical follow-up assessments [5]. The peri-implant tissue acts as a barrier and protects against oral bacteria. Contamination of the implant surface by microorganisms can be prevented, thus avoiding inflammation of the surrounding tissue and bone [6,7].

There are multiple options to enhance cell proliferation and ultimately osseointegration, including etching, coating, and sandblasting [8–10]. Several coating options have been described in the literature. There are nanoparticle coatings with osteo-integrative activity, such as Al_2O_3 nanoparticles, hydroxyapatite, and calcium phosphate, nanoparticle coatings with osteo-integrative and antibacterial activities, such as TiO_2 nanoparticles and nano-crystalline diamond, and nanoparticles with antimicrobial activity such as dental coating materials such as Ag nanoparticles, ZnO nanoparticles, CuO nanoparticle, Quercitrin and Chlorhexidine [11].

Previous research has demonstrated the effects of cold atmospheric plasma pre-treatment of titanium surfaces on the initial attachment of primary human fibroblasts [6]. Over the years, several studies have focused on the effect of treating different titanium surfaces with different plasma jets on osteoblasts [12].

Both fibroblasts and osteoblasts show improved spreading on titanium surfaces when treated with cold atmospheric plasma. Faster proliferation and adhesion of immortal-ized human gingival fibroblasts (HGF-1) and osteoblast-like cells (MG-63) on coated and uncoated titanium and zirconia discs were observed after treatment [13].

However, studies investigating the impact on the initial attachment of primary human osteoblasts are currently lacking. Therefore, the aim of this study was to investigate the attachment and biological activity of primary human osteoblasts on micro-structured titanium surfaces after treatment with cold atmospheric plasma during the first 24 h.

2. Materials and Methods

2.1. Setting

2.1.1. Titanium Discs

The test specimens consisted of 144 titanium discs (titanium grade 2, Friadent, Mannheim, Germany) with a diameter of 5 mm and a height of 1 mm. The surfaces of the discs were micro-structured through sandblasting and etching to a surface roughness value (Ra) of 2 μm. Of these, 54 discs were left untreated as controls.

2.1.2. Plasma Device

This study used a plasma source developed at the Leibniz Institute for Surface Mod-ification located in Leipzig, Germany. The miniaturized plasma source equipped with a coaxial electrode system was microwave-excited and operated with helium gas. The inner electrode was formed by an inner conductor consisting of a 0.3 mm thick steel tube. The second electrode, which was made of 3.4 mm thick steel and served to shield the generated microwaves, was formed by the outer body of the plasma source. The process gases were added through the inner conductor, and their flow rate was regulated by a mass flow controller. To activate the plasma source, a pulsed microwave generator was used generating at a frequency of 2.45 GHz. The generator allowed for a peak power range of 100–300 W, an average power range of 1–9 W, and a pulse width of 1–10 μs.

The plasma treatment settings, including the pulse width, pulse power, and average microwave power, could be adjusted as desired. The plasma source was mounted on a computer-controlled 3-axis movement system (Steinmeyer MC-G047, Feinmess Dresden GmbH, Dresden, Germany) to ensure uniformity of treatment. The parameters used for processing comprised a peak power of 300 W and a pulse width of 5 μs, resulting in a mean power of 5 W and a scanning speed of 8 mm/s. The plasma jet was used to process the surface in a meandering line scan. The working distance was set at 2 mm, and a gas flow of 2 L/min was used. Helium was selected as the carrier gas and was mixed with 5 sccm of oxygen. These parameters limited the surface temperature of the titanium to about 40 °C during the plasma treatment.

2.1.3. Primary Human Osteoblasts

Primary human osteoblasts from femoral trabecular bone tissue from the knee or hip joint region were obtained from PromoCell (Heidelberg, Germany). Cryopreserved cells were cultured in DMEM/10% FBS in culture flasks, with third passage cells being used in the experiments.

2.1.4. Sample Treatment

The samples were treated in a controlled environment in an S2 laboratory where the air temperature was $22 \pm 2\ °C$ and the air was changed eight times per hour at a slight positive pressure. After the treatment, primary human osteoblasts were immediately applied to a well plate under a sterile bench. Both the treated samples and the untreated controls received 10^4 cells. The titanium specimens were kept moist with DMEM-liquid medium for 30 min before the wells were filled with 200 µL of the same medium. The samples were then incubated in an incubator (Hera Cell 240, Thermo Scientific, Waltham, MA, USA) at $37\ °C$ in a 5% CO_2 atmosphere for 4, 12, and 24 h.

2.2. Sample Analysis

2.2.1. Surface Coverage, Images

Fluorescence microscopy was used to visualize the colonization of the primary human osteoblasts on the titanium discs. Staining was performed with Vinculin, Phalloidin, and DAPI. The cells were fixed to the specimens with a 2.5% glutaraldehyde solution (Carl Roth GmbH, Karlsruhe, Germany) for 30 min. After the glutaraldehyde treatment, the specimens were washed five times with phosphate buffered saline (PBS) to remove glutaraldehyde. Permeabilization of the cell wall was initiated by exposure to 0.1% Triton-X (Roche, Grenzach-Wyhlen, Germany) in PBS. A double rinse step was performed using PBS, and the non-specific background was minimized by using 1% bovine serum albumin (BSA in PBS, Gibco, Invitrogen, Carlsbad, CA, USA). The samples were placed in a humidified chamber and treated with a primary antibody directed against vinculin (Anti-Vinculin, mouse monoclonal antibodies; Sigma-Aldrich Chemie GmbH, Steinheim, Germany) diluted in 5% bovine serum albumin. The samples were then washed twice with 0.1% Tween 20 (Roche, Grenzach-Wyhlen, Germany) in phosphate-buffered saline and the Tween 20 was removed with phosphate-buffered saline. The samples were then treated with a secondary antibody (anti-mouse IgG, RD Systems, Minneapolis, MN, USA) for 1 h while maintaining the humidity in the chamber. This was followed by another rinse with 0.1% Tween 20 and PBS. As a final step the samples were immersed in a phalloidin stock solution (1 mg/mL MetOH) for 25 min at room temperature, in the dark, followed by the application of DAPI solution (Roche, Grenzach-Wyhlen, Germany) for 10 min, also in the dark.

The titanium discs were fixed onto Superfrost slides (Menzel-glasses, Braunschweig, Germany) before being thoroughly examined using a reflected-light microscope (Axio Scope A1, Carl Zeiss, Göttingen, Germany) with $2.5\times$, $10\times$, and $20\times$ magnification objectives. A general view was then taken for each sample using a $25\times$ magnification (2.5 (lens) $\times$ 10 (ocular)). In addition, high magnification images were taken to provide representative illustrations.

2.2.2. Alkaline Phosphatase Activity

Colorimetric determination of alkaline phosphatase activity was performed using a Biovision kit obtained from Mountain View, California, USA. After extracting 80 µL of medium from each well of culture plates containing titanium discs, the medium was transferred to a 96-well plate manufactured by Greiner bio-one in Frickenhausen, Germany. This was followed by the addition of 50 µL of para-nitro-phenyl-phosphate (pNPP) solution, and the plate was left in the dark at an ambient temperature for 60 min. At the same time, a blank color sample was prepared. The standard solution consisted of 40 µL of 5 mM pNPP solution in 160 µL of assay buffer. A range of increasing concentrations was used to construct the standard curve. The experiments were performed in duplicate. After a one-hour reaction time, 20 µL of stop solution was added, and the absorbance was measured at a wavelength of 405 nm using an ELISA reader (Tecan Infinite 200, Magellan V6.6, Tecan, Groedig, Austria).

2.2.3. Cell Proliferation/Cell Viability Assay

After incubation for 4, 12, and 24 h, each well of the culture plates was treated with 20 μL of WST-1 reagent (Cell Proliferation Kit WST-1, Water-Soluble Tetrazolium). In addition, a control well was prepared containing only medium, without WST-1 reagent. A blank was also prepared containing only the WST-1 reagent. After two hours of incubation under the above culture conditions, the well plate was vortexed for one minute. Light absorption was measured at a wavelength of 420 nm using an ELISA reader, and a reference value was determined at a wavelength of 600 nm.

2.3. Statistics

The Mann–Whitney U test was performed to compare control and test samples statistically. Statistical significance was considered when the p-value was $p < 0.05$.

3. Results

3.1. Fluorescence Microscopy

Fluorescence microscopy conducted after 4, 12, and 24 h demonstrated the attachment of the primary human osteoblasts on the titanium surfaces. After 4 h of cultivation, the primary human osteoblasts present on the micro-structured titanium surface within the test and control groups adhered to the titanium surfaces with comparable areas of focal contacts (Figure 1A,B). After 12 h, however, differences between the test and control groups were visible. Osteoblasts covered the entire surface of the micro-structured titanium discs treated with cold atmospheric plasma. They were dispersed from the application site, whereas cells on untreated titanium remained in the area of application without extensive spreading (Figure 2A,B). However, the attachment areas of the individual cells on the treated and untreated titanium were comparable. After 24 h, there were no differences in the pattern of cell distribution on the micro-structured titanium surfaces of both the test and control groups. In addition, focal attachment areas remained comparable (Figure 3A,B).

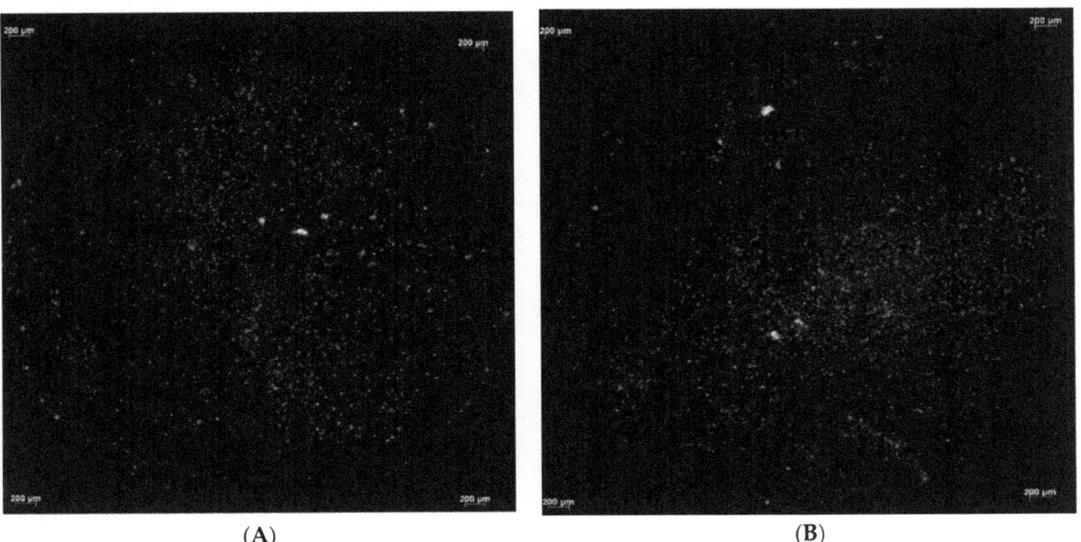

(**A**)   (**B**)

Figure 1. Fluorescence micrographs of primary human osteoblasts on micro-structured titanium discs (**A**) and controls (**B**) after 4 h of cultivation (vinculin, phalloidin, and DAPI staining). The attachment areas of individual cells were comparable (**A,B**).

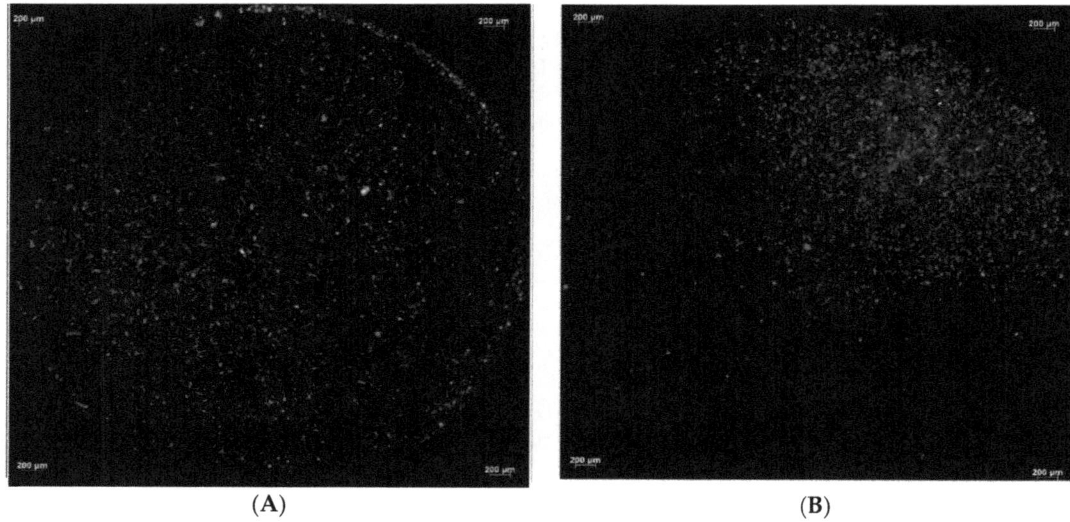

(A)　　　　　　　　　　　　(B)

Figure 2. Fluorescence micrographs of primary human osteoblasts on micro-structured titanium discs treated with cold atmospheric plasma (**A**) and control samples (**B**) after 12 h of cultivation (Vinculin, Phalloidin, DAPI stains). The cells are spread homogeneously on the treated titanium surfaces (**A**), whereas the cells on the untreated titanium remained around the application area (**B**). The attachment pattern of the individual cells is not different (**A**,**B**).

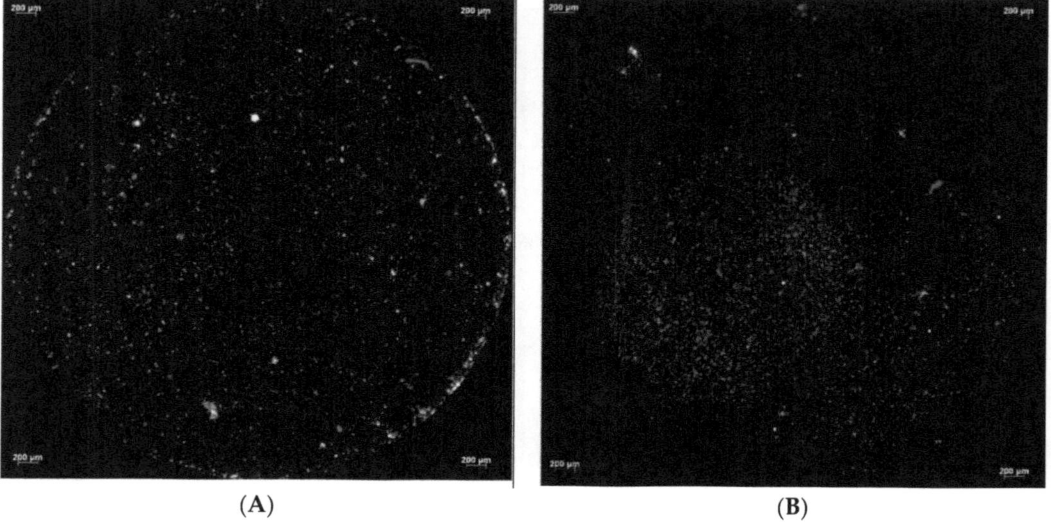

(A)　　　　　　　　　　　　(B)

Figure 3. Fluorescent micrographs of primary human osteoblasts on cold atmospheric plasma-treated (**A**) and untreated (**B**) micro-structured titanium discs after 24 h cultivation (vinculin, phalloidin, and DAPI staining). The distribution of cells on the titanium surfaces and their adherence areas show no apparent differences.

3.2. Cell Activity

Statistically significant differences in the alkaline phosphatase activity were observed between the test and control groups at 4 and 12 h (4 h: p = 0.013, 12 h: p= 0.015), although the differences in the mean values and standard deviations between the two groups were small. No statistically significant differences were observed between the plasma-treated and control samples after 24 h. After 4 h alkaline phosphatase activity was 8.1 ($\pm$0.2) U/mL in the control group and 8.5 ($\pm$0.4) U/mL in the test group, after 12 h it was 17.4 U/mL ($\pm$1.2) in the control group and 19.2 ($\pm$1.9) U/mL in the test group, and after 24 h it was 37.2 U/mL ($\pm$1.2) in the control group and 36.4 ($\pm$4.5) U/mL in the test group (Figure 4).

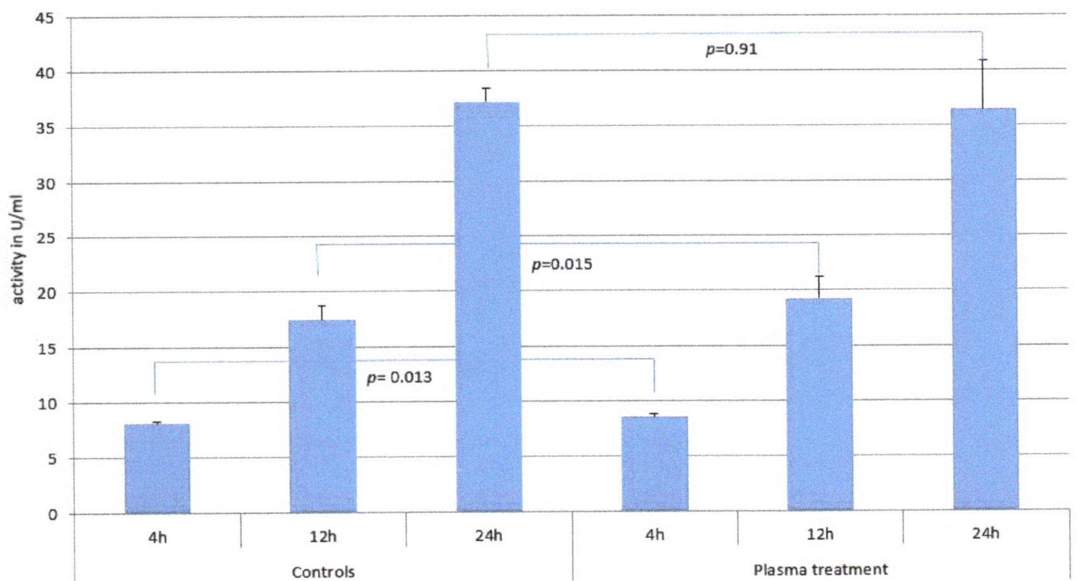

Figure 4. Comparison of alkaline phosphatase activity after 4, 12, and 24 h between micro-structured titanium (plasma treated) samples and controls. The bars correspond to the mean value and the line on top to the $\pm$standard deviation. Statistically significant differences were found for the 4 h and 12 h results (4 h: p = 0.013, 12 h: p = 0.015). No statistically significant differences were found for the 24 h results (p = 0.91).

When the cell proliferation was compared using the WST-1 assay, statistically significant differences were observed between the test and control groups at 4, 12, and 24 h (4 h: p = 0.022, 12 h: p= 0.036, 24 h: p= 0.006). In particular, the OD value after 4 h in the test group of 0.21 ($\pm$0.07) compared to 0.11 ($\pm$0.01) in the control group showed a clear difference (p = 0.022). After 12 h of cultivation, the OD value was 0.31 ($\pm$0.07) in the test group and 0.24 ($\pm$0.05) in the control group, a difference that was also statistically significant (p = 0.036). Even after culturing the osteoblasts for 24 h, there were still differences in cell proliferation with values of 0.32 ($\pm$0.03) for the test group and 0.26 ($\pm$0.03) for the control group. This difference was also statistically significant with a p-value of 0.006 (Figure 5).

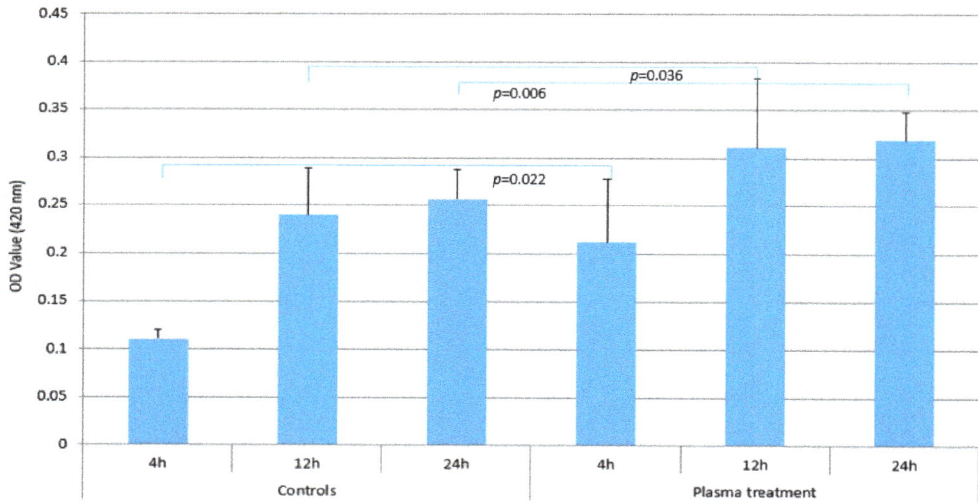

Figure 5. Comparison of cell proliferation using WST-1 assay after 4, 12, and 24 h between plasma treated samples and untreated controls. The bars correspond to the mean value and the line on top to the ±standard deviation. Statistically significant differences were found for the 4 h, 12 h, and 24 h results (4 h: $p = 0.022$, 12 h: $p = 0.036$, 24 h: $p = 0.006$).

4. Discussion

Previous studies have demonstrated the favorable response of osteoblasts to biofilm-covered titanium treated with cold atmospheric plasma and mechanical cleaning [1]. Plasma-treated titanium surfaces showed reduced contact angles, larger cell sizes, and better spreading of osteoblastic cells compared to non-treated controls [3]. The beneficial effects of cold atmospheric plasma treatment on titanium surfaces and osteoblasts have been demonstrated on smooth, moderately rough, and rough implant surfaces [14]. In addition, studies have shown the potential to remove biofilms from micro-structured surfaces of titanium implants [15]. In addition, cold atmospheric plasma facilitates osteoblastic cell differentiation, which could serve as an exciting tool for bone regeneration in the future [16].

The effects have previously been studied on primary human gingival fibroblasts and have shown that pre-treatment of titanium surfaces leads to improved initial adhesion of these cells [6]. To the best of our knowledge, this is the first study to demonstrate beneficial effects of cold atmospheric plasma to condition micro-structured titanium on the early attachment of primary human osteoblasts.

After 4 h, the cells of both groups showed a comparable attachment area on the treated and untreated titanium surfaces, with cell proliferation being higher in the test group after 4 h, 12 h, and 24 h, as measured by the WST-1 test. However, after 12 h, the cells on the cold atmospheric plasma-treated titanium surfaces had covered the entire surface of the titanium disc, whereas the cells on untreated titanium remained in the application area. However, the attachment areas of individual cells were not visually different in the fluorescence microscopic images. After 24 h, the distribution of cells on both titanium surfaces and their respective attachment areas were identical. Again however, the WST-1 kit indicated higher cell proliferation on the plasma-treated titanium surfaces. The adhesion of human osteoblast-like cells (MG-63) was not altered on titanium grade IV implant surfaces that had been treated cold atmospheric plasma treatment within an observation period of 24 h [13].

On air plasma pre-treated titanium MC3T3-E1 cells covered a larger surface area after 2 h. In addition, better cell proliferation and migration were observed with a more developed cellular network. The authors point out that the effects of plasma on cells cannot be maintained for a long time [17]. The results are consistent with ours. It may be necessary to treat implants with cold atmospheric plasma immediately before implantation to improve the success rate [17].

Oxygen-plasma treatment of titanium surfaces had no effect on the metabolic activity and proliferation of primary human alveolar bone osteoblasts from day 1 to day 7. Plasma pre-treated surfaces showed comparable cell densities to controls at day 1, 3, and 7. No differences in the organization of actin and vinculin were observed between treated and untreated surfaces [18]. However, Henningsen et al. also describe a higher initial cell attachment (argon plasma) after 2 h in comparison to non-plasma-treated titanium discs. They used murine osteoblast-like cells MC3T3-E1 in vitro. They describe better results (number of cells) for plasma-treated surfaces (argon and O_2 plasma) compared to controls during a 72 h incubation. They also found better cell proliferation after 24 h, which is in line with our results [19]. Other studies have also confirmed these results [20]. Larger average cell areas (human osteoblasts, MG-63) were described 24 h after the argon-plasma treatment of titanium surfaces. Cells on these surfaces were enlarged, and cells on the untreated surfaces were less adapted. In summary, higher levels of cell proliferation and adhesion were described [21]; the penultimate of which is also consistent with our findings.

Titanium surfaces treated with argon-plasma showed improved adhesion of rat bone marrow cells after 6 h [22]. Lee et al. describe higher number of attached cells (human osteosarcoma cell line) on plasma-treated titanium using a dielectric barrier discharge (DBD) plasma after 2 h. After 5 days the number of cells in the test group was 40.2% higher than in the control group. After 24 h, cells in the test group were more uniformly attached to the implant surface and cell density was much higher [23]. These results are confirmed by Long et al. who found that after 24 h up to 30% more cells (MC3T3-E1 mouse pre-osteoblasts) adhered to the plasma-treated surface [24]. A significant increase in the number of osteoblasts (MC3T3-E1 and MG-63) adhering to argon-plasma-treated titanium surfaces is also described for 10 min [14]. After 12 h, Wang et al. described excellent adhesion and elongation of osteoblast rat cells. The cell area increased after titanium plasma treatment, even after 24 h [25]. Better adhesion of rat bone marrow cells for plasma-treated titanium disks at 1, 3, 6, and 24 h was demonstrated by Ujino [26]. These results are confirmed by Hayashi who investigated the effect of plasma treatment on titanium surfaces and rat bone marrow cells [27]. These results contradict our own. There may be several reasons for the conflicting results described. Firstly, the noble gas and the exposure time of the beam were different. The exposure time plays a crucial role. A higher cell density is described for plasma-treated surfaces after 1 min of surface treatment compared to 12 and 16 min [28]. These results are confirmed by Swart et al. The highest cell adhesion was observed on surfaces that had been treated with argon plasma for one minute. The presence of inorganic contaminants was observed with longer treatment times. This may have influenced the behavior of the cells [29]. Other exposure times have also been investigated in the literature [30]. Han et al. describe a weak proportionality of the number of osteoblasts to the treatment time for the same culture period. A significant increase in the number of cells was observed after 4 min of plasma treatment [20].

Secondly, the surfaces were different. It has been described several times in the literature that bone-related cells prefer rough surfaces. In this case, better initial cell proliferation and adhesion have been demonstrated [31]. The use of rough, micro-structured surfaces in this study may explain why the adhesion of individual osteoblasts was not really improved after plasma treatment compared to the control group. Thirdly, different types of studies were carried out. The study presented was an in vitro study. However, animal studies were also included in the discussion. Also, different cells were used. In this study, primary human osteoblasts were used. The use of primary human osteoblasts provides a more realistic representation than immortalized cells. These cannot replace primary human cells [32]. Similarities in mineralization and cell proliferation between primary human osteoblasts and MC3T3-E1 cells are reported. Furthermore, SaOs2 and

MG-63 cells demonstrated a higher proliferation rate than primary human osteoblasts. In addition, SaOs2, but not MG-63 cells demonstrated similar mineralization potential, gene regulation and alkaline phosphatase activity compared to primary human osteoblasts [32].

For alkaline phosphatase activity, the absolute mean values differed only slightly between the test and control groups. Although the values for the plasma-treated titanium samples were statistically significantly higher (at 4 and 12 h), the difference in the mean values was small. The statistical significance here is probably due to the low standard deviations. This indicates well-controlled and reproducible conditions but should not be overinterpreted. The values for the test and control groups after 4 h, 12 h, and 24 h indicate that the primary human osteoblasts accepted the micro-structured titanium as a surface for settling. The alkaline phosphatase activity of rat bone marrow cells was higher at 7 and 14 days in the plasma-treated titanium group than in the control group. Plasma treatment therefore increases alkaline phosphatase activity [26]. These results are also confirmed by Hayashi using rat bone marrow cells, with the best achieved results being obtained using the argon plasma jet [27]. Lee et al. also investigated the alkaline phosphatase activity of the human osteosarcoma cell line used. After 7 days, it was 81.5 % higher than the control [23].

To the best of the authors' knowledge, there is currently no study that has also examined the alkaline phosphatase activity after 4, 12, and 24 h of primary human osteoblasts. A direct comparison with the results in the literature is therefore difficult. In principle, the results obtained so far indicate that the alkaline phosphatase activity increases after plasma treatment of the surface. This is supported by the fact that the literature also describes an increase in cell proliferation compared to the controls. The increase in cell proliferation is consistent with our findings. In this study, cell proliferation did not correspond to the alkaline phosphatase activity, which is a marker of osteoblast activity, early bone differentiation, and bone formation [26]. It is possible that the alkaline phosphatase is comparable because the cells find favorable conditions on both the plasma-treated and the untreated titanium surfaces. The different cell proliferation values indicate that the cells spread more actively on the plasma-treated titanium.

The titanium test specimens were fabricated from pure, medical-grade titanium (titanium grade 2) and had surface configurations typical of intra-osseous implants. Treatment was performed using a miniaturized plasma source that has been previously used and investigated [26]. The experiments were performed under standard environmental conditions. Parameters were carefully selected to ensure that biologically acceptable temperatures were maintained in the treatment area. This allowed for the most realistic possible approximation of real conditions in practice. The cultivation time was limited to a maximum of 24 h in order to avoid the need for a prophylactic antibiotic addition to the medium. The analytical methods used focused on visualizing the attachment, morphology, and biological activity of primary human osteoblasts. Both plasma-treated and untreated control samples showed cell proliferation for up to 24 h. This supports previous research indicating the exceptional biocompatibility of titanium [33].

However, there are clear limitations to this study. This is an in vitro study, so it is not possible to assess the extent to which the effects described might also be observed in a clinical setting (e.g., influence of saliva and blood). The cells were derived from the trabecular bone, not from the mandible, and were therefore purchased rather than harvested from the patient. Studies investigating the effect of primary human osteoblasts of the mandible are needed. The results also have limitations: after 24 h, the beneficial effect of the plasma pre-treatment of titanium on the fibroblasts (cell distribution and alkaline phosphatase activity) appears to be minimal compared to the control. It would be interesting to see if similar results could be demonstrated in vivo. If the results are similar, in the overall assessment, no decisive effect of the plasma pre-treatment of titanium on primary human osteoblasts can be determined. Then this would only be the case initially.

Nevertheless, the results highlight the wide range of potential applications for cold atmospheric plasma in a variety of dental fields. The results of this study may be of value in dental practice. The application of cold atmospheric plasma to titanium increases the wettability of the surface and thus improves the distribution of biological substances and cells applied to the surface. These results are consistent with our own and with other researchers' findings.

The present in vitro study shows that primary human osteoblasts on micro-structured titanium exhibit a faster cell proliferation and initial homogeneous distribution as well as a higher alkaline phosphatase activity after 4 and 12 h, which opens up the application of cold atmospheric plasmas in implant dentistry. The technique can be used in pre-implantation conditioning and to regenerate lost attachment caused by peri-implantitis [17,34]. Further studies are needed to evaluate the potential and opportunities of cold atmospheric plasma in dentistry, especially in a clinical setting [34]. However, it is undeniable that its implementation offers exciting possibilities, particularly for improving the success of implants.

5. Conclusions

The results of the in vitro study show that cold atmospheric plasma treatment improves the initial attachment of primary human osteoblasts. After 12 h, osteoblasts had covered the entire surface of the pre-treated titanium, whereas cells on the untreated titanium remained in the area of application. However, the areas of attachment of individual cells were not visually different in the fluorescence microscopic images. After 24 h, the distribution of cells on both titanium surfaces and their respective attachment areas were identical. After 4 h of culture, slight advantages were observed for alkaline phosphatase activity and clear advantages for cell proliferation as assessed by the cell-proliferation WST-1-assay. After a longer culture period of 24 h, the benefits of alkaline phosphatase activity diminished, demonstrating the exceptional biocompatibility of the micro-structured titanium used. In terms of cell proliferation, the experimental group showed improved performance, which was also statistically significant. These results suggest that cold atmospheric plasma may have clinical applications in peri-implantitis treatment and in pre-implantation. Future studies in a clinical setting are required, as the results are not immediately applicable to practice.

Author Contributions: The study was planned by S.R., M.H. and A.S. The experimental work was conducted by S.R. and M.L. The data analysis and interpretation were conducted by M.P.G., J.N., S.R., A.L., M.H., A.S. and M.L. The manuscript draft was written by M.P.G., S.R. and J.N. The manuscript revision was conducted by S.R., J.N., M.P.G., M.H., A.L., M.L. and A.S. All authors have read and agreed to the published version of the manuscript.

Funding: This study was funded partly by the German Federal Ministry of Education and Research (BMBF FKZ 01 EZ 0730/0731).

Institutional Review Board Statement: Not applicable.

Informed Consent Statement: Not applicable.

Data Availability Statement: The data supporting the findings of this study are available from the corresponding author upon reasonable request.

Acknowledgments: The authors thank Andreas Schubert for the technical support.

Conflicts of Interest: Author Antje Lehmann was employed by the company ADMEDES GmbH. The remaining authors declare that the research was conducted in the absence of any commercial or financial relationships that could be construed as a potential conflict of interest. The company ADMEDES GmbH had no role in the design of the study; in the collection, analyses, or interpretation of data; in the writing of the manuscript, or in the decision to publish the results.

Abbreviations

CAP, cold atmospheric plasma; WST-1 4, water soluble tetrazolium 4.

References

1. Jiao, Y.; Tay, F.R.; Niu, L.N.; Chen, J.H. Advancing antimicrobial strategies for managing oral biofilm infections. *Int. J. Oral Sci.* **2019**, *11*, 28. [CrossRef]
2. Kamionka, J.; Matthes, R.; Holtfreter, B.; Pink, C.; Schlüter, R.; von Woedtke, T.; Kocher, T.; Jablonowski, L. Efficiency of cold atmospheric plasma, cleaning powders and their combination for biofilm removal on two different titanium implant surfaces. *Clin. Oral Investig.* **2022**, *26*, 3179–3187. [CrossRef]
3. Duske, K.; Koban, I.; Kindel, E.; Schröder, K.; Nebe, B.; Holtfreter, B.; Jablonowski, L.; Weltmann, K.D.; Kocher, T. Atmospheric plasma enhances wettability and cell spreading on dental implant metals. *J. Clin. Periodontol.* **2012**, *39*, 400–407. [CrossRef] [PubMed]
4. Walter, N.; Stich, T.; Docheva, D.; Alt, V.; Rupp, M. Evolution of implants and advancements for osseointegration: A narrative review. *Injury* **2022**, *53*, 369–373. [CrossRef] [PubMed]
5. Souza, J.C.M.; Sordi, M.B.; Kanazawa, M.; Ravindran, S.; Henriques, B.; Silva, F.S.; Aparicio, C.; Cooper, L.F. Nano- scale modification of titanium implant surfaces to enhance osseointegration. *Acta Biomater.* **2019**, *94*, 112–131. [CrossRef] [PubMed]
6. Gund, M.P.; Naim, J.; Lehmann, A.; Hannig, M.; Linsenmann, C.; Schindler, A.; Rupf, S. Effects of Cold Atmospheric Plasma Pre-Treatment of Titanium on the Biological Activity of Primary Human Gingival Fibroblasts. *Biomedicines* **2023**, *11*, 1185. [CrossRef] [PubMed]
7. Afshari, M.; Amini, S.; Hashemibeni, B. Effect of low frequency ultrasound waves on the morphology and viability of cultured human gingival fibroblasts. *J. Taibah Univ. Med. Sci.* **2023**, *18*, 1406–1416. [CrossRef] [PubMed]
8. Smeets, R.; Stadlinger, B.; Schwarz, F.; Beck-Broichsitter, B.; Jung, O.; Precht, C.; Kloss, F.; Gröbe, A.; Heiland, M.; Ebker, T. Impact of Dental Implant Surface Modifications on Osseointegration. *BioMed Res.* **2016**, *2016*, 217–221. [CrossRef]
9. Jiang, X.; Yao, Y.; Tang, W.; Han, D.; Zhang, L.; Zhao, K.; Wang, S.; Meng, Y. Design of dental implants at materials level: An overview. *J. Biomed. Mater.* **2020**, *108*, 1634–1661. [CrossRef] [PubMed]
10. Jemat, A.; Ghazali, M.J.; Razali, M.; Otsuka, Y. Surface Modifications and Their Effects on Titanium Dental Implants. *BioMed Res.* **2015**, *2015*, 1634–1661. [CrossRef]
11. Parnia, F.; Yazdani, J.; Javaherzadeh, V.; Maleki Dizaj, S. Overview of Nanoparticle Coating of Dental Implants for Enhanced Osseointegration and Antimicrobial Purposes. *J. Pharm. Pharm. Sci.* **2017**, *20*, 148–160. [CrossRef] [PubMed]
12. Carossa, M.; Cavagnetto, D.; Mancini, F.; Mosca Balma, A.; Mussano, F. Plasma of Argon Treatment of the Implant Surface, Systematic Review of In Vitro Studies. *Biomolecules* **2022**, *12*, 1219. [CrossRef] [PubMed]
13. Wagner, G.; Eggers, B.; Duddeck, D.; Kramer, F.J.; Bourauel, C.; Jepsen, S.; Deschner, J.; Nokhbehsaim, M. Influence of cold atmospheric plasma on dental implant materials—An in vitro analysis. *Clin. Oral Investig.* **2022**, *26*, 2949–2963. [CrossRef]
14. Canullo, L.; Genova, T.; Mandracci, P.; Mussano, F.; Abundo, R.; Fiorellini, J.P. Morphometric Changes Induced by Cold Argon Plasma Treatment on Osteoblasts Grown on Different Dental Implant Surfaces. *Int. J. Periodontics Restor. Dent.* **2017**, *37*, 541–548. [CrossRef] [PubMed]
15. Matthes, R.; Duske, K.; Kebede, T.G.; Pink, C.; Schlüter, R.; von Woedtke, T.; Weltmann, K.D.; Kocher, T.; Jablonowski, L. Osteoblast growth, after cleaning of biofilm-covered titanium discs with air-polishing and cold plasma. *J. Clin. Periodontol.* **2017**, *44*, 672–680. [CrossRef]
16. Tominami, K.; Kanetaka, H.; Sasaki, S.; Mokudai, T.; Kaneko, T.; Niwano, Y. Cold atmospheric plasma enhances osteoblast differentiation. *PLoS ONE* **2017**, *12*, e0180507. [CrossRef] [PubMed]
17. Xie, Y.T.; Wang, Q.; Lin, Z.; Li, S.J.; He, J.; Zhang, X.W.; Deng, C.F.; Jiang, L.L.; Zhao, B.H. The Effects of Air Cold Atmospheric Plasma on Cellular Early Attachment, Proliferation and Migration on Pure Titanium Surfaces. *Sci. Adv. Mater.* **2019**, *11*, 1392–1401. [CrossRef]
18. Rabel, K.; Kohal, R.J.; Steinberg, T.; Rolauffs, B.; Adolfsson, E.; Altmann, B. Human osteoblast and fibroblast response to oral implant biomaterials functionalized with non-thermal oxygen plasma. *Sci. Rep.* **2021**, *11*, 17302. [CrossRef]
19. Henningsen, A.; Smeets, R.; Hartjen, P.; Heinrich, O.; Heuberger, R.; Heiland, M.; Precht, C.; Cacaci, C. Photofunctionalization and non-thermal plasma activation of titanium surfaces. *Clin. Oral Investig.* **2018**, *22*, 1045–1054. [CrossRef]
20. Han, I.; Vagaska, B.; Seo, H.J.; Kang, J.K.; Kwon, B.J.; Lee, M.H.; Park, J.C. Promoted cell and material interaction on atmospheric pressure plasma treated titanium. *Appl. Surf. Sci.* **2012**, *258*, 4718–4723. [CrossRef]
21. González-Blanco, C.; Rizo-Gorrita, M.; Luna-Oliva, I.; Serrera-Figallo, M.Á.; Torres-Lagares, D.; Gutiérrez-Pérez, J.L. Human Osteoblast Cell Behaviour on Titanium Discs Treated with Argon Plasma. *Materials* **2019**, *12*, 1735. [CrossRef]
22. Komasa, S.; Kusumoto, T.; Hayashi, R.; Takao, S.; Li, M.; Yan, S.; Zeng, Y.; Yang, Y.; Hu, H.; Kobayashi, Y.; et al. Effect of Argon-Based Atmospheric Pressure Plasma Treatment on Hard Tissue Formation on Titanium Surface. *Int. J. Mol. Sci.* **2021**, *22*, 7617. [CrossRef] [PubMed]
23. Lee, H.; Jeon, H.J.; Jung, A.; Kim, J.; Kim, J.Y.; Lee, S.H.; Kim, H.; Yeom, M.S.; Choe, W.; Gweon, B.; et al. Improvement of osseointegration efficacy of titanium implant through plasma surface treatment. *Biomed. Eng. Lett.* **2022**, *12*, 421–432. [CrossRef]
24. Long, L.; Zhang, M.; Gan, S.; Zheng, Z.; He, Y.; Xu, J.; Fu, R.; Guo, Q.; Yu, D.; Chen, W. Comparison of early osseointegration of non-thermal atmospheric plasma-functionalized/ SLActive titanium implant surfaces in beagle dogs. *Front. Bioeng. Biotechnol.* **2022**, *10*, 965248. [CrossRef] [PubMed]
25. Wang, L.; Wang, W.; Zhao, H.; Liu, Y.; Liu, J.; Bai, N. Bioactive Effects of Low-Temperature Argon-Oxygen Plasma on a Titanium Implant Surface. *ACS Omega* **2020**, *5*, 3996–4003. [CrossRef] [PubMed]

26. Ujino, D.; Nishizaki, H.; Higuchi, S.; Komasa, S.; Okazaki, J. Effect of Plasma Treatment of Titanium Surface on Biocompatibility. *Appl. Sci.* **2019**, *9*, 2257. [CrossRef]
27. Hayashi, R.; Takao, S.; Komasa, S.; Sekino, T.; Kusumoto, T.; Maekawa, K. Effects of Argon Gas Plasma Treatment on Biocompatibility of Nanostructured Titanium. *Int. J. Mol. Sci.* **2024**, *25*, 149. [CrossRef]
28. Guo, L.; Zou, Z.; Smeets, R.; Kluwe, L.; Hartjen, P.; Cacaci, C.; Gosau, M.; Henningsen, A. Time Dependency of Non-Thermal Oxygen Plasma and Ultraviolet Irradiation on Cellular Attachment and mRNA Expression of Growth Factors in Osteoblasts on Titanium and Zirconia Surfaces. *Int. J. Mol. Sci.* **2020**, *21*, 8598. [CrossRef]
29. Swart, K.M.; Keller, J.C.; Wightman, J.P.; Draughn, R.A.; Stanford, C.M.; Michaels, C.M. Short-term plasma-cleaning treatments enhance in vitro osteoblast attachment to titanium. *J. Oral Implant.* **1992**, *18*, 130–137.
30. Seon, G.M.; Seo, H.J.; Kwon, S.Y.; Lee, M.H.; Kwon, B.J.; Kim, M.S.; Koo, M.A.; Park, B.J.; Park, J.C. Titanium surface modification by using microwave-induced argon plasma in various conditions to enhance osteoblast biocompatibility. *Biomater. Res.* **2015**, *19*, 13. [CrossRef]
31. Tsujita, H.; Nishizaki, H.; Miyake, A.; Takao, S.; Komasa, S. Effect of plasma treatment on titanium surface on the tissue surrounding implant material. *Int. J. Mol. Sci.* **2021**, *22*, 6931. [CrossRef] [PubMed]
32. Czekanska, E.M.; Stoddart, M.J.; Ralphs, J.R.; Richards, R.G.; Hayes, J.S. A phenotypic comparison of osteoblast cell lines versus human primary osteoblasts for biomaterials testing. *J. Biomed. Mater. Res. A* **2014**, *102*, 2636–2643. [CrossRef] [PubMed]
33. Zhang, W.S.; Liu, Y.; Shao, S.Y.; Shu, C.Q.; Zhou, Y.H.; Zhang, S.M.; Qiu, J. Surface characteristics and in vitro biocompatibility of titanium preserved in a vitamin C-containing saline storage solution. *J. Mater. Sci. Mater. Med.* **2024**, *35*, 3. [CrossRef] [PubMed]
34. Alqutaibi, A.Y.; Aljohani, A.; Alduri, A.; Masoudi, A.; Alsaedi, A.M.; Al-Sharani, H.M.; Farghal, A.E.; Alnazzawi, A.A.; Aboalrejal, A.N.; Mohamed, A.H.; et al. The Effectiveness of Cold Atmospheric Plasma (CAP) on Bacterial Reduction in Dental Implants: A Systematic Review. *Biomolecules* **2023**, *13*, 1528. [CrossRef]

 *biomedicines*

Article

Selective Effects of Cold Atmospheric Plasma on Bone Sarcoma Cells and Human Osteoblasts

Andreas Nitsch [1], Konrad F. Sieb [1], Sara Qarqash [1], Janosch Schoon [1], Axel Ekkernkamp [1,2], Georgi I. Wassilew [1], Maya Niethard [1,3] and Lyubomir Haralambiev [1,2,*]

[1] Center for Orthopedics, Trauma Surgery and Rehabilitation Medicine, University Medicine Greifswald, Ferdinand-Sauerbruch-Straße, 17475 Greifswald, Germany
[2] Department of Trauma and Orthopaedic Surgery, BG Klinikum Unfallkrankenhaus Berlin, Warener Straße 7, 12683 Berlin, Germany
[3] Sarcoma Centre, HELIOS-Klinikum Berlin-Buch, Schwanebecker Chaussee 50, 13125 Berlin, Germany
[*] Correspondence: lyubomir.haralambiev@med.uni-greifswald.de; Tel.: +49-3834-8622541

Abstract: Background: The use of cold atmospheric plasma (CAP) in oncology has been intensively investigated over the past 15 years as it inhibits the growth of many tumor cells. It is known that reactive oxidative species (ROS) produced in CAP are responsible for this effect. However, to translate the use of CAP into medical practice, it is essential to know how CAP treatment affects non-malignant cells. Thus, the current in vitro study deals with the effect of CAP on human bone cancer cells and human osteoblasts. Here, identical CAP treatment regimens were applied to the malignant and non-malignant bone cells and their impact was compared. Methods: Two different human bone cancer cell types, U2-OS (osteosarcoma) and A673 (Ewing's sarcoma), and non-malignant primary osteoblasts (HOB) were used. The CAP treatment was performed with the clinically approved kINPen MED. After CAP treatment, growth kinetics and a viability assay were performed. For detecting apoptosis, a caspase-3/7 assay and a TUNEL assay were used. Accumulated ROS was measured in cell culture medium and intracellular. To investigate the influence of CAP on cell motility, a scratch assay was carried out. Results: The CAP treatment showed strong inhibition of cell growth and viability in bone cancer cells. Apoptotic processes were enhanced in the malignant cells. Osteoblasts showed a higher potential for ROS resistance in comparison to malignant cells. There was no difference in cell motility between benign and malignant cells following CAP treatment. Conclusions: Osteoblasts show better tolerance to CAP treatment, indicated by less affected viability compared to CAP-treated bone cancer cells. This points toward the selective effect of CAP on sarcoma cells and represents a further step toward the clinical application of CAP.

Keywords: cold atmospheric plasma; human osteoblast cells; bone cancer; osteosarcoma cells; Ewing's sarcoma; apoptosis; reactive oxygen species; hydrogen peroxide

Citation: Nitsch, A.; Sieb, K.F.; Qarqash, S.; Schoon, J.; Ekkernkamp, A.; Wassilew, G.I.; Niethard, M.; Haralambiev, L. Selective Effects of Cold Atmospheric Plasma on Bone Sarcoma Cells and Human Osteoblasts. *Biomedicines* **2023**, *11*, 601. https://doi.org/10.3390/biomedicines11020601

Academic Editor: Christoph Viktor Suschek

Received: 26 January 2023
Revised: 14 February 2023
Accepted: 15 February 2023
Published: 17 February 2023

1. Introduction

Tumors affecting the skeleton are a great medical and economical challenge. Independent of their origin—primary bone cancer or metastases—these tumors destroy the bone structure and function [1–3]. The treatment and prognosis of bone tumors depend on the one hand on their entity, localization, and spread, on the other hand, on patient characteristics such as age, morbidity, etc. [4]. Currently, surgery for radical tumor excision, adjuvant and neoadjuvant chemotherapy, and radiation as a single therapy or their combinations are state of the art in bone cancer treatment [5–7]. Local treatments such as laser ablation, thermal coagulation [8], or cryosurgery [9] serve as additional methods; however, they are not sufficient to be used as the sole therapy. The current treatment methods, especially chemotherapy, have plenty of undesirable side effects such as changes in bone development, increased bone resorption, and thus an increased risk of bone fractures [10–12]. Alternative techniques for intraoperative bone cancer therapy are needed.

In the last 10 years, cold atmospheric plasma (CAP) was introduced as a promising local anti-cancer therapy [13]. CAP is an ionized gas that appears very suitable for use on human tissue due to its body-like temperature (approx. 40 °C). The biological effects of CAP are associated with the numerous reactive species, ions, free electrons, electromagnetic fields, and the low level of UV radiation it contains. The reactive oxygen and nitrogen species (RONS), such as H_2O_2 [14,15], NO_2^- [16], and $ONOO^-$ [17,18] generated in CAP are also thought to be responsible for the mechanisms of action of CAP treatment in malignant cells. The cytotoxic effects of CAP are related to an increase in intracellular ROS [19,20], DNA damage [21,22], the disabling of antioxidant defenses [23,24], but also a direct influence on the cell life cycle such as cell arrest [25–27] and apoptosis [28,29] or necrotic cell death [30]. In numerous types of cancer such as glioblastoma, ovarian [31,32], gastric [33], pancreatic [34–36], lung [37], or colorectal carcinoma [35] breast cancer [38–40], and melanoma [41,42]—which also preferentially metastasize to the bones [43,44]— but also in primary bone tumors such as osteosarcoma, chondrosarcoma, and Ewing's sarcoma, the anti-cancer effect of CAP has been proven in vitro [45–47]. CAP also induced growth inhibition in bone sarcoma cells, the stimulation of cell apoptosis, and the impairment of cell membrane functions [45–52]. Previous studies indicate that the plasma source and dose can have apoptotic effects on cancer cells while not affecting healthy cells [53–55]. However, a comparison of malignant bone cells and primary non-malignant bone cells with regard to possible selective in vitro effects of CAP remains elusive. Thus, the aim of this study is to treat osteosarcoma cells, Ewing's sarcoma cells, and primary human osteoblasts with CAP and to compare the in vitro effects with respect to ROS accumulation, cell viability, and apoptosis.

2. Materials and Methods

2.1. Cell Culture

Two different human bone sarcoma cell lines were used: U2-OS (Osteosarcoma; Cell Lines Service, Eppelheim, Germany) and A673 (Ewing's sarcoma; American Type Culture Collection, Manassas, VA, USA). In addition, non-malignant bone cells were used: human osteoblasts HOB (PromoCell, Heidelberg, Germany). U2-OS and A673 cells were cultured in Dulbecco's modified Eagle's medium (DMEM) containing 1.0 g/L glucose, 10% fetal bovine serum, 1 mM sodium pyruvate, and 1% penicillin/streptomycin (all reagents from PAN Biotech, Aidenbach, Germany). Human osteoblasts (HOB) were cultured in Human Osteoblast Growth Medium with Supplement Mix (HOBM; obtained from PromoCell, Heidelberg, Germany). All cells were incubated at 37 °C and 5% CO_2.

2.2. Cold Atmospheric Plasma Treatment

CAP treatment was performed with the kIN-Pen® MED plasma jet (Neoplas tools, Greifswald, Germany). The flow rate of the carrier gas argon was adjusted to 3 slm. Control cells were treated analogously but without igniting the gas plasma, i.e., only by argon gas flow. Cell suspensions were treated for different durations (specified in the corresponding methods section).

2.3. Growth Kinetics

In total, 2×10^3 cells were suspended in 200 μL cell culture medium and were transferred to a 24-well plate and treated with CAP or carrier gas argon for 10 s. Immediately after the treatment, 800 μL cell culture medium was added to the wells. The cells were harvested with gentle trypsinization after 24, 48, 72, 96, and 120 h. The number of cells was determined using the CASY cell counter and analyzer model TT (OLS OMNI Life Science, Bremen, Germany). At least three independent experiments were performed.

2.4. Viability Assay

In total, 5×10^4 (24 h and 48 h) or 1×10^4 (120h) cells were suspended in 200 μL cell culture medium and were transferred to a 24-well plate and treated with CAP or carrier gas argon for 5 s, 10 s, and 20 s (24 h and 48 h) or 5 s, 10 s, 20 s, 30 s, and 60 s (120 h). The cell suspensions were transferred in 96-plate and incubated at 37 °C and 5% CO_2 over 24 h, 48 h, or 120 h. Cell viability was determined using CellTiter-Blue® Cell Viability Assay (Promega GmbH, Walldorf, Germany). The assay was carried out according to the manufacturer's instructions [56,57]. The incubation period of the reagent was 1 h. The fluorescence intensity at 560/590 nm was recorded with a TECAN m200 multiplate reader (TECAN, Männedorf, Switzerland). At least three independent experiments were performed, each with triplicates. The fluorescence intensities of the CAP-treated samples were normalized to those of the carrier gas-treated controls.

2.5. Caspase-3/7-Assay

In total, 5.0×10^4 cells were treated with CAP or argon for 10 s and incubated for 24 h and 48 h. To normalize the quantified fluorescence intensity to the cell number, a second plate was carried out parallel. After the incubation period, the used medium was removed and 100 µL of Caspase 3/7 detection solution DPBS with 2 µM CellEvent™ Caspase 3/7 Green Detection Reagent (Thermo Fisher Scientific, Waltham, MA, USA) was incubated for 45 min. The fluorescence at 495/535 nm was recorded with a TECAN m200 multiplate reader (TECAN, Männedorf, Switzerland).

2.6. TUNEL-Assay

In total, 5.0×10^4 cells were treated with CAP or argon for 10 s and incubated for 24 h and 48 h. The TiterTACS™ Colorimetric Apoptosis Detection Kit (Trevigen, Gaithersburg, MD, USA) was used according to the manufacturer's instructions [58]. Absorption at 450 nm was quantified using the Infinite M200 plate reader (Tecan, Männedorf, Switzerland). The absorption of the samples was normalized to cell numbers using a parallel second plate.

2.7. Live-Dead Staining

The cells were treated with CAP or carrier gas argon for 10 s and were transferred to 96-well plates. After 24 h incubation, cells were stained with a live/dead cell imaging kit (Thermo Fisher Scientific, Waltham, MA, USA).

2.8. Intracellular Oxidative Stress Quantification

The cells were harvested, and the suspension was diluted to 1.0×10^6 cells per milliliter in medium. The cell suspension (200 µL) was transferred in wells of a 24-well plate and treated with CAP or argon for 10 s. As a positive control, the cells were treated with 500 µM H_2O_2. After 1 h incubation, the cells were stained with CellROX deep red (Thermo Fisher Scientific, Waltham, MA, USA) and incubated for 30 min. After centrifugation, the labeled cells were resuspended in measuring buffer and analyzed in an Attunue™ Flow Cytometer (Thermo Fisher Scientific, Waltham, MA, USA) and evaluated with FlowJo Software Version 10 (Tree Star Inc., Ashland, OR, USA). The gating strategy is shown in Figure 1. At least three independent experiments were performed. The mean fluorescence intensity (MFI) was normalized to the MFI of the argon-treated control cells.

2.9. Quantification of Hydrogen Peroxide Formation

The cells were harvested and diluted to 1.0×10^6 cells per milliliter in HOBM und DMEM. Further, 200 µL cell suspension was treated with CAP for 0 s, 5 s, 10 s, 20 s, and 40 s. In addition, 200 µL of medium without cells was treated identically. Immediately after the treatment the medium or cell-free supernatant cells were transferred in a 96-well plate and diluted to 1:100, and the Amplex Red hydrogen peroxide assay (Thermo Fisher Scientific, Waltham, MA, USA) was carried out according to the manufacturer's instructions [59]. After 6 h of incubation, the H_2O_2 concentration was quantified again.

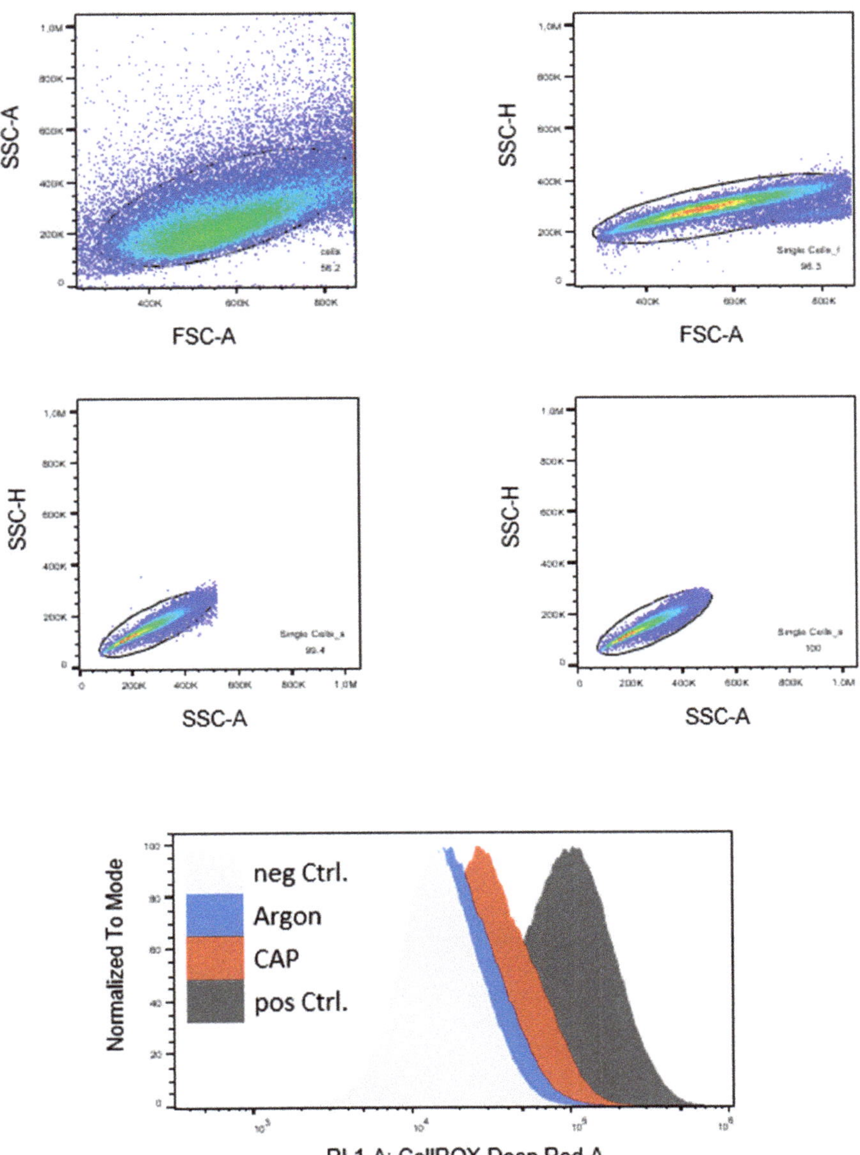

Figure 1. Human osteoblasts (HOB) and bone cancer cells (U2-OS and A673) were treated with cold atmospheric plasma (CAP) or carrier gas argon for 10 s. Controls were performed without treatment (neg Ctrl.) and with 5500 µm H_2O_2. Here, U2-OS is shown as a representative example. Gating strategy: Debris and doublets were excluded by forward- and side-scatter characteristics. The mean fluorescence intensity (MFI) of CellROX deep red was compared. SSC-A: side-scatter area, SSC-H: side-scatter height, FSC-W: forward-scatter width, FSC-H: forward-scatter height.

2.10. Scratch-Assay

In total, 1×10^5 cells were seeded in the wells of 2-well cell culture inserts (ibidi, Gräfeling, Germany) 24 h before the start of the assay. The insert was removed, and the cells were washed twice with DPBS, and 200 µL CAP- or argon-treated medium was added. To prevent the scratch from being closed by proliferation, the assays were performed under low serum conditions. With the software Zen 2012 pro, the cell-free area was recorded over 24 h in intervals of 2 h. The evaluation of the cell-free area was determined with ImageJ software. The quantified cell-free areas were normalized to the cell-free area at the beginning of the experiment.

2.11. Statistic

Unless otherwise stated, all data were depicted as mean values with standard deviation. At least three independent experiments were performed. The differences between the groups were evaluated using the t-test, ANOVA, and two-way ANOVA with a post hoc Tukey test. The software GraphPad Prism 9.1.2 was used for the evaluation and the graphic processing.

3. Results

3.1. Effects of Cell Growth and Viability

The cells were treated with CAP in a medium suspension for 10 s, to investigate the effects of CAP treatment on non-malignant human osteoblast and bone cancer cells. The number of viable cells was determined after 24 h, 48 h, 72 h, 96 h, and 120 h. Whereas cell proliferation of malignant cells is significantly inhibited after a single CAP treatment, the effects on non-malignant cells are less pronounced (Figure 2A–C). Complementary to the live cell count analysis, the cell viability was also examined. At 24 h and 48 h after treatment, the malignant cells showed a treatment time-dependent reduction in cell viability. In contrast, there was no reduction in cell viability in the non-malignant human osteoblast cells (Figure 2D–F). The different CAP treatment times were examined for up to 60 s, and the cells were incubated for up to 120 h. Compared to the control cells, treated with the carrier gas argon, the cell viability of the human osteoblasts was not significantly reduced by any of the tested CAP treatment times after 120 h. In contrast, a 10 s treatment in both malignant bone cancer cell lines led to a significant reduction in cell viability (Figure 2G).

Although the viability of HOB cells decreased slightly with increasing treatment duration, these effects were not as pronounced as they were for the malignant cell lines. The cell viability of the HOB is less impaired after 60 s CAP treatment than that of the two malignant bone cancer cell lines after 5 s of treatment.

Both the treatment time and the type of cells used had a significant impact on cell viability after 120 h incubation.

The viability of HOBs was less affected by CAP treatment and is significantly different from A673 cells after only 5 s and 10 s of treatment. The differences in cell viability between non-malignant and cancer cells increased with the prolongation of CAP treatment times (Figure 2G).

3.2. Induction von Apoptosis

Furthermore, studies on apoptosis induction in the context of CAP were performed using Caspase-3/7 and TUNEL assays. At 24 h and 48 h after 10 s of CAP treatment, there was significantly increased caspase-3/7 activity in the bone cancer cell lines. In the U2-OS cells, the activity increased 1.5-fold compared to controls. In the A673 cells, the caspase-3/7 activity was increased 2.5-fold after 24 h and almost 4-fold after 48 h. No significant change in caspase activity was found in the human osteoblasts 24 h after treatment. After 48 h, an increase in caspase activity was also observed in the non-malignant cells (Figure 3A–C).

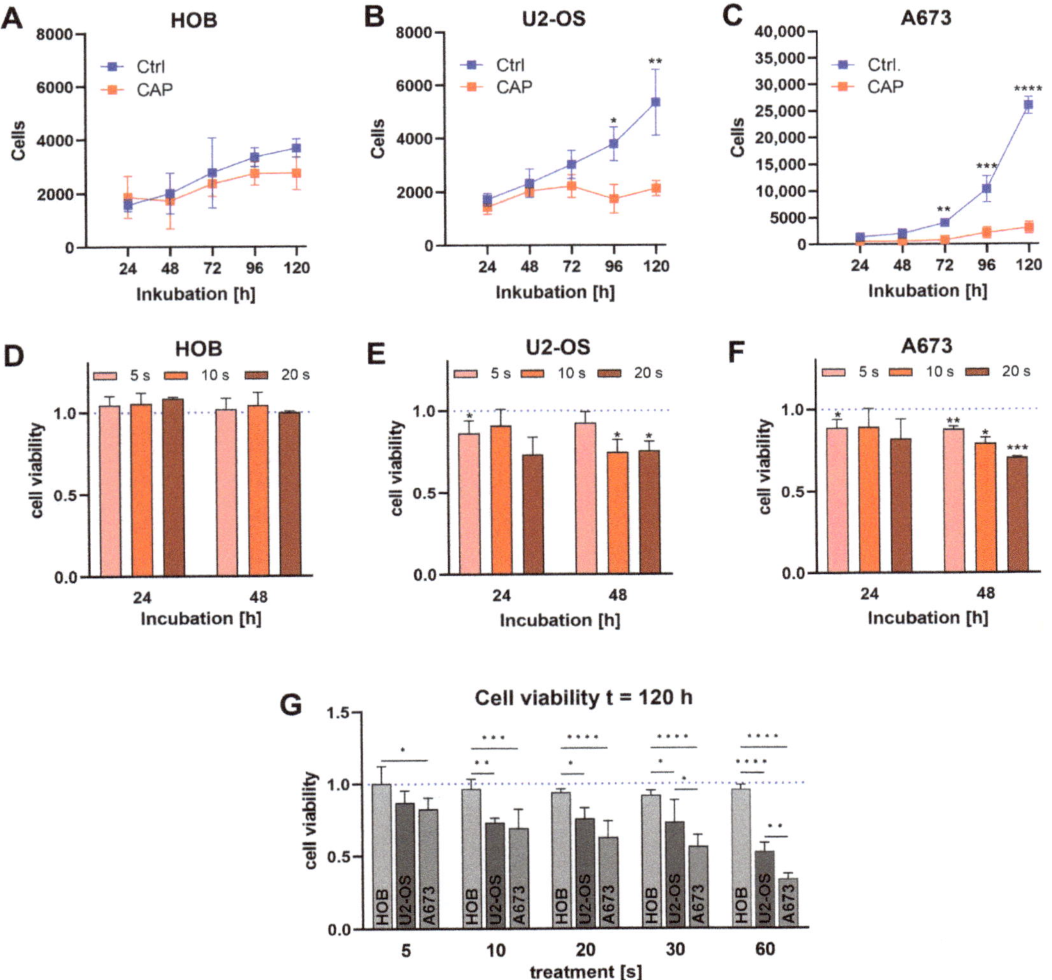

Figure 2. Human osteoblasts (HOB) and bone cancer cells (U2-OS and A673) were treated with cold atmospheric plasma (CAP). During the 120 h incubation, the cell count was determined every 24 h using a CASY Cell Counter and Analyzer (**A–C**). After CAP treatment, the non-malignant and the cancer cells were treated for 24 h, 48 h (**D–F**), and 120 h (**G**). Cell viability was determined using a titer blue assay. Mean values ± SD were normalized to the control treatments. Significant differences are indicated as follows * $p < 0.05$, ** $p < 0.01$, *** $p < 0.001$, **** $p < 0.0001$.

Apoptosis detection using the TUNEL method confirmed the results of the Caspase-3/7 assays. The malignant cells showed significantly increased TUNEL signals both after 24 h and after 48 h. Here, the effect was also more pronounced in the A673 in comparison with the other cells. Although after 48 h showed a tendency to increase the apoptosis rate, HOBs did not show any significant increase in the TUNEL signal (Figure 3D–E).

Live/dead staining confirmed that CAP treatments lead to a reduction in viability and an increase in dead cells. This effect was most evident in the A673 cell line, but no relevant difference was observed between the CAP-treated and the untreated cells in the HOB cells (Figure 3G–I).

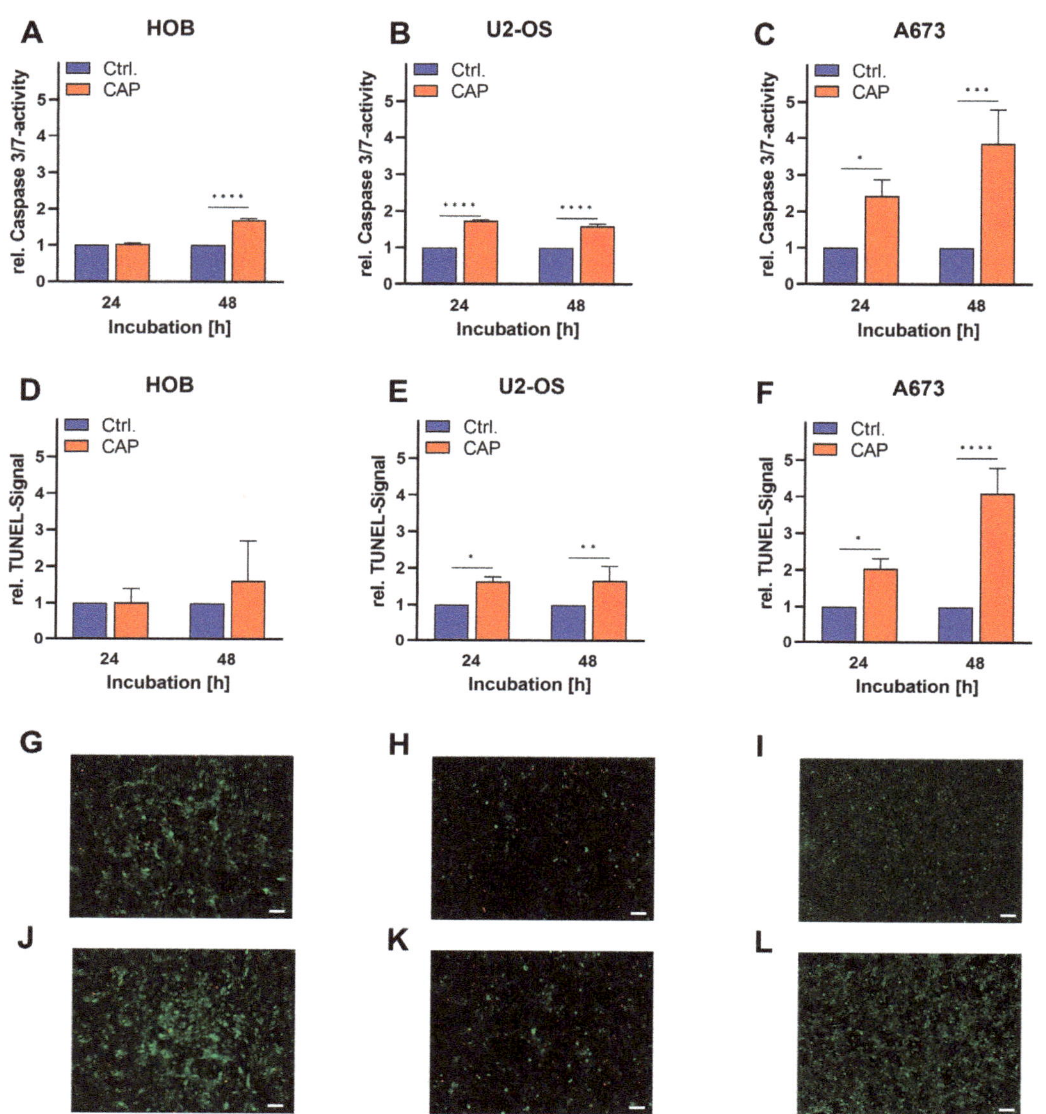

Figure 3. Human osteoblasts (HOB) and bone cancer cells (U2-OS and A673) were treated with cold atmospheric plasma (CAP) or argon for 10 s. After 24 h and 48 h, the apoptosis assays caspase-3/7 (**A–C**) and TUNEL (**D–F**) were performed. Cells were stained live (green)/dead (red) cell imaging kit (**G–L**), first row (**G–I**): argon treated, second row (**J–L**): CAP treated cells; representative images were shown. Data were given as mean ± SD of relative fluorescence (**A–C**) or absorption (**D–F**). Scale indicator is 100 µm. Significant differences are indicated as follows: * $p < 0.05$, ** $p < 0.01$, *** $p < 0.001$, **** $p < 0.0001$.

3.3. Alteration of Extra- and Intracellular Oxidative Levels

One of the most important effects of CAP is the production of various reactive species. Therefore, the cellular oxidative stress after CAP exposure was investigated. CAP treatment led to an increase in oxidative stress in HOB and A673 (Figure 4A). The production of hydrogen peroxide by CAP was investigated by treating both cell culture media with CAP for up to 40 s. A treatment time-dependent formation of hydrogen peroxide was found. A longer treatment time generated a higher concentration of hydrogen peroxide (HOBM: r = 0.994, R^2 = 0.988; DMEM: r = 0.995; R^2 = 0.990). There was no significant difference between the two media types concerning the formation of hydrogen peroxide (Figure 4B).

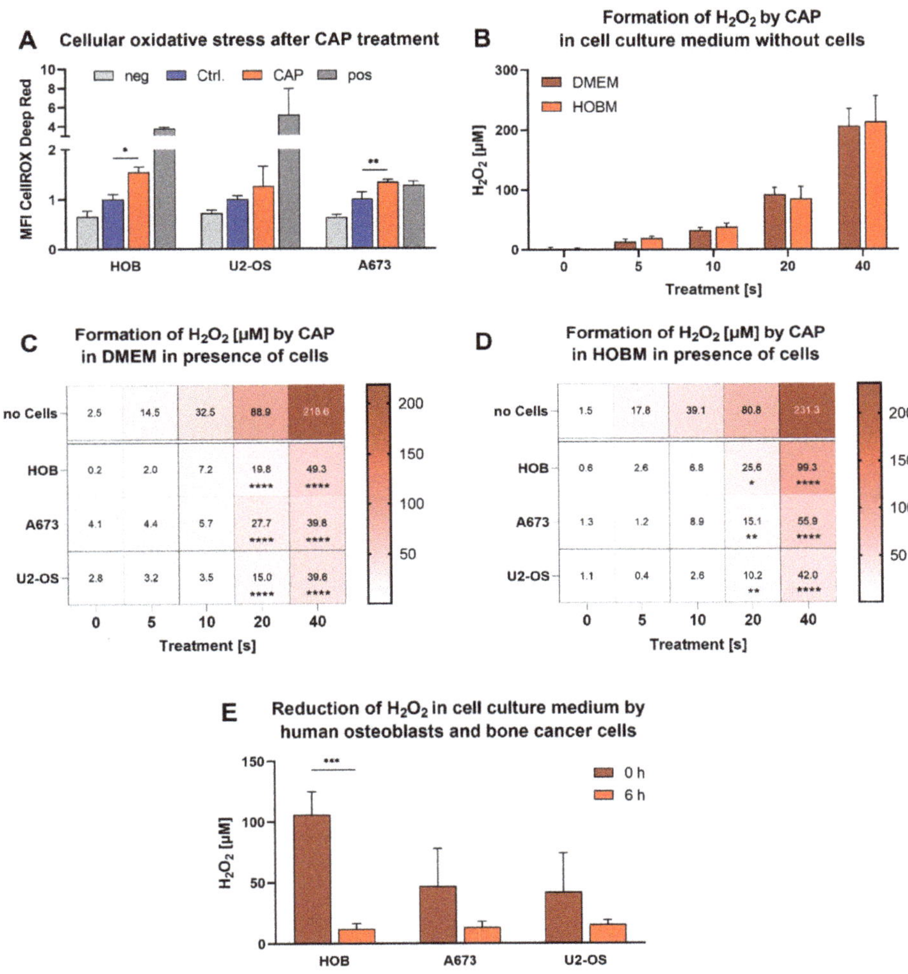

Figure 4. Intracellular oxidative stress increased significantly in human osteoblasts (HOB) and bone cancer cells (U2-OS and A673) after cold atmospheric plasma (CAP) treatment (**A**). Hydrogen peroxide (H_2O_2) was formed after CAP treatment. The amount of hydrogen peroxide formed differed depending on the treatment time, type of cell-free medium (DMEM and human osteoblast medium (HOBM) (**B–D**)) or cell suspension (**C–E**). Data were given as mean fluorescence intensity ± SD (**A**) or as mean concentration ± SD of H_2O_2 determined by a standard curve (**D–E**). Significant differences are indicated as follows: * $p < 0.05$, ** $p < 0.01$, *** $p < 0.001$, **** $p < 0.0001$.

Cell culture supernatants were analyzed to examine the buffering properties of the cells. The treatment time-dependent increase in hydrogen peroxide concentration by the cells was found to be <4-fold reduced (Figure 4C,D). This effect occurred in supernatants of all cell lines and all media used. To investigate the decomposition of hydrogen peroxide, the concentrations were quantified after 6 h of incubation. Here, the concentration declined in the supernatants of all cell lines (Figure 4E).

Wound healing assays were performed to investigate the influence of CAP treatment on cell motility. All cell types were slightly reduced in their motility following CAP treatment, although this effect was not statistically significant. In general, the baseline motility of the cells used differed considerably. Whereas the U2-OS cells had almost completely closed the cell-free area after 18 h, the A673 cells showed no tendency to move within 24 h (Figure 5).

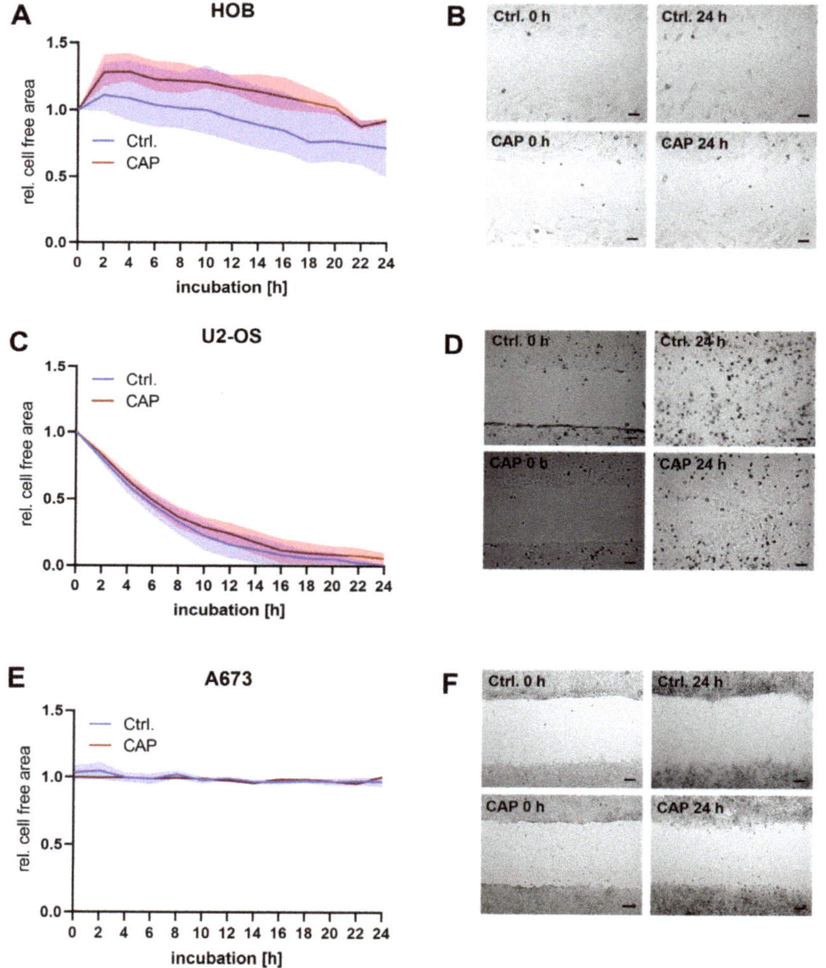

Figure 5. Human osteoblasts (HOB) and bone cancer cells (U2-OS and A673) were incubated in two-well cell culture inserts. After the removal of the insert, cells were treated with cold atmospheric plasma (CAP) or argon-treated medium. The cell-free area was recorded over 24 h in intervals of 2 h (**A,C,E**). The quantified cell-free areas were normalized to the cell-free area at the beginning of the experiment. (**B,D,F**): show representative images. Scale indicator is 100 μm. Data were shown as mean with range.

4. Discussion

The current study was able to find clear differences in the cellular response of bone tumor cells and osteoblasts to the exposure of CAP. Although the CAP-induced formation of ROS was comparably high, the malignant cells showed a higher sensitivity to CAP.

Although many studies have discussed CAP as a promising anti-cancer therapy, its selectivity has rarely been reported [60–63]. In dermatological studies, an increased tolerance to the CAP exposure of skin fibroblasts (non-cancer skin cells) compared to melanoma cells was found. This differential susceptibility of non-malignant skin cells and melanoma cells to CAP exposure underscores the applicability of CAP in the clinical setting [64]. In this study, the main selective effects of CAP on malignant cells are attributed to ROS mechanisms. ROS has a strong impact on numerous biological processes in cells. In normal cells, ROS is controlled by regulation between the silver lining of low and high ROS concentration. ROS has a concentration-dependent influence on tumor cells and their environment. At moderate concentrations, signal cascades such as mitogen-activated protein kinase, c-Jun N-terminal kinase, and vascular endothelial growth factor (VEGF) are stimulated. At high concentrations of ROS, inhibitory effects on the angiogenesis, metastasis, and survival of cancer cells, primarily via the induction of apoptosis [65].

Altered survival signaling was indicated as the driving selective effect of CAP in glioblastoma cells [66]. The current study has shown that CAP treatment inhibits the growth of osteosarcoma and Ewing's sarcoma cells confirming the findings of our previous studies [45,46,51], as well as those by other groups [48,67]. As CA-based reactive species are generated in the extracellular micro-environment, they not only specifically interact with the tumor cells but also with non-malignant cells, such as HOB. However, our in vitro study indicates that the non-malignant HOB shows lower susceptibility to CAP treatments. The loss of viability following CAP treatments is less pronounced in HOB in comparison to bone cancer cells. The possible selective effect of CAP in tumor cells versus non-malignant cells is believed to be due to their biological differences in terms of antioxidant resistance [68]. The production and metabolism of RONS by cancer cells are exploited therapeutically to unleash the effects of oxygen radicals. Thus, the increased ROS concentration exceeds the antioxidant resistance mechanisms of the cancer cell and leads to its apoptosis [68]. In the presented study, this hypothesis was confirmed by showing that apoptosis is enhanced in osteosarcoma and Ewing's sarcoma in comparison to HOB.

The generation of ROS and its impact on the different cell types was quantified since the effects of CAP can be distinctively influenced by the type of cell culture medium [24,69]. Ingredients such as N-acetylcysteine, ascorbic acid, and other antioxidants can largely buffer ROS production by CAP and its effects [16,70–72]. Both DMEM and HOB medium showed a proportional increase in hydrogen peroxide depending on the CAP exposure time. The comparison of equivalent CAP treatment times in cell culture supernatant allowed for evaluating the individual level of ROS exposure. The quantification of the ROS levels 6 h after CAP exposure indicates cell type-specific differences regarding ROS tolerance. The non-malignant cells (HOB) showed a more pronounced tolerance to H_2O_2 than the bone cancer cell lines, although the osteoblasts were exposed to higher ROS concentrations. Although a high concentration of H_2O_2 is produced by CAP treatment, which is harmful to various cell types [73,74], in vitro studies show that the effects of CAP are not simply inducible by H_2O_2 treatment. Rather, the synergistic effect of numerous other reactive species such as NO_2^-/NO_3^- together with hydrogen peroxide is held responsible for the cytotoxic and anti-proliferative effect in cancer cells [18,75–77].

The selectivity of the CAP treatment of osteosarcoma cells was previously reported by others [48]. Hamouda et al. [78] showed that HOB is more resistant to CAP treatment than to the osteosarcoma cell line SaOS-2. Two research groups have also compared the effects of CAP on mesenchymal stromal cells, another cell type from healthy bone, with the CAP effect on osteosarcoma cells. Mesenchymal stromal cells showed better CAP resistance than the bone cancer cells [48,79]. Ermakov et al. found not only a CAP-related increased proliferation activity but also an increased expression of osteogenic differentiation markers in human mesenchymal stem cells [79]. However, another study that focused on the effect of CAP at biocompatible doses on primary mesenchymal stromal cells of the bone marrow from various donors indicates no enhanced osteogenic differentiation potential [80].

Morphological changes in the surface and shape of numerous cancer cells after CAP treatment such as the loss of cytoplasmic protrusions and formation of tiny protuberances have been observed [81–85]. These changes are accompanied by architectural changes of the cytoskeleton such as F-actin [81,84,86], the reduced expression of integrin [81,87,88], and focal adhesion kinase [81], which is often related to reduced migration rates of CAP-treated cancer cells [87,89]. Changes in the cytoskeleton of bone cancer cells have been shown in our previous works [46,49]. The current results indicate a discrete reduction in cell motility; however, no differences were found between malignant and non-malignant bone cells.

Although the effects of CAP produced by different sources of osteosarcoma cells are comparable [51,90] the physical parameters such as voltage, current, power, and electromagnetic field are very individual for different plasma devices and have a direct influence on the plasma parameters and thus on the concentrations of the individual reactive species [62,91–94]. The quantification of the hydrogen peroxide concentration after different CAP exposure times clearly demonstrates a time-dependent increase. The CAP treatment is dosed over the exposure time and must be evaluated differently for each device and cell type [51,79]. Thus, it is necessary to perform studies investigating the effects of CAP from different sources on different bone cancer cells. The current study shows the effect of CAP not only on osteosarcoma cells but also on Ewing's sarcoma cells in direct comparison to the effect on non-malignant bone cells.

5. Conclusions

The results of the presented study indicate a dose-dependent cytotoxic effect of CAP on osteosarcoma and Ewing's sarcoma cells. The ROS-related apoptosis processes in the bone sarcoma cells appear to be largely absent in non-malignant HOBs. The HOBs show a significantly increased CAP resistance and thus better CAP tolerability. Our study provides important information on the possible selectivity of a CAP application for anti-cancer treatment in a pre-clinical in vitro setup.

Author Contributions: Conceptualization, L.H., A.N. and M.N.; methodology, A.N., K.F.S. and L.H.; software, K.F.S. and S.Q.; validation, J.S., K.F.S. and S.Q.; formal analysis, A.N. and M.N.; investigation, L.H. and M.N.; resources, J.S. and L.H.; data curation, L.H., A.E. and G.I.W.; writing—original draft preparation, A.N., M.N. and L.H.; writing—review and editing, J.S., A.E. and G.I.W.; visualization, K.F.S. and A.N.; supervision, A.E. and G.I.W.; project administration, L.H., A.E. and G.I.W.; funding acquisition, A.E. and G.I.W.; All authors have read and agreed to the published version of the manuscript.

Funding: We acknowledge support for the Article Processing Charge from the DFG (German Research Foundation, 393148499) and the Open Access Publication Fund of the University of Greifswald. J.S. receives the Domagk Master Class (DMC) scholarship funded by the University Medicine Greifswald. S.Q. receives the Domagk scholarship funded by the University Medicine Greifswald.

Institutional Review Board Statement: Not applicable.

Informed Consent Statement: Not applicable.

Data Availability Statement: The data that support the findings of this study are available from the corresponding author upon reasonable request.

Acknowledgments: The authors thank Damián Muzzio and Jens Ehrhardt for their technical support and for providing laboratory equipment.

Conflicts of Interest: The authors declare no conflict of interest.

References

1. Croucher, P.I.; McDonald, M.M.; Martin, T.J. Bone metastasis: The importance of the neighbourhood. *Nat. Rev. Cancer* **2016**, *16*, 373–386. [CrossRef] [PubMed]
2. Zhang, Y.; Zhang, L.; Zhang, G.; Li, S.; Duan, J.; Cheng, J.; Ding, G.; Zhou, C.; Zhang, J.; Luo, P.; et al. Osteosarcoma metastasis: Prospective role of ezrin. *Tumor Biol.* **2014**, *35*, 5055–5059. [CrossRef] [PubMed]
3. Ban, J.; Fock, V.; Aryee, D.N.T.; Kovar, H. Mechanisms, Diagnosis and Treatment of Bone Metastases. *Cells* **2021**, *10*, 2944. [CrossRef] [PubMed]
4. Coleman, R.E. Clinical features of metastatic bone disease and risk of skeletal morbidity. *Clin. Cancer Res.* **2006**, *12*, 6243s–6249s. [CrossRef] [PubMed]
5. Andreou, D.; Henrichs, M.P.; Gosheger, G.; Nottrott, M.; Streitbürger, A.; Hardes, J. New surgical treatment options for bone tumors. *Der Pathol.* **2014**, *35* (Suppl. 2), 232–236. [CrossRef]
6. Brown, H.K.; Schiavone, K.; Gouin, F.; Heymann, M.-F.; Heymann, D. Biology of Bone Sarcomas and New Therapeutic Developments. *Calcif. Tissue Int.* **2018**, *102*, 174–195. [CrossRef] [PubMed]
7. Casas-Ganem, J.; Healey, J.H. Advances that are changing the diagnosis and treatment of malignant bone tumors. *Curr. Opin. Rheumatol.* **2005**, *17*, 79–85. [CrossRef]
8. Ringe, K.I.; Panzica, M.; von Falck, C. Thermoablation of Bone Tumors. *Rofo* **2016**, *188*, 539–550. [CrossRef]
9. Deschamps, F.; Farouil, G.; de Baere, T. Percutaneous ablation of bone tumors. *Diagn. Interv. Imaging* **2014**, *95*, 659–663. [CrossRef]
10. Skjødt, M.K.; Frost, M.; Abrahamsen, B. Side effects of drugs for osteoporosis and metastatic bone disease. *Br. J. Clin. Pharmacol.* **2019**, *85*, 1063–1071. [CrossRef]
11. Walczak, B.E.; Irwin, R.B. Sarcoma chemotherapy. *J. Am. Acad. Orthop. Surg* **2013**, *21*, 480–491. [CrossRef] [PubMed]
12. Mavrogenis, A.F.; Papagelopoulos, P.J.; Romantini, M.; Angelini, A.; Ruggieri, P. Side Effects of Chemotherapy in Musculoskeletal Oncology. *J. Long-Term Eff. Med. Implant.* **2010**, *20*, 1–12. [CrossRef]
13. Keidar, M.; Walk, R.; Shashurin, A.; Srinivasan, P.; Sandler, A.; Dasgupta, S.; Ravi, R.; Guerrero-Preston, R.; Trink, B. Cold plasma selectivity and the possibility of a paradigm shift in cancer therapy. *Br. J. Cancer* **2011**, *105*, 1295–1301. [CrossRef] [PubMed]
14. Panngom, K.; Baik, K.Y.; Nam, M.K.; Han, J.H.; Rhim, H.; Choi, E.H. Preferential killing of human lung cancer cell lines with mitochondrial dysfunction by nonthermal dielectric barrier discharge plasma. *Cell Death Dis.* **2013**, *4*, e642. [CrossRef]
15. Ahn, H.J.; Kim, K.I.; Hoan, N.N.; Kim, C.H.; Moon, E.; Choi, K.S.; Yang, S.S.; Lee, J.-S. Targeting Cancer Cells with Reactive Oxygen and Nitrogen Species Generated by Atmospheric-Pressure Air Plasma. *PLoS ONE* **2014**, *9*, e86173. [CrossRef] [PubMed]
16. Ma, Y.; Ha, C.S.; Hwang, S.W.; Lee, H.J.; Kim, G.C.; Lee, K.-W.; Song, K. Non-Thermal Atmospheric Pressure Plasma Preferentially Induces Apoptosis in p53-Mutated Cancer Cells by Activating ROS Stress-Response Pathways. *PLoS ONE* **2014**, *9*, e91947. [CrossRef]
17. Lukes, P.; Dolezalova, E.; Sisrova, I.; Clupek, M. Aqueous-phase chemistry and bactericidal effects from an air discharge plasma in contact with water: Evidence for the formation of peroxynitrite through a pseudo-second-order post-discharge reaction of H_2O_2 and HNO_2. *Plasma Sources Sci. Technol.* **2014**, *23*, 015019. [CrossRef]
18. Graves, D.B. The emerging role of reactive oxygen and nitrogen species in redox biology and some implications for plasma applications to medicine and biology. *J. Phys. D Appl. Phys.* **2012**, *45*, 263001. [CrossRef]
19. Vandamme, M.; Robert, E.; Lerondel, S.; Sarron, V.; Ries, D.; Dozias, S.; Sobilo, J.; Gosset, D.; Kieda, C.; Legrain, B.; et al. ROS implication in a new antitumor strategy based on non-thermal plasma. *Int. J. Cancer* **2012**, *130*, 2185–2194. [CrossRef]
20. Ahn, H.J.; Kim, K.I.; Kim, G.; Moon, E.; Yang, S.S.; Lee, J.S. Atmospheric-pressure plasma jet induces apoptosis involving mitochondria via generation of free radicals. *PLoS ONE* **2011**, *6*, e28154. [CrossRef]
21. Köritzer, J.; Boxhammer, V.; Schäfer, A.; Shimizu, T.; Klämpfl, T.G.; Li, Y.-F.; Welz, C.; Schwenk-Zieger, S.; Morfill, G.E.; Zimmermann, J.L.; et al. Restoration of Sensitivity in Chemo—Resistant Glioma Cells by Cold Atmospheric Plasma. *PLoS ONE* **2013**, *8*, e64498. [CrossRef] [PubMed]
22. Kalghatgi, S.; Kelly, C.M.; Cerchar, E.; Torabi, B.; Alekseev, O.; Fridman, A.; Friedman, G.; Azizkhan-Clifford, J. Effects of Non-Thermal Plasma on Mammalian Cells. *PLoS ONE* **2011**, *6*, e16270. [CrossRef]
23. Zhao, S.; Xiong, Z.; Mao, X.; Meng, D.; Lei, Q.; Li, Y.; Deng, P.; Chen, M.; Tu, M.; Lu, X.; et al. Atmospheric pressure room temperature plasma jets facilitate oxidative and nitrative stress and lead to endoplasmic reticulum stress dependent apoptosis in HepG2 cells. *PLoS ONE* **2013**, *8*, e73665. [CrossRef] [PubMed]
24. Kaushik, N.K.; Kaushik, N.; Park, D.; Choi, E.H. Altered Antioxidant System Stimulates Dielectric Barrier Discharge Plasma-Induced Cell Death for Solid Tumor Cell Treatment. *PLoS ONE* **2014**, *9*, e103349. [CrossRef] [PubMed]
25. Siu, A.; Volotskova, O.; Cheng, X.; Khalsa, S.S.; Bian, K.; Murad, F.; Keidar, M.; Sherman, J.H. Differential Effects of Cold Atmospheric Plasma in the Treatment of Malignant Glioma. *PLoS ONE* **2015**, *10*, e0126313. [CrossRef] [PubMed]
26. Volotskova, O.; Hawley, T.S.; Stepp, M.A.; Keidar, M. Targeting the cancer cell cycle by cold atmospheric plasma. *Sci. Rep.* **2012**, *2*, 636. [CrossRef] [PubMed]
27. Hua, D.; Cai, D.; Ning, M.; Yu, L.; Zhang, Z.; Han, P.; Dai, X. Cold atmospheric plasma selectively induces G(0)/G(1) cell cycle arrest and apoptosis in AR-independent prostate cancer cells. *J. Cancer* **2021**, *12*, 5977–5986. [CrossRef]
28. Turrini, E.; Laurita, R.; Stancampiano, A.; Catanzaro, E.; Calcabrini, C.; Maffei, F.; Gherardi, M.; Colombo, V.; Fimognari, C. Cold Atmospheric Plasma Induces Apoptosis and Oxidative Stress Pathway Regulation in T-Lymphoblastoid Leukemia Cells. *Oxidative Med. Cell. Longev.* **2017**, *2017*, 4271065. [CrossRef]

29. Bauer, G.; Sersenová, D.; Graves, D.B.; Machala, Z. Cold Atmospheric Plasma and Plasma-Activated Medium Trigger RONS-Based Tumor Cell Apoptosis. *Sci. Rep.* **2019**, *9*, 14210. [CrossRef]

30. Virard, F.; Cousty, S.; Cambus, J.P.; Valentin, A.; Kémoun, P.; Clément, F. Cold Atmospheric Plasma Induces a Predominantly Necrotic Cell Death via the Microenvironment. *PLoS ONE* **2015**, *10*, e0133120. [CrossRef]

31. Utsumi, F.; Kajiyama, H.; Nakamura, K.; Tanaka, H.; Mizuno, M.; Ishikawa, K.; Kondo, H.; Kano, H.; Hori, M.; Kikkawa, F. Effect of Indirect Nonequilibrium Atmospheric Pressure Plasma on Anti-Proliferative Activity against Chronic Chemo-Resistant Ovarian Cancer Cells In Vitro and In Vivo. *PLoS ONE* **2013**, *8*, e81576. [CrossRef] [PubMed]

32. Koensgen, D.; Besic, I.; Gumbel, D.; Kaul, A.; Weiss, M.; Diesing, K.; Kramer, A.; Bekeschus, S.; Mustea, A.; Stope, M.B. Cold Atmospheric Plasma (CAP) and CAP-Stimulated Cell Culture Media Suppress Ovarian Cancer Cell Growth—A Putative Treatment Option in Ovarian Cancer Therapy. *Anticancer. Res.* **2017**, *37*, 6739–6744. [CrossRef] [PubMed]

33. Torii, K.; Yamada, S.; Nakamura, K.; Tanaka, H.; Kajiyama, H.; Tanahashi, K.; Iwata, N.; Kanda, M.; Kobayashi, D.; Tanaka, C.; et al. Effectiveness of plasma treatment on gastric cancer cells. *Gastric Cancer* **2015**, *18*, 635–643. [CrossRef]

34. Partecke, L.I.; Evert, K.; Haugk, J.; Doering, F.; Normann, L.; Diedrich, S.; Weiss, F.-U.; Evert, M.; Huebner, N.O.; Guenther, C.; et al. Tissue tolerable plasma (TTP) induces apoptosis in pancreatic cancer cells in vitro and in vivo. *BMC Cancer* **2012**, *12*, 473. [CrossRef] [PubMed]

35. Liedtke, K.R.; Diedrich, S.; Pati, O.; Freund, E.; Flieger, R.; Heidecke, C.D.; Partecke, L.I.; Bekeschus, S. Cold Physical Plasma Selectively Elicits Apoptosis in Murine Pancreatic Cancer Cells In Vitro and In Ovo. *Anticancer. Res.* **2018**, *38*, 5655–5663. [CrossRef]

36. Bekeschus, S.; Käding, A.; Schröder, T.; Wende, K.; Hackbarth, C.; Liedtke, K.R.; van der Linde, J.; von Woedtke, T.; Heidecke, C.-D.; Partecke, L.-I. Cold Physical Plasma-Treated Buffered Saline Solution as Effective Agent Against Pancreatic Cancer Cells. *Anti-Cancer Agents Med. Chem.* **2018**, *18*, 824–831. [CrossRef] [PubMed]

37. Joh, H.M.; Choi, J.Y.; Kim, S.J.; Chung, T.H.; Kang, T.-H. Effect of additive oxygen gas on cellular response of lung cancer cells induced by atmospheric pressure helium plasma jet. *Sci. Rep.* **2014**, *4*, 6638. [CrossRef] [PubMed]

38. Aggelopoulos, C.A.; Christodoulou, A.M.; Tachliabouri, M.; Meropoulis, S.; Christopoulou, M.E.; Karalis, T.T.; Chatzopoulos, A.; Skandalis, S.S. Cold Atmospheric Plasma Attenuates Breast Cancer Cell Growth Through Regulation of Cell Microenvironment Effectors. *Front. Oncol.* **2021**, *11*, 826865. [CrossRef]

39. Chupradit, S.; Widjaja, G.; Radhi Majeed, B.; Kuznetsova, M.; Ansari, M.J.; Suksatan, W.; Turki Jalil, A.; Ghazi Esfahani, B. Recent advances in cold atmospheric plasma (CAP) for breast cancer therapy. *Cell Biol. Int.* **2023**, *47*, 327–340. [CrossRef]

40. Cheng, X.; Murthy, S.R.K.; Zhuang, T.; Ly, L.; Jones, O.; Basadonna, G.; Keidar, M.; Kanaan, Y.; Canady, J. Canady Helios Cold Plasma Induces Breast Cancer Cell Death by Oxidation of Histone mRNA. *Int. J. Mol. Sci.* **2021**, *22*, 9578. [CrossRef]

41. Arndt, S.; Fadil, F.; Dettmer, K.; Unger, P.; Boskovic, M.; Samol, C.; Bosserhoff, A.K.; Zimmermann, J.L.; Gruber, M.; Gronwald, W.; et al. Cold Atmospheric Plasma Changes the Amino Acid Composition of Solutions and Influences the Anti-Tumor Effect on Melanoma Cells. *Int. J. Mol. Sci.* **2021**, *22*, 7886. [CrossRef] [PubMed]

42. Zimmermann, T.; Gebhardt, L.A.; Kreiss, L.; Schneider, C.; Arndt, S.; Karrer, S.; Friedrich, O.; Fischer, M.J.M.; Bosserhoff, A.-K. Acidified Nitrite Contributes to the Antitumor Effect of Cold Atmospheric Plasma on Melanoma Cells. *Int. J. Mol. Sci.* **2021**, *22*, 3757. [CrossRef] [PubMed]

43. Huang, J.F.; Shen, J.; Li, X.; Rengan, R.; Silvestris, N.; Wang, M.; Derosa, L.; Zheng, X.; Belli, A.; Zhang, X.L.; et al. Incidence of patients with bone metastases at diagnosis of solid tumors in adults: A large population-based study. *Ann. Transl. Med.* **2020**, *8*, 482. [CrossRef] [PubMed]

44. Tjensvoll, K.; Oltedal, S.; Heikkilä, R.; Kvaløy, J.T.; Gilje, B.; Reuben, J.M.; Smaaland, R.; Nordgård, O. Persistent tumor cells in bone marrow of non-metastatic breast cancer patients after primary surgery are associated with inferior outcome. *BMC Cancer* **2012**, *12*, 190. [CrossRef] [PubMed]

45. Haralambiev, L.; Nitsch, A.; Einenkel, R.; Muzzio, D.O.; Gelbrich, N.; Burchardt, M.; Zygmunt, M.; Ekkernkamp, A.; Stope, M.B.; Gümbel, D. The Effect of Cold Atmospheric Plasma on the Membrane Permeability of Human Osteosarcoma Cells. *Anticancer. Res.* **2020**, *40*, 841–846. [CrossRef]

46. Jacoby, J.M.; Strakeljahn, S.; Nitsch, A.; Bekeschus, S.; Hinz, P.; Mustea, A.; Ekkernkamp, A.; Tzvetkov, M.V.; Haralambiev, L.; Stope, M.B. An Innovative Therapeutic Option for the Treatment of Skeletal Sarcomas: Elimination of Osteo-and Ewing's Sarcoma Cells Using Physical Gas Plasma. *Int. J. Mol. Sci.* **2020**, *21*, 4460. [CrossRef]

47. Nitsch, A.; Strakeljahn, S.; Jacoby, J.M.; Sieb, K.F.; Mustea, A.; Bekeschus, S.; Ekkernkamp, A.; Stope, M.B.; Haralambiev, L. New Approach against Chondrosoma Cells—Cold Plasma Treatment Inhibits Cell Motility and Metabolism, and Leads to Apoptosis. *Biomedicines* **2022**, *10*, 688.

48. Canal, C.; Fontelo, R.; Hamouda, I.; Guillem-Marti, J.; Cvelbar, U.; Ginebra, M.-P. Plasma-induced selectivity in bone cancer cells death. *Free. Radic. Biol. Med.* **2017**, *110*, 72–80. [CrossRef]

49. Haralambiev, L.; Nitsch, A.; Jacoby, J.M.; Strakeljahn, S.; Bekeschus, S.; Mustea, A.; Ekkernkamp, A.; Stope, M.B. Cold Atmospheric Plasma Treatment of Chondrosarcoma Cells Affects Proliferation and Cell Membrane Permeability. *Int. J. Mol. Sci.* **2020**, *21*, 2291. [CrossRef]

50. Haralambiev, L.; Wien, L.; Gelbrich, N.; Kramer, A.; Mustea, A.; Burchardt, M.; Ekkernkamp, A.; Stope, M.B.; Gümbel, D. Effects of Cold Atmospheric Plasma on the Expression of Chemokines, Growth Factors, TNF Superfamily Members, Interleukins, and Cytokines in Human Osteosarcoma Cells. *Anticancer. Res.* **2019**, *39*, 151–157. [CrossRef]

51. Haralambiev, L.; Wien, L.; Gelbrich, N.; Lange, J.; Bakir, S.; Kramer, A.; Burchardt, M.; Ekkernkamp, A.; Gümbel, D.; Stope, M.B. Cold atmospheric plasma inhibits the growth of osteosarcoma cells by inducing apoptosis, independent of the device used. *Oncol. Lett.* **2020**, *19*, 283–290. [CrossRef] [PubMed]
52. Tornin, J.; Mateu-Sanz, M.; Rodríguez, A.; Labay, C.; Rodríguez, R.; Canal, C. Pyruvate Plays a Main Role in the Antitumoral Selectivity of Cold Atmospheric Plasma in Osteosarcoma. *Sci. Rep.* **2019**, *9*, 10681. [CrossRef]
53. Yan, D.; Talbot, A.; Nourmohammadi, N.; Sherman, J.; Cheng, X.; Keidar, M. Toward understanding the selective anticancer capacity of cold atmospheric plasma—A model based on aquaporins (Review). *Biointerphases* **2015**, *10*, 040801. [CrossRef] [PubMed]
54. Ratovitski, E.A.; Cheng, X.; Yan, D.; Sherman, J.H.; Canady, J.; Trink, B.; Keidar, M. Anti-Cancer Therapies of 21st Century: Novel Approach to Treat Human Cancers Using Cold Atmospheric Plasma. *Plasma Process. Polym.* **2014**, *11*, 1128–1137. [CrossRef]
55. Morfill, G.E.; Kong, M.G.; Zimmermann, J.L. Focus on Plasma Medicine. *New J. Phys.* **2009**, *11*, 115011. [CrossRef]
56. Promega Corporation. *CellTiter-Blue®Cell Viability Assay Technical Bulletin #TB317*; Promega Corporation: Madison, WI, USA, 2016.
57. O'Brien, J.; Wilson, I.; Orton, T.; Pognan, F. Investigation of the Alamar Blue (resazurin) fluorescent dye for the assessment of mammalian cell cytotoxicity. *Eur. J. Biochem.* **2000**, *267*, 5421–5426. [CrossRef] [PubMed]
58. R&D Systems Inc. *HT TiterTACS™Apoptosis Detection Kit User Manual*; R&D Systems Inc.: Minneapolis, MN, USA, 2019.
59. Molecular Probes Inc. *Amplex® Red Hydrogen Peroxide/Peroxidase Assay Kit*; Molecular Probes Inc.: Eugene, OR, USA, 2009.
60. Yan, D.; Talbot, A.; Nourmohammadi, N.; Cheng, X.; Canady, J.; Sherman, J.; Keidar, M. Principles of using Cold Atmospheric Plasma Stimulated Media for Cancer Treatment. *Sci. Rep.* **2015**, *5*, 18339. [CrossRef]
61. Dubuc, A.; Monsarrat, P.; Virard, F.; Merbahi, N.; Sarrette, J.-P.; Laurencin-Dalicieux, S.; Cousty, S. Use of cold-atmospheric plasma in oncology: A concise systematic review. *Adv. Med. Oncol.* **2018**, *10*, 1758835918786475. [CrossRef]
62. Brány, D.; Dvorská, D.; Halašová, E.; Škovierová, H. Cold Atmospheric Plasma: A Powerful Tool for Modern Medicine. *Int. J. Mol. Sci.* **2020**, *21*, 2932. [CrossRef]
63. Dai, X.; Bazaka, K.; Thompson, E.W.; Ostrikov, K.K. Cold Atmospheric Plasma: A Promising Controller of Cancer Cell States. *Cancers* **2020**, *12*, 3360. [CrossRef]
64. Kim, S.-J.; Seong, M.-J.; Mun, J.-J.; Bae, J.-H.; Joh, H.-M.; Chung, T.-H. Differential Sensitivity of Melanoma Cells and Their Non-Cancerous Counterpart to Cold Atmospheric Plasma-Induced Reactive Oxygen and Nitrogen Species. *Int. J. Mol. Sci.* **2022**, *23*, 14092. [CrossRef] [PubMed]
65. Aggarwal, V.; Tuli, H.S.; Varol, A.; Thakral, F.; Yerer, M.B.; Sak, K.; Varol, M.; Jain, A.; Khan, M.A.; Sethi, G. Role of Reactive Oxygen Species in Cancer Progression: Molecular Mechanisms and Recent Advancements. *Biomolecules* **2019**, *9*, 735. [CrossRef] [PubMed]
66. Tanaka, H.; Mizuno, M.; Ishikawa, K.; Nakamura, K.; Kajiyama, H.; Kano, H.; Kikkawa, F.; Hori, M. Plasma-Activated Medium Selectively Kills Glioblastoma Brain Tumor Cells by Down-Regulating a Survival Signaling Molecule, AKT Kinase. *Plasma Med.* **2011**, *1*, 265–277. [CrossRef]
67. Mateu-Sanz, M.; Tornín, J.; Brulin, B.; Khlyustova, A.; Ginebra, M.-P.; Layrolle, P.; Canal, C. Cold Plasma-Treated Ringer's Saline: A Weapon to Target Osteosarcoma. *Cancers* **2020**, *12*, 227. [CrossRef] [PubMed]
68. Wang, J.; Yi, J. Cancer cell killing via ROS: To increase or decrease, that is the question. *Cancer Biol. Ther.* **2008**, *7*, 1875–1884. [CrossRef]
69. Li, Y.; Tang, T.; Lee, H.; Song, K. Cold atmospheric pressure plasma-activated medium induces selective cell death in human hepatocellular carcinoma cells independently of singlet oxygen, hydrogen peroxide, nitric oxide and nitrite/nitrate. *Int. J. Mol. Sci.* **2021**, *22*, 5548. [CrossRef]
70. Halasi, M.; Wang, M.; Chavan, T.S.; Gaponenko, V.; Hay, N.; Gartel, A.L. ROS inhibitor N-acetyl-L-cysteine antagonizes the activity of proteasome inhibitors. *Biochem. J.* **2013**, *454*, 201–208. [CrossRef]
71. Moniruzzaman, R.; Rehman, M.U.; Zhao, Q.-L.; Jawaid, P.; Mitsuhashi, Y.; Imaue, S.; Fujiwara, K.; Ogawa, R.; Tomihara, K.; Saitoh, J.-I. Roles of intracellular and extracellular ROS formation in apoptosis induced by cold atmospheric helium plasma and X-irradiation in the presence of sulfasalazine. *Free. Radic. Biol. Med.* **2018**, *129*, 537–547. [CrossRef]
72. Gurzov, E.N.; Tran, M.; Fernandez-Rojo, M.A.; Merry, T.L.; Zhang, X.; Xu, Y.; Fukushima, A.; Waters, M.J.; Watt, M.J.; Andrikopoulos, S. Hepatic oxidative stress promotes insulin-STAT-5 signaling and obesity by inactivating protein tyrosine phosphatase N2. *Cell Metab.* **2014**, *20*, 85–102. [CrossRef]
73. Nathan, C.F.; Cohn, Z.A. Antitumor effects of hydrogen peroxide in vivo. *J. Exp. Med.* **1981**, *154*, 1539–1553. [CrossRef]
74. Lopez-Lazaro, M. Dual role of hydrogen peroxide in cancer: Possible relevance to cancer chemoprevention and therapy. *Cancer Lett.* **2007**, *252*, 1–8. [CrossRef] [PubMed]
75. Garibaldi, A.; Gullino, M.L.; Minuto, G. Diseases of basil and their management. *Plant. Dis.* **1997**, *81*, 124–132. [CrossRef] [PubMed]
76. Mohades, S.; Laroussi, M.; Sears, J.; Barekzi, N.; Razavi, H. Evaluation of the effects of a plasma activated medium on cancer cells. *Phys. Plasmas* **2015**, *22*, 122001. [CrossRef]
77. Tabuchi, Y.; Uchiyama, H.; Zhao, Q.-L.; Yunoki, T.; Andocs, G.; Nojima, N.; Takeda, K.; Ishikawa, K.; Hori, M.; Kondo, T. Effects of nitrogen on the apoptosis of and changes in gene expression in human lymphoma U937 cells exposed to argon-based cold atmospheric pressure plasma. *Int. J. Mol. Med.* **2016**, *37*, 1706–1714. [CrossRef]

78. Hamouda, I.; Labay, C.; Cvelbar, U.; Ginebra, M.P.; Canal, C. Selectivity of direct plasma treatment and plasma-conditioned media in bone cancer cell lines. *Sci. Rep.* **2021**, *11*, 17521. [CrossRef]

79. Ermakov, A.M.; Ermakova, O.N.; Afanasyeva, V.A.; Popov, A.L. Dose-dependent effects of cold atmospheric argon plasma on the mesenchymal stem and osteosarcoma cells in vitro. *Int. J. Mol. Sci.* **2021**, *22*, 6797. [CrossRef]

80. Fischer, M.; Schoon, J.; Freund, E.; Miebach, L.; Weltmann, K.-D.; Bekeschus, S.; Wassilew, G.I. Biocompatible Gas Plasma Treatment Affects Secretion Profiles but Not Osteogenic Differentiation in Patient-Derived Mesenchymal Stromal Cells. *Int. J. Mol. Sci.* **2022**, *23*, 2038. [CrossRef] [PubMed]

81. Lee, H.; Shon, C.; Kim, Y.; Kim, S.; Kim, G.; Kong, M.G. Degradation of adhesion molecules of G361 melanoma cells by a non-thermal atmospheric pressure microplasma. *New J. Phys.* **2009**, *11*, 115026. [CrossRef]

82. Kim, G.; Kim, W.; Kim, K.; Lee, J. DNA damage and mitochondria dysfunction in cell apoptosis induced by nonthermal air plasma. *Appl. Phys. Lett.* **2010**, *96*, 021502. [CrossRef]

83. Kim, G.; Kim, G.; Park, S.; Jeon, S.; Seo, H.; Iza, F.; Lee, J.K. Air plasma coupled with antibody-conjugated nanoparticles: A new weapon against cancer. *J. Phys. D Appl. Phys.* **2008**, *42*, 032005. [CrossRef]

84. Chang, J.W.; Kang, S.U.; Shin, Y.S.; Kim, K.I.; Seo, S.J.; Yang, S.S.; Lee, J.-S.; Moon, E.; Lee, K.; Kim, C.-H. Non-thermal atmospheric pressure plasma inhibits thyroid papillary cancer cell invasion via cytoskeletal modulation, altered MMP-2/-9/uPA activity. *PLoS ONE* **2014**, *9*, e92198. [CrossRef]

85. Recek, N.; Cheng, X.; Keidar, M.; Cvelbar, U.; Vesel, A.; Mozetic, M.; Sherman, J. Effect of cold plasma on glial cell morphology studied by atomic force microscopy. *PLoS ONE* **2015**, *10*, e0119111. [CrossRef]

86. Naciri, M.; Dowling, D.; Al-Rubeai, M. Differential Sensitivity of Mammalian Cell Lines to Non-Thermal Atmospheric Plasma. *Plasma Process. Polym.* **2014**, *11*, 391–400. [CrossRef]

87. Shashurin, A.; Stepp, M.A.; Hawley, T.S.; Pal-Ghosh, S.; Brieda, L.; Bronnikov, S.; Jurjus, R.A.; Keidar, M. Influence of cold plasma atmospheric jet on surface integrin expression of living cells. *Plasma Process. Polym.* **2010**, *7*, 294–300. [CrossRef]

88. Ishaq, M.; Evans, M.D.; Ostrikov, K.K. Atmospheric pressure gas plasma-induced colorectal cancer cell death is mediated by Nox2–ASK1 apoptosis pathways and oxidative stress is mitigated by Srx–Nrf2 anti-oxidant system. *Biochim. Et Biophys. Acta (BBA)-Mol. Cell Res.* **2014**, *1843*, 2827–2837. [CrossRef] [PubMed]

89. Wang, M.; Holmes, B.; Cheng, X.; Zhu, W.; Keidar, M.; Zhang, L.G. Cold atmospheric plasma for selectively ablating metastatic breast cancer cells. *PLoS ONE* **2013**, *8*, e73741. [CrossRef] [PubMed]

90. Haralambiev, L.; Neuffer, O.; Nitsch, A.; Kross, N.C.; Bekeschus, S.; Hinz, P.; Mustea, A.; Ekkernkamp, A.; Gümbel, D.; Stope, M.B. Inhibition of Angiogenesis by Treatment with Cold Atmospheric Plasma as a Promising Therapeutic Approach in Oncology. *Int. J. Mol. Sci.* **2020**, *21*, 7098. [CrossRef] [PubMed]

91. Isbary, G.; Shimizu, T.; Li, Y.-F.; Stolz, W.; Thomas, H.M.; Morfill, G.E.; Zimmermann, J.L. Cold atmospheric plasma devices for medical issues. *Expert Rev. Med. Devices* **2013**, *10*, 367–377. [CrossRef] [PubMed]

92. Tabares, F.L.; Junkar, I. Cold plasma systems and their application in surface treatments for medicine. *Molecules* **2021**, *26*, 1903. [CrossRef] [PubMed]

93. Martusevich, A.K.; Surovegina, A.V.; Bocharin, I.V.; Nazarov, V.V.; Minenko, I.A.; Artamonov, M.Y. Cold argon athmospheric plasma for biomedicine: Biological effects, applications and possibilities. *Antioxidants* **2022**, *11*, 1262. [CrossRef]

94. Barekzi, N.; Laroussi, M. Dose-dependent killing of leukemia cells by low-temperature plasma. *J. Phys. D Appl. Phys.* **2012**, *45*, 422002. [CrossRef]

Article

Enrichment of Bone Tissue with Antibacterially Effective Amounts of Nitric Oxide Derivatives by Treatment with Dielectric Barrier Discharge Plasmas Optimized for Nitrogen Oxide Chemistry

Dennis Feibel [1], Alexander Kwiatkowski [1], Christian Opländer [2], Gerrit Grieb [3], Joachim Windolf [1] and Christoph V. Suschek [1,*]

[1] Department for Orthopedics and Trauma Surgery, Medical Faculty, Heinrich-Heine-University Düsseldorf, Moorenstraße 5, 40225 Düsseldorf, Germany

[2] Institute for Research in Operative Medicine (IFOM), Cologne-Merheim Medical Center, University Witten/Herdecke, 58455 Witten-Herdecke, Germany

[3] Department of Plastic Surgery and Hand Surgery, Burn Centre, Medical Faculty, RWTH Aachen University, 52074 Aachen, Germany

* Correspondence: suschek@hhu.de

Abstract: Cold atmospheric plasmas (CAPs) generated by dielectric barrier discharge (DBD), particularly those containing higher amounts of nitric oxide (NO) or NO derivates (NOD), are attracting increasing interest in medical fields. In the present study, we, for the first time, evaluated DBD-CAP-induced NOD accumulation and therapeutically relevant NO release in calcified bone tissue. This knowledge is of great importance for the development of new therapies against bacterial-infectious complications during bone healing, such as osteitis or osteomyelitis. We found that by modulating the power dissipation in the discharge, it is possible (1) to significantly increase the uptake of NODs in bone tissue, even into deeper regions, (2) to significantly decrease the pH in CAP-exposed bone tissue, (3) to induce a long-lasting and modulable NO production in the bone samples as well as (4) to significantly protect the treated bone tissue against bacterial contaminations, and to induce a strong bactericidal effect in bacterially infected bone samples. Our results strongly suggest that the current DBD technology opens up effective NO-based therapy options in the treatment of local bacterial infections of the bone tissue through the possibility of a targeted modulation of the NOD content in the generated CAPs.

Keywords: dielectric barrier discharge (DBD); cold atmospheric plasma (CAP); nitric oxide radical (NO)

Citation: Feibel, D.; Kwiatkowski, A.; Opländer, C.; Grieb, G.; Windolf, J.; Suschek, C.V. Enrichment of Bone Tissue with Antibacterially Effective Amounts of Nitric Oxide Derivatives by Treatment with Dielectric Barrier Discharge Plasmas Optimized for Nitrogen Oxide Chemistry. *Biomedicines* **2023**, *11*, 244. https://doi.org/10.3390/biomedicines11020244

Academic Editor: Mike Barbeck

Received: 15 December 2022
Revised: 13 January 2023
Accepted: 14 January 2023
Published: 17 January 2023

1. Introduction

Cold atmospheric plasmas (CAP) generated with the help of modern plasma technology under atmospheric pressure conditions allow for interaction with living biological tissues with few side effects due to the low process temperatures [1]. This results in manifold promising therapy options for the treatment of various diseases in humans and animals [2,3]. With the CAPs, a distinction is made between "direct" and "indirect" plasma [4]. Put simply, an indirect plasma is created by ionizing a gas that flows through an electric field. The resulting outflowing plasma can then be applied to the area to be treated in the form of a "plasma jet". Direct plasma can be generated with the help of dielectric barrier discharge (DBD) technology, with the sample to be treated serving as a counter electrode to the DBD device electrode. In the gap between the electrode of the DBD source and the object to be treated, the ambient atmosphere is ionized, and the generated CAP directly exerts its effect on the treated tissue area [4].

CAPs contain a mixture of different oxygen, nitric oxide and other radical species, UV radiation and a strong flow of charges. All components individually or in combination have the potential ability to influence biological functions if they are applied directly to tissue or cells [1]. Plasma generated with ambient air includes reactive nitrogen species (RNS) such as nitric oxide (NO), nitrogen dioxide (NO_2) and reactive oxygen species (ROS) such as hydrogen peroxide (H_2O_2), ozone (O_3), superoxide radicals (O_2^-), hydroxyl radicals (OH), and many other products in different concentrations and compositions [5,6]. The content of NO and other NO derivatives (NODs) in DBD-generated CAPs are of particular importance. In the context of tissue homeostasis, NO plays a decisive physiological role, e.g., in the regulation of vasodilation, thrombogenesis, the immune and inflammatory response, angiogenesis, cell proliferation and cell differentiation, collagen metabolism, cell death via apoptosis or necrosis, as well as antibacterial defense [7]. In the human organism, NO can be generated enzymatically from the amino acid L-arginine by at least three isoenzymes from the family of NO synthases in probably every cell type [8]. Due to its pivotal role in the regulation of tissue homeostasis, insufficient physiological local NO production or NO availability correlates, for example, with the well-known clinical picture of chronic, bacterially infected and poorly healing wounds [9]. With regard to the therapy of chronic wounds, the positive effect of NO-based therapies could be demonstrated by exogenously applied NO gas, NO donors or NO-containing plasmas which all exerted significant improvement in the wound healing status, including a clear and strong reduction in the bacterial load on the treated wounds [10–12]. Since the bacterial infestation of wounds is apparently a decisive driver for the delay in wound healing, it is not surprising that other NO-independent measures that lead to the reduction in bacterial infestation can also positively support wound healing, such as plasma compositions, which are more characterized by a dominant H_2O_2 generation and less NO chemistry [13]. It must be noted, however, that in addition to its pronounced antibacterial effect, NO can also induce numerous other wound healing-supporting mechanisms and thus have additional positive properties that H_2O_2 and other ROS do not have.

As part of a previous study to evaluate the therapeutic significance of NO-containing DBD-CAPs, we observed that the treatment of human skin with CAPs, which were generated with a power dissipated in the discharge of 122 mW by the same DBD source as we did for the current study, led to an accumulation of considerable amounts of NO derivatives and a decrease in the pH in the treated skin tissue. [14]. Both parameters led to a release of NO in the skin and the development of physiological NO effects, such as enhanced local vasodilation [15]. Interestingly, such a situation did not lead to the occurrence of increased cell toxicity in the treated skin sample, either in vitro or in vivo [16]. In the treated skin tissue, an NO-releasing system could thus be built up with NO-containing CAPs, in which the duration of the NO release and the level of NO production were dependent on the amount of accumulated NO derivatives and thus could be varied solely by the duration of the plasma treatment [15].

Due to the necessity of a free anatomical access of a plasma source to the surface to be treated, plasma-oriented therapies appear to be predestined for the treatment of diseases from the dermatological or stomatological context. In cases of surgical interventions, however, this technique could be used if an otherwise not freely accessible area was opened as part of a medical procedure, like a revision. This would be the case, for example, in the context of supportive local therapy for bacterial complications in bone tissue, e.g., in the clinical picture of osteitis or osteomyelitis. These are often dramatic infectious and inflammatory complications of bone healing caused by bacteria [17], accompanied by the formation of biofilms, overactivation of immune competent cells and irreversible destruction of the bone substance [18]. In addition to systemic antibiosis, the treatment of these complications often consists of surgical debridement of the focus of infection and local use of antibiotic measures [17,18]. Due to the bactericidal properties of NO against free-living bacteria as well as against bacteria organized in a biofilm [19,20], its anti-inflammatory and immunomodulating effects [21] and positive properties in the regulation of osteogenesis [22,23], the use of NO-containing CAPs could represent an effective therapy option in the treatment of osteitis or osteomyelitis.

To date, there are no meaningful studies on whether and to what extent bone tissue can interact with CAP-induced NO derivatives and what bactericidal potential such an interaction can cause. Therefore, in the present study, we used a DBD plasma device which enabled us to achieve varying degrees of accumulation of NODs in exposed bone tissue. We investigated the depth of penetration of the NODs into the bone tissue, the degree of NO release and the antibacterial properties of the NO-optimized CAPs used in different models of bacterial infection of bone tissue. Our results clearly show that the use of plasma technologies, with the help of which a pharmacologically relevant NO release can be induced in exposed bone tissue, represents an effective supportive therapy option in the treatment of bacterial complications of the bone.

2. Materials and Methods

2.1. Materials

If not indicated otherwise, all chemicals were obtained from Sigma-Aldrich, Merck KGaA (Darmstadt, Germany).

2.2. DBD Plasma Source

We used a prototype DBD device which consists of one driven, cylindrical copper electrode covered with aluminum oxide and with a total diameter of 10 mm, exactly as described and characterized recently [15]. CAP was ignited in the gap (1 mm) between this dielectric covered electrode and grounded samples (water, agar, bone tissue), which represented the counter electrode. We have evaluated the effect of CAPs generated under four different device settings. CAPs were generated in ambient air by applying 13.5 kV or 17 kV maximum voltage pulses (kV) combined with trigger frequencies of 300 or 600 Hz (13.5 kV/300 Hz, 13.5 kV/600 Hz, 17 kV/300 Hz, 17 kV/600 Hz) which, with an air gap of 1 mm and positive polarity, corresponded to the power dissipated in the discharge of 122 mW, 275 mW, 221 mW and 410 mW, respectively.

2.3. Determination of pH Values

At the corresponding times of the respective experiments, the pH values were quantified by means of the Calimatic 766 pH meter (Knick, Berlin, Germany), both before and after the treatment of the samples with the respective CAPs. A combination pH electrode with a flat membrane (InLab-surface, Mettler-Toledo, Giessen, Germany) was used to measure the pH on solid objects (agar, bone surfaces).

2.4. Detection and Quantification of the NO Derivatives

The concentrations of nitrite and other chemically (iodine) reducible nitric oxide derivates in plasma treated aqueous solutions or aqueous exudates of plasma-exposed bone samples were quantified by an iodine/iodide-based assay, and using the NO-analyzer CLD 88 (Ecophysics) exactly as described previously [24]. Additionally, in order to determine nitrate, samples were incubated with 0.1 mol/l vanadium (III) chloride in 1 M hydrochloric acid refluxing at 95 °C under nitrogen [25]. The NOD quantities were quantified using nitrite/nitrate standards in standardized calibration solutions. The nitrate concentrations were calculated by subtracting the nitrite values obtained from the total NOD values.

To quantify the NO gas release from DBD plasma-treated aqueous solution, the respective solution was transferred to a quartz glass cylinder after the CAP exposure (20 mL) and permanently flushed through with a constant gas flow with N_2 (100 mL N_2/min). The resulting N_2/NO mixture was evacuated at the same flow rate and fed into the CLD for quantitative analysis [24]. The integral calculation of NO release from the solution was carried out using a specially designed program using the Matlab software (The MathWorks, Inc., Natick, MA, USA).

In order to quantify the release of native NO gas from human bone samples treated with DBD plasma, the bone preparations were transferred to an airtight chamber (35.2 cm^3) immediately after the plasma exposure. The chamber loaded with the bone specimen was flushed with an inert gas (N_2) for 1 min to remove oxygen. Then the chamber was hermetically sealed using two closing valves and NO emanating from the bone samples was accumulated in the chamber for 120 s. After opening the valves, the contents of the chamber were then fed into the CLD analyzer, and the NO content was determined [24].

2.5. Characterization of Accumulation of Nitric Oxide Derivates in CAP-exposed Aqueous Samples

As part of preliminary tests, we evaluated the power dissipated in the discharge at which the greatest accumulation of NO derivatives into the exposed solution could be achieved. To do this, we treated 20 mL of H_2O_{dest} in 50 mL centrifuge tubes with CAP generated under conditions of the device settings mentioned above for a period of 1, 5, 10 or 15 min. We then quantified the pH of the solution and the CAP-dependent accumulation of iodine-reducible NO derivates into the exposed solution. The time-resolved release of NO from the plasma-treated sample were quantified by using the chemiluminescence based NO-analyzer CLD 88 as mentioned above.

2.6. Bone Samples

We used pre-machined bovine bone chips (IDS, Boldon, UK), consisting exclusively of cortical tissue with a diameter of approx. 7 mm and 0.5 mm thick, self-prepared bone punches made from pig shoulder blades (10 mm in diameter) which, in addition to the outer cortex (1–2 mm), also contained parts of the cancellous bone of different thicknesses. In addition, we made 10×10 mm specimens from human femoral neck (*Collum femoris*). The samples were used with the consent of the donors and in accordance with the guidelines of the Ethics Committee of the University Hospital Düsseldorf (study number 3634) and in compliance with the Declaration of Helsinki principles (revision of October 2013). With these human specimens, we were able to produce bone samples of various thicknesses by removing bone tissue using a medical circular bone saw.

2.7. CAP Treatment of Bone Samples

The cortical tissue chips were equilibrated in a buffered aqueous solution for 10 min before use. The bone chips were either treated individually with plasma or, in order to characterize the penetration depth of NO derivatives (NODs) into the bone samples, several such chips were stacked on top of one another and treated with the CAP from above. In order to prevent the lateral entry of plasma components into this bone stack, the stack was sealed with wax.

The native porcine bone punches, which in cross-section form a "sandwich structure" consisting of cancellous tissue enclosed upwards and downwards by cortical bone tissue, were exposed to CAPs on one side from above. Alternatively, this "sandwich structure" was cut in half across the cancellous tissue using a sharp scalpel or a saw and the specimens were treated with plasma from the cortical side. The 10×10 mm specimens of the human femoral neck were treated identically. The bone samples were treated with CAPs for 1, 5, 10, 15, 20 or 30 min in various experimental conditions described above. The power dissipated in the discharge was 122 to 410 mW, so that the dissipated electrical work was calculated at 0.122 to 12.3 W min. At the indicated time points after plasma exposure, the bone samples were transferred to 5 mL PBS.

In order to characterize the penetration depth of NODs into the bone tissue after CAP treatment (5 min, 410 mW power dissipated in the discharge), tissue layers were removed millimeter by millimeter from the apical side of the CAP-exposed bone samples using cutting tools and coarse files and the removed tissue layers were then each transferred to 2 mL of PBS.

The NODs contained in plasma-exposed bone tissue were washed out as completely as possible by gently shaking the PBS solution for 30 min, into which the bone preparations had previously been transferred. The NOD content of these aqueous exudates was quantified using the CLD as described above.

In order to detect and quantify a release of native nitric oxide from DBD plasma-treated (10 min 275 mW power dissipated in the discharge) human bone preparations of the femur neck (10 × 10 mm), these preparations were transferred to the gastight chamber immediately after plasma treatment and NO release was quantified by CLD as described above.

For all bone tissue samples, the NOD values obtained after DBD plasma exposure were normalized to the weight of the samples.

2.8. Bacterial Culture and Evaluation of the Bactericidal Effect of DBD Plasmas

(1) Test bacterium

Staphylococcus epidermidis stock solution with a McFarland value of 0.5 mcf contained a bacterial concentration of 1.5×10^8 cfu/mL. The bacterial culture was carried out on 9 cm bacterial culture plates with Mueller Hinton 2 agar + 5% sheep blood (MHS, from bioMérieux SA, Marcy l'Etoile, France.

(2) Bactericidal impact of plasma treatment of sterile bone samples on the subsequently inoculated bacteria.

This experimental set-up was intended to check whether a CAP pretreatment of bone tissue can develop a bacteriostatic or bactericidal effect during a subsequent inoculation of the plasma treated samples. For this purpose, the bone samples were exposed to CAPs (10 min 275 mW power dissipated in the discharge) and approximately 20 min later, these bone samples were inoculated on the plasma exposed surface of the sample or alternatively on the opposite, non-plasma treated, sample side with 1.5×10^6 bacteria. After 24 h of incubation in a moisture 37 °C temperate chamber, the infected samples were transferred to 600 mL of sterile PBS. In order to detach the bacteria, the bone samples were shaken intensively for 30 min with an orbital shaker (vortex) and then wiped off intensively with a cotton swab. Following this procedure, the tip of the cotton swab was cut off and taken up in the existing sample volume and vortexed for a further 5 min. Subsequently, 200 μL of this bacterial solution was plated in different dilutions to agar plates and after a further 24 h, the number of visible bacterial colonies was documented and the reduction in the bacterial count ($\log_{10}$ reduction of cfu/mL) was calculated. Bone samples inoculated with bacteria but not treated with plasma served as positive controls. As a negative control, we used plasma treated bone samples that were not inoculated with bacteria.

(3) Bactericidal impact of plasma treatment of bone samples inoculated with bacteria.

With this second experimental setup, we wanted to evaluate the effect of DBD CAPs on bone samples that were already inoculated by bacteria. For this purpose, the bone preparations were inoculated with 1.5×10^6 bacteria and the bacterially infected surface was treated with the DBD plasma (10 min 275 mW power dissipated in the discharge), either 20 min or 24 h after the bacterial inoculation. In order to record bactericidal in-depth effects of the plasma, we treated, in alternative test approaches, the non-infected sample side with plasma. After 24 h since the last plasma exposure, the samples were transferred to 600 mL of PBS and the reduction in the bacterial count ($\log_{10}$ reduction of cfu/mL) was calculated identically as described above in the last chapter. Again, bone samples inoculated with bacteria, but not treated with plasma, served as positive controls, whereas sterile bone samples that were otherwise treated identically served as negative controls.

(4) Characterization of the bactericidal effect of plasma pretreatment of agar plates followed by bacterial inoculation.

In order to further elaborate the therapeutic relevance of plasma pretreatment of a tissue as a preventive and protective option against bacterial infections, we treated selected areas of agar plates for 5 to 30 min with DBD CAPs (275 mW power dissipated in the discharge). After 24 h, we inoculated the plates with bacteria and characterized the bacterial growth after a further 24 h of incubation.

2.9. Statistical Analysis

For statistical analysis, we used GraphPad Prism 8 (San Diego, CA, USA). Significant differences were evaluated using either paired two-tailed Student's *t*-test or ANOVA followed by an appropriate post hoc multiple comparison test (Tukey method). A *p*-value < 0.05 was considered significant.

3. Results

3.1. Exposure Time- and Performance-Dependent Accumulation of NO Derivatives into CAP-Exposed Aqueous Solutions and Bone Samples

CAP treatment of the aqueous solution led to an exponential increase in the accumulated amounts of nitrite, which was dependent on the exposure time and electrical work dissipated in the discharge (Figure 1A). In parallel, we observed a decrease in the solution pH, which aimed for a limit value of approx. pH 3.5 (Figure 1B). The increase in nitrite accumulation and the corresponding decrease in the solution pH correlated positively with a significant generation of gaseous NO from plasma exposed samples, as quantified by CLD (Figure 1C).

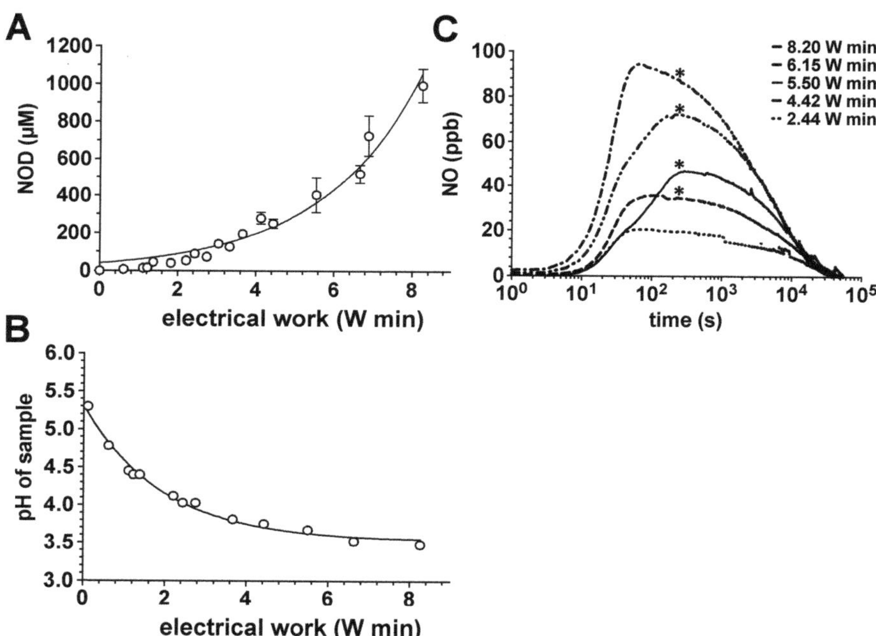

Figure 1. Impact of the power dissipation in the discharge on the pH, NOD accumulation and NO release in aqueous solutions exposed to the CAPs. Using the DBD plasma source, 20 mL of water samples were treated with CAPs generated with the power dissipated in the discharge of 122, 221, 275 or 410 mW for 5, 10, 15 or 20 min. The nitrite accumulation in the treated solutions was then quantified by CLD, the pH value of the solution was assessed with a pH meter and the course and the amount of NO release from these solutions was also characterized by CLD. (**A**) Nitrite content of the plasma-treated samples as a function of the electrical work dissipated in the discharge. Open circles represent the mean ± S.D. of four individual experiments. (**B**) pH values of the plasma-treated samples as a function of the electrical work dissipated in the discharge. (**C**) NO release from samples treated with plasma. A representative result of six individual tests is shown. *, $p < 0.05$, as compared to NO release from solutions treated with plasmas generated under conditions of less electrical work dissipated in the discharge.

Additionally, we treated bone punch preparations from the pigs shoulder with DBD plasmas of different power dissipated in the discharge. As shown in Figure 2A–C, bone tissue can be "loaded" with NODs. The amount of accumulated nitrite correlated positively with exposure time and the increasing power dissipated in the discharge. As part of the differential analysis of the accumulated NOD species, we found with human bone preparations that the majority of the NODs represent nitrate (Figure 2E), here detected in mM amounts, and about 1/20th to 1/50th of the NOD was stored in the form of nitrite (Figure 2D).

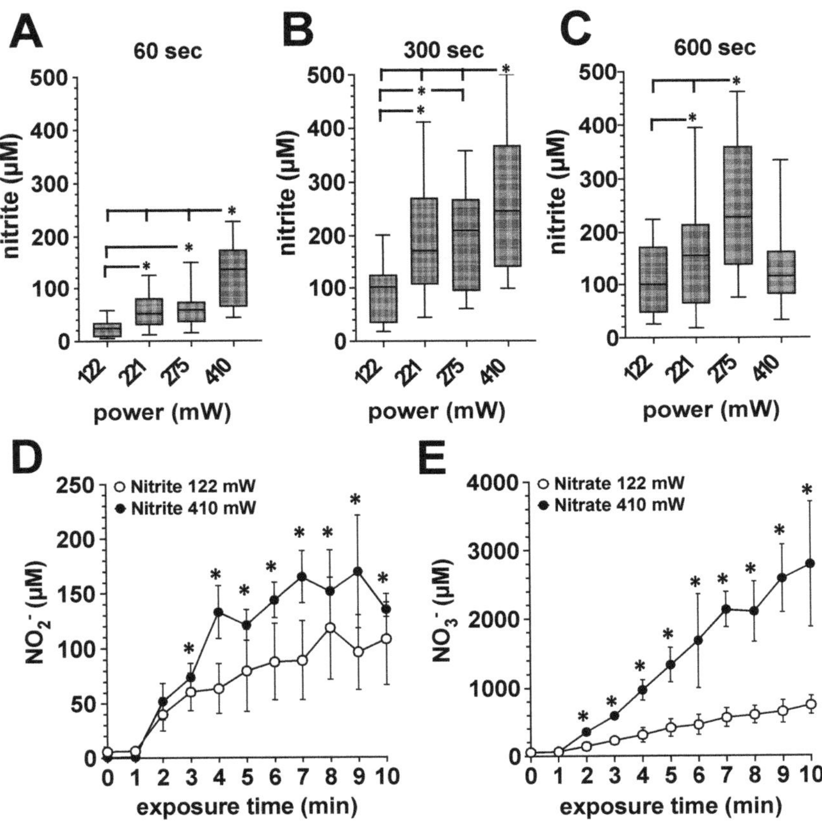

Figure 2. Influence of the power dissipation in the discharge on the plasma-induced accumulation of nitrite and nitrate in exposed bone tissue. Using the DBD plasma source, bone tissue samples were treated with plasma generated with the power dissipated in the discharge of 122, 221, 275 or 410 mW for for 1, 5 or 10 min. The nitrite and nitrate accumulation in the bone tissue was quantified by CLD, as described in the Material and Methods section. A–C, Porcine bone tissue treated 60 s (**A**), 300 s (**B**) or 600 s (**C**) with plasma generated with the power dissipated in the discharge of 122, 221, 275 or 410 mW. Values of 9 individual ($n = 9$) experiments are shown as boxplot with median value and with whiskers with minimum and maximum. *, $p < 0.05$. (**D**), Accumulation of nitrite or nitrate (**E**) in human bone tissue from the femoral head shaft after exposure to plasma generated with the power dissipated in the discharge of 122 mW (white circles) or 410 mW (black circles). Values represent the mean ± S.D. of four individual experiments. *, $p < 0.05$ as compared to the corresponding values obtained with the power dissipated in the discharge of 122 mW (white circles).

3.2. Quantification of NO Emanation from CAP-Exposed Bone Samples

In order to quantify NO release from plasma-treated bone samples, we transferred the samples after the corresponding plasma exposure into a gas-tight chamber which was filled with an inert carrier gas (N_2) (Figure 3A). After every two minutes, we quantified the accumulated NO amount by CLD. We observed a significant release of NO from human bone samples exposed to DBD plasma. The amount of NO released from the bone tissue (500–800 mg) was in the range between 15 and 22 pmol/min, immediately after the plasma exposure (Figure 3B). In the further course, the NO emanation from the bone tissue constantly decreased, but eight hours after the plasma treatment, it still averaged approx. 10 pmol/min. In parallel, we observed a significant decrease in the pH of DBD plasma exposed bone sample (Figure 3D).

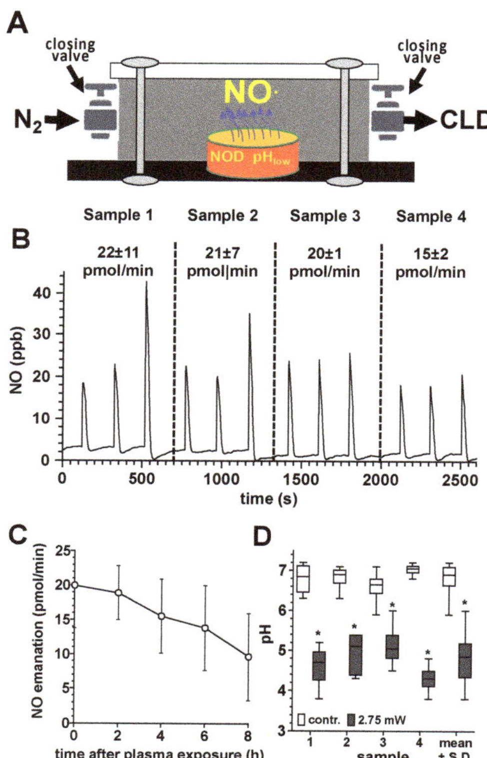

Figure 3. Quantification of NO release from plasma-treated human bone samples. (**A**), sketch of the experimental setup for quantifying the NO release of plasma-exposed bone tissue. DBD plasma-treated human bone samples (10 min at 275 mW power dissipation in the discharge) of the femur neck (500–800 mg) were transferred to a hermetically sealed chamber (35.2 cm^3), flushed with nitrogen to remove oxygen immediately after the plasma exposure. NO liberated from the bone samples was accumulated in the chamber for 120 s and the NO content was quantified by flushing the gas contents of the chamber into the CLD analyzer. (**B**), NO release (pmol/min) from the treated bone tissue (mean values ± S.D. of three measurements (triplicates) of each of the four samples (samples 1–4). (**C**), Characterization of the NO emanation (pmol/min) from the tissue of the four plasma-treated bone samples at different time points after the plasma treatment. (**D**), pH values of the treated surface of the four plasma treated bone samples immediately after plasma exposure, as detected by using a flat membrane pH electrode. *, $p < 0.05$ as compared to the control samples.

3.3. Characterization of the Penetration Depth and Distribution of NO Derivatives in Bone Tissue Exposed to DBD Plasma

Knowledge of the depth of penetration of NODs into the treated bone tissue is of great relevance in a possible plasma-based therapy. We, therefore, treated native human bone samples of the femur neck with DBD plasma (5 min 410 mW) and quantified the concentration of nitrite at different depths and time points after the plasma treatment (Figure 4). In the depth intervals studied and under the conditions described, we could observe a high accumulation of nitrite in the upper layers of the treated bones, 30 min after plasma exposure. However, with an increasing depth of the bone tissue, the nitrite concentration decreased sharply. Interestingly, 90 and 180 min after the plasma treatment, we were able to detect a steadily increasing nitrite concentration also in the deeper regions of the plasma-exposed bone, which indicates a diffusion-controlled distribution of the NODs transferred into the bone tissue (Figure 4).

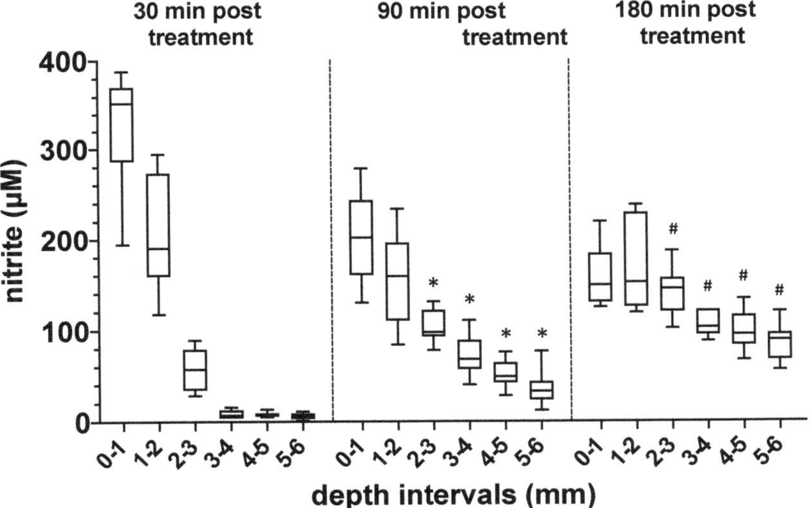

Figure 4. Evaluation of the depth of penetration of nitrogen oxide derivatives into bone tissue after exposure to plasma. Quantification of the nitrite content (µM) at different depths of human bone samples treated with DBD plasma (5 min at 410 mW power dissipation in the discharge). Values from 5 individual (n = 5) experiments are presented as a boxplot with median and with whiskers with minimum and maximum. The nitrite content in the bone samples was detected 20, 90 and 180 min after plasma exposure using CLD technology. *, $p < 0.05$ as compared to corresponding values detected after 20 min. #, $p < 0.05$ as compared to corresponding values detected at 20 or 90 min.

3.4. Bactericidal Impact of Plasma Treatment of Bone Samples Inoculated with Bacteria

In order to be able to better assess the bactericidal effect of plasma treatment on bone tissue, we have inoculated the bone preparations (punch preparations from the pork shoulder blade) with a *Staphylococcus epidermidis* strain (1.5×10^6). After 24 h, we treated the bacterially inoculated side (Figure 5A) or the non-inoculated opposite side of the specimens (Figure 5B) with CAPs (power dissipated in the discharge was 275 mW) for 10 min. After a further 24 h, we dissolved the bacteria from the plasma-treated samples and the untreated bone preparations, applied the resulting bacterial solutions in different dilutions to agar plates, and determined the reduction in the bacterial count (log10 reduction of cfu/mL). As shown in Figure 5A, plasma treatment of the bacterially inoculated side led to a $\log_{10}$ 3.92 ± 0.13 reduction and of the non-inoculated opposite side of the specimens, to a $\log_{10}$ 3.73 ± 0.27 reduction in cfu/mL (Figure 5B).

Mode 1: Inocculation ➔ 24 h incubation ➔ plasma treatment of the inocculated bone surface ➔ 24 h incubation ➔ agar culture

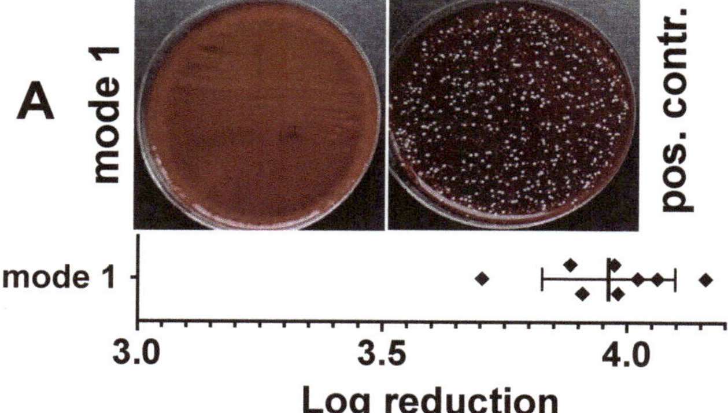

Mode 2: Inocculation ➔ 24 h incubation ➔ plasma treatment of the non-inocculated bone surface ➔ 24 h incubation ➔ agar culture

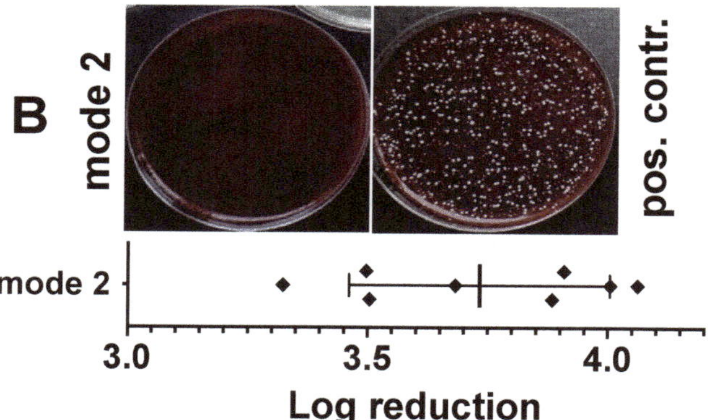

Figure 5. Characterization of the antibacterial effect of DBD plasma in a model of a bacterially infected bone preparation. Punch preparations from the pork shoulder blade were inoculated with the *Staphylococcus epidermidis* strain (1.5×10^6). After 24 h of incubation, the bacterially infected area (**A**, mode 1) or the non-infected side (**B**, mode 2) of the specimens were exposed to the DBD CAPs (5 min at 410 mW power dissipation in the discharge). After a further 24 h, the bacteria were detached and the reduction in the bacterial count ($\log_{10}$ reduction of cfu/mL) was calculated. As positive controls (pos. contr.), the corresponding bone samples were treated identically with the bacteria, but were not exposed to plasma. Bars shown represent the mean ± S.D. of six individual experiments.

In an alternative experimental setup, the results of which are shown in Figure 6, we treated bacterially inoculated bone preparations with DBD plasma identically as described above, but immediately, about 10 min after inoculation. Plasma treatment of the bacterially inoculated side (Figure 6A) led to a $\log_{10}$ 3.76 ± 0.34 reduction and of the non-inoculated opposite side of the specimens to a $\log_{10}$ 3.71 ± 0.30 reduction in cfu/mL (Figure 6B).

3.5. Bactericidal Effect of Plasma Pretreatment of Agar Plates Followed by Bacterial Inoculation

In order to check to what extent plasma exposure could also serve as a preventive or prophylactic bactericidal therapy option, we treated agar plates with CAPs generated with the power dissipated in the discharge of 122 mW, 221 mW or 275 mW (**C**), for 5 min, 10 min, 20 min or 30 min. After 24 h, agar plates were inoculated with the bacteria. As shown in Figure 7, bacterial growth was apparently inhibited on the plasma exposed agar areas (Figure 7A–C). The growth inhibition of bacteria apparently correlated with the exposure time and the power dissipated in the discharge (Figure 7D).

Mode 3: Inocculation ➜ plasma-treatment of the inoculated bone surface ➜ 24 h incubation ➜ agar culture

Mode 4: Inocculation ➜ plasma-treatment of the non-inoculated bone surface ➜ 24 h incubation ➜ agar culture

Figure 6. Characterization of the antibacterial effect of DBD CAPs in a model of a bacterially contaminated bone preparation. Punch preparations from the pork shoulder blade were inoculated with the Staphylococcus epidermidis strain (1.5×10^6). Immediately (approximately 10–15 min) after bacterial inoculation, the non-contaminated area (**A**, mode 3) or the bacterially contaminated side (**B**, mode 4) of the specimens were exposed to the DBD plasma (10 min at 275 mW power dissipation in the discharge). After 24 h, the bacteria were detached from the bone preparations and the resulting bacterial solutions of mode 3 or mode 4 were spread out on agar plates (photographic images of mode 3 or mode 4). As a positive control (pos. contr.), the corresponding bone samples were treated identically, but not exposed to plasma. The quantitative evaluation was carried out by counting the photographically recorded colonies on the agar plate using the *imageJ* software. Bars shown represent the mean ± S.D. of six individual experiments. *, $p < 0.05$ as compared to the positive control (pos. contr.).

3.6. Bactericidal Impact of Plasma Treatment of Sterile Bone Samples on the Subsequently Inoculated Bacteria

In a modified experiment with punch preparations from the pork shoulder blade, we treated the bone samples from one side with the CAPs (10 min 275 mW) and after 24 h we inoculated either the treated or the untreated side of the bone sample with bacteria. After a further 24 h, we determined the reduction in the bacterial count (log10 reduction of cfu/mL) as described above. In Figure 8, we show that bacterial inoculation on the plasma treated side (Figure 8A) led to a $\log_{10}$ 3.68 ± 0.26 reduction and of the non-inoculated opposite side of the specimens, to a $\log_{10}$ 3.46 ± 0.27 reduction in cfu/mL (Figure 8B).

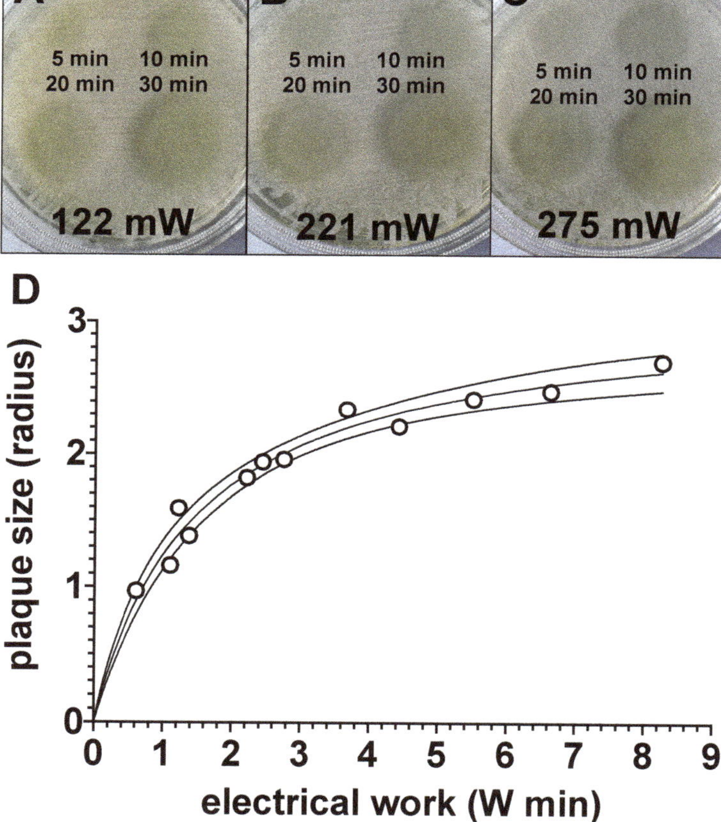

Figure 7. Evaluation of the effect of preventive plasma treatment of bacterial growth agar plates prior to bacterial inoculation. Four quadrants of bacterial agar growth plates were treated with plasma, generated with the power dissipated in the discharge of 122 mW (**A**), 221 mW (**B**) or 275 mW (**C**), for 5 min (top left), 10 min (top right) 20 min (bottom left) or 30 min (bottom right). The used electrode had a total diameter of 10 mm. A total of 24 h after plasma exposure, agar plates were inoculated with bacteria. After a further 24 h, the radii of the bacteria-free areas, as shown in (**A–C**), were measured with a ruler and these values were plotted graphically in relation to the electrical work dissipated in the discharge (**D**). Values shown in (**D**) represent the mean of two individual experiments. The solid line represents the interpolations line; the area between the solid lines represents 95%-asymptotic confidence interval.

Mode 5: Plasma treatment ➔ inocculation of the plasma-treated bone surface ➔ 24 h incubation ➔ agar culture

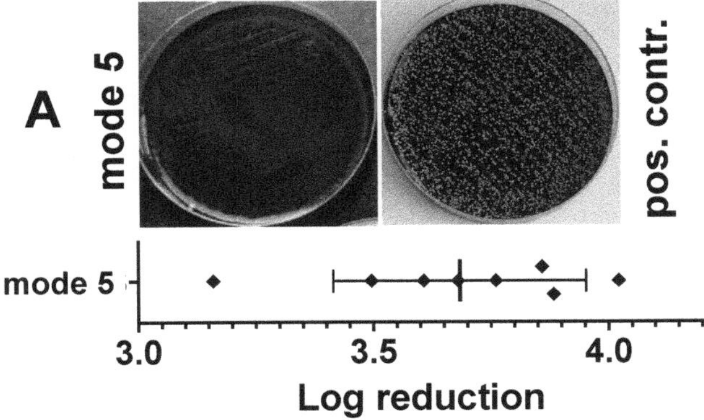

Mode 6: Plasma treatment ➔ inocculation of the non-treated bone surface ➔ 24 h incubation ➔ agar culture

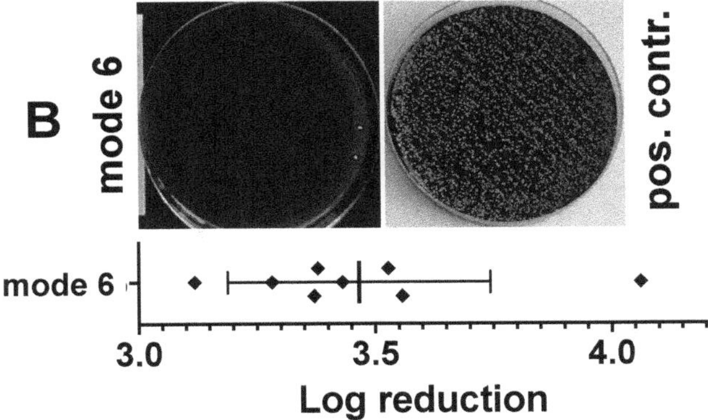

Figure 8. Characterization of the antibacterial effect of a preventive DBD CAP treatment in a model of a bacterially infected bone samples. Punch preparations from the pork shoulder blade were exposed to the DBD plasma (10 min at 275 mW power dissipation in the discharge). Ten minutes after plasma exposure, the non-treated side of the sample (**A**, mode 5) or the plasma exposed bone area of the samples (**B**, mode 6) were inoculated with the Staphylococcus epidermidis strain (1.5×10^6). After 24 h, the bacteria were detached from the bone preparations and the resulting bacterial solutions of mode 5 or mode 6 were spread out on agar plates (photographic images of A; mode 5 or B; mode 6). As a positive control (pos. contr.), the corresponding bone samples were treated identically, but not exposed to plasma. The quantitative evaluation was carried out by counting the photographically recorded colonies on the agar plate using the *imageJ* software. Bars shown represent the mean $\pm$ S.D. of six individual experiments. *, $p < 0.05$ as compared to the positive control (pos. contr.).

4. Discussion

Bacterial infections of the bone tissue, such as osteitis or osteomyelitis, are very serious and unfortunately not uncommon complications and major challenges in trauma and orthopedic surgery. Larger orthopedic interventions or fractures with severe tissue injuries are often associated with a disturbed local immune response, which promotes implant-associated infections due to opportunistic pathogens. A total of 60% of all implant-associated infections are caused almost equally by two strains of staphylococci, *S. aureus*

(34%) and *S. epidermidis* (32%) [26,27]. Microorganisms that have penetrated the wound can then settle on the surface of implants and devitalize areas of bone tissue, where they form a biofilm, a hydrated matrix of extracellular components. In the biofilm, dividing bacteria can be found, which in this planktonic form are potentially sensitive to antibiotics, but most of the bacteria in the biofilm are in their sessile form, which is characterized by a reduced metabolism and reproduction rate, and are therefore far less sensitive to bactericides [28]. The free-swimming planktonic form of bacteria are mainly found in the expansion phase necessary for the colonization of new habitats. Compared to the planktonic form, the biofilm variant offers up to four orders of magnitude higher resistance and tolerance to the immune system, biocides and antibiotics [29,30]. This fact shows how difficult it can be to eradicate biofilm infections. Additionally, the matrix of the biofilm represents an effective barrier between bacteria and immunocompetent cells of the innate or acquired immune response, as well as an effective barrier against antibiotics [31].

The biofilm is the predominant form of bacterial growth on implants and the driving pathological source of chronic ostitis [32,33]. Almost all implant-associated infections caused by biofilm-forming bacteria lead to the development of subacute to chronic ostitis, which is accompanied by progressive chronic bone degeneration with impaired and delayed healing. In order to avoid life-threatening complications, such as sepsis, adequate therapy with an attempt to completely eliminate the pathogen causing ostitis is necessary [34]. Radical debridement, lavage, and removal of implants in combination with systemic or oral antibiotic therapy are, therefore, the gold standard of treatment [35]. However, it should not go unmentioned here that intravenous or oral antibiotic therapies can lead to severe systemic side effects, and often only have minor effects due to insufficient antibiotic concentrations at the local infection site [36].

To date, some molecular factors have been identified which are able to induce a transformation of the antibiotic-resistant sessile and surface-associated biofilm phenotype to the antibiotic-susceptible planktonically spreading phenotype in a biofilm. One of these factors is NO [19,20]. Even very low NO concentrations in the nano molar range lead to an increased dispersion of the biofilm and at the same time, to an increased sensitivity of the biofilm and the dispersed bacteria to several classes of antimicrobial agents, as well as a strong increase in the effectiveness of these antibiotics [37,38]. The dispersion properties of NO seem to be well conserved across bacterial species [19,20]. The ability of NO to increase the dispersion of biofilms represents a great opportunity for the development of new and more efficient therapeutics to control biofilm-related infections and to overcome biofilm resistance. The bacteria would not be directly killed by the low doses of NO, but the released planktonic cells would have a decidedly increased susceptibility to antibiotics and other antimicrobial agents and could thus be effectively eliminated [38,39]. Moreover, NO-based strategies to reduce or eliminate biofilms would greatly benefit from combined treatments with standard antibiotics to prevent or eliminate bacterial infections. In the case of higher NO concentrations, the development of further and significantly more reactive and pathogen-killing nitrogen species (RNS) would be favored, whereby a damaging effect of these RNSs on host cells would also be expected [40]. RNSs can significantly change the functionality of nucleic acids and proteins through nitrosylation or nitration of amine, thiol and tyrosine residues as well as metal centers and, under certain circumstances, irreversibly damage it [41]. Due to the good diffusion capacity of RNS and the aforementioned diverse modes of action, RNSs are effective broad spectrum antimicrobial agents at higher concentrations, which could kill biofilms of gram-negative and gram-positive bacteria [19,20].

At this point, we would like to point out that we here deliberately omit the entire topic of the role, regulation and support of local enzymatic NO production and focus exclusively on therapy options addressed by exogenous application of NO. Interestingly, regarding the clinical picture of ostitis, we could not find any evidence of experimental or clinical therapeutical use of NO in the current literature databases. The reason for this is undoubtedly the lack of availability of suitable NO sources that could be used therapeutically meaningfully after removing the implants, debridement and lavage. The two essential exogenous NO sources that have already been successfully used, e.g., in the treatment of bacterially infected wounds in vivo, are gaseous NO and NO donors [42]. Both are, however, for the local treatment of osteitis or osteomyelitis unsuitable. The long therapeutic duration of use of gaseous NO in high concentrations represents a considerable source of danger and stress for the health of all parties, regardless of the high costs, and therefore is clinically impractical. The use of NO donors for pharmacological therapy of an already existing ostitis appears unsuitable as well, as the degradation of these substances, i.e., the release of NO, varies greatly depending on the humidity, the pH value and the temperature of the environment [43]. Therefore, with NO donors, the desired release kinetics and effective NO concentration cannot be specifically estimated or predicted. It should not go unmentioned, however, that materials scientists in particular, when functionalizing implants with spontaneously NO-releasing systems, pursue the idea of counteracting bacterial infection of the bone as a preventive measure [44].

In contrast to the two NO sources mentioned, a NOD-plasma-based therapeutic approach would be technically and pharmacologically meaningful. By using flexible DBD electrode mats, which can adapt to the anatomical topography of a bone in comparably large areas, treatment could occur in one step. Such a technology, adapted to DBD technology, is already available and could easily be optimized for the situation of an exposed infected bone. In addition, NOD-accumulating CAPs generated by other plasma technologies could of course also be used, such as the indirectly generated CAPs in the plasma-jet variant.

With the help of NOD-plasma-application, a multitude of therapy-specific goals on non-infected, as well as already bacterially infected bone tissue, could be addressed by combining the duration of application and the power dissipated in the discharge. In the case of non-infected bones, as a preventive treatment, particularly endangered areas could be treated differently depending on the existing risk of bacterial infection. We are thinking, for example, of the areas of the bone fracture after the bone has been repositioned and retained. In the case of open fractures and environmental exposure of the bone stumps, a preventive NOD-plasma-exposure of the bone stumps prior to repositioning could help to prevent infections. In addition, preventive treatment of bone areas that are particularly susceptible to infection would be justified. Of particular importance here are the screw holes, which represent a relevant source of risk for pathogens to penetrate deeper areas of the bone tissue. In each of the scenarios mentioned, the NOD-plasma-exposed bone tissue could be "charged" to different degrees with acidified NO derivatives, depending on the duration of exposure and discharge capacity, analogous to our results that we carried out with human skin tissue [15]. The resulting release of NO in an acidic environment, alone or synergistically with antibiotics, would help to prevent a possible bacterial contamination and infection of the operating area during a surgical procedure.

At the time of our investigations with human skin tissue [15], however, it was not yet clear whether the observed accumulation of NODs in the plasma-exposed samples was directly due to the NO contained in the plasma, or was mediated by nitrogen oxides other than NO. The data presented here strongly suggest that the NO contained in the plasma is not primarily responsible for the effects observed in our bone model. The DBD source used here generated, at the lowest discharge power (125 mW), plasmas with a NO concentration of approximately 200 ppb, but an approximately 500-fold higher concentration of NO_2 (100 ppm) [15]. We therefore postulate that the NO_2 contained in the CAP in the moist environment of the exposed biological tissue essentially leads to the generation of nitrous acid (HNO_2) and nitric acid (HNO_3) (Equation (1)). Nitrous acid (HNO_2) generates consequently gaseous NO (Equation (2), Equation (3)), as well as the formation of nitrite (Equation (4)) that under acidic conditions supports HNO_2 formation and therefore additional release of NO (Equation (5)).

$$2NO_2 + H_2O \rightleftharpoons HNO_3 + HNO_2 \tag{1}$$

$$3HNO_2 \rightleftharpoons HNO_3 + 2NO(g) + H_2O \tag{2}$$

$$HNO_2 \rightleftharpoons NO + NO_2 + H_2O \tag{3}$$

$$2NO_2(g) + H_2O \rightleftharpoons 2H^+ + NO_2^- + NO_3^-. \tag{4}$$

$$H^+ + NO_2^- \rightleftharpoons HNO_2(aq) \; HNO_2(g) \tag{5}$$

These assumptions are supported by our observations that the NOD-plasma treatment of the aqueous solutions led to a significant pH decrease and aimed for the pH value of 3.29 that is characteristic of HNO_2, and at the same time induced a long-lasting release of NO from the exposed solution.

In the concept of the NOD-plasma-based therapy option against biofilms, however, another aspect would play the essential role. Under acidic conditions, nitrite reacts to form nitrous acid and other nitrogenous metabolites [45] and shows a pronounced antimicrobial effect against bacteria, fungi and the common oral and skin pathogens, as well as bacterial spores [46,47]. Interestingly, acidified nitrite also enhances the antibacterial and fungicidal effects of hydrogen peroxide [48]. In this context, it is important to mention that DBD CAPs, in addition to NODs, also lead to accumulation of a not inconsiderable amount of hydrogen peroxide into treated biological samples [49]. It is therefore very likely that these two factors are largely responsible for the antimicrobial effect of CAPs, as was already postulated by Naitali et al. [50]. In particular, the generation of peroxynitrite from the reaction of nitrite and H_2O_2 under acidic conditions could be the reason behind the increased antimicrobial effect of CAPs [51].

Of course, serious concerns about possible risks and side effects could be raised with regard to CAP-based therapy on bone tissue. With intensive use of NOD-plasma, accompanied by a temporary strong acidification of the bone tissue and the entry of high amounts of NODs, as would be the case in the treatment of severe biofilm contamination of bone tissue, cytotoxic events on cells of the bone tissue could not be completely avoided. Previous data show [52,53] that ROS/RNS-containing plasmas can have a pronounced cytotoxic character, depending on the duration of application or ROS/RNS concentration. The cytotoxic mechanism includes a structural change in proteins, nucleic acids and lipid membranes caused by oxidative and nitrosative stress leading to the loss of the biological function of these structures, up to the induction of apoptotic or necrotic cell death. It is also such a cytotoxic aspect of CAPs that could limit potential therapeutic applications based on the intensive use of highly reactive plasmas. However, the cytotoxic character of potentially cytotoxic plasmas in the treatment of bacterially infected bone tissue would occupy a special position. In such a case, even a partial devitalization of the treated bone tissue would be tolerable to a large extent, as it can be assumed that such a devitalization would probably only be a temporary side effect of plasma therapy. As we show here, even with the highest rate of accumulation of NO derivatives, we could no longer detect any relevant amounts of NO activity in bone tissue, at the latest 72 h after exposure to plasma. This would enable the remaining sterile bone matrix tissue during the subsequent post-therapy regeneration phase to be repopulated by cells immigrating from the biologically intact edges and a revitalized intact bone tissue would be created again.

5. Conclusions

DBD-generated CAPs, energetically optimized for nitric oxide chemistry, can "charge" bone tissue with appropriate amounts of NO derivatives and lead to the establishment of a therapeutically relevant NO-releasing system in bone tissue. The quantity of NO equivalents introduced into the bone tissue appears to be solely a function of the exposure time and the amount of electrical work used to generate the plasma. It is thus possible, depending on the therapeutic needs required, to treat bone tissue in a targeted manner on its surface or also in the depth of the tissue with therapeutically relevant amounts of NOD. In the near future, depending on the amount of enriched NODs in the bone tissue, NOD-plasma-assisted therapy of bacterial osteitis could achieve both NO-dependent spread and NOD-induced destruction of a biofilm. Thus, this technology could be used successfully as an accompanying therapy option to other forms of therapy, but also as the sole therapy option in the fight against bacterial bone infections.

Author Contributions: D.F., collection and assembly of data, study design, data analysis and interpretation, manuscript writing; A.K., collection and assembly of data, data analysis and interpretation; C.O., assembly of data, data analysis, manuscript writing; G.G., assembly of data, data analysis; J.W., final approval of manuscript, provision of study material or patients, data analysis and interpretation; C.V.S., conception and design, data analysis and interpretation, manuscript writing. All authors have read and agreed to the published version of the manuscript.

Funding: This work was supported by the Faculty of Medicine of the Heinrich-Heine-University Düsseldorf.

Institutional Review Board Statement: The experimental protocol and the use of human material have been approved by the local Ethics Committee of the Medical Faculty of the Heinrich-Heine-University Düsseldorf (study number: 3634). All experiments were conducted in compliance with the Declaration of Helsinki principles.

Informed Consent Statement: Not applicable.

Data Availability Statement: The datasets used and/or analyzed during the current study are available from the corresponding author on reasonable request.

Acknowledgments: We thank Christa Maria Wilkens, Samira Seghrouchni, Jutta Schneider, and Sabine Lensing-Hoehn for technical assistance.

Conflicts of Interest: The authors state no conflict of interest.

References

1. Fridman, G.; Friedman, G.; Gutsol, A.; Shekhter, A.; Vasilets, V.; Fridman, A. Applied plasma medicine. *Plasma Process. Polym.* **2008**, *5*, 503–533. [CrossRef]
2. Ermolaeva, S.; Sysolyatina, E.; Gintsburg, A. Atmospheric pressure nonthermal plasmas for bacterial biofilm prevention and eradication. *Biointerphases* **2015**, *10*, 029404. [CrossRef] [PubMed]
3. Yousfi, M.; Merbahi, N.; Pathak, A.; Eichwald, O. Low-temperature plasmas at atmospheric pressure: Toward new pharmaceutical treatments in medicine. *Fundam. Clin. Pharmacol.* **2014**, *28*, 123–135. [CrossRef]
4. Kong, M.; Kroesen, G.; Morfill, G.; Nosenko, T.; Shimizu, T.; van Dijk, J.; Zimmermann, J.L. Plasma medicine: An introductory review. *New J. Phys.* **2009**, *11*, 115012. [CrossRef]
5. Laroussi, M.; Leipold, F. Evaluation of the roles of reactive species, heat, and UV radiation in the inactivation of bacterial cells by air plasmas at atmospheric pressure. *Int. J. Mass Spectrom.* **2004**, *233*, 81–86. [CrossRef]
6. Nosenko, T.; Shimizu, T.; Steffes, B.; Zimmermann, J.; Stolz, W.; Schmidt, H.; Isbary, G.; Pompl, R.; Bunk, W.; Fujii, S.; et al. Low-Temperature Atmospheric-Pressure Plasmas as a Source of Reactive Oxygen and Nitrogen Species for Chronic Wound Disinfection. *Free Radic. Bio. Med.* **2009**, *47*, S128.
7. Bruch-Gerharz, D.; Ruzicka, T.; Kolb-Bachofen, V. Nitric oxide and its implications in skin homeostasis and disease—A review. *Arch. Dermatol. Res.* **1998**, *290*, 643–651. [CrossRef]
8. Stuehr, D.J. Mammalian nitric oxide synthases. *Biochim. Biophys. Acta* **1999**, *1411*, 217–230. [CrossRef]
9. Cals-Grierson, M.-M.; Ormerod, A. Nitric oxide function in the skin. *Nitric Oxide* **2004**, *10*, 179–193. [CrossRef] [PubMed]
10. Kandhwal, M.; Behl, T.; Kumar, A.; Arora, S. Understanding the Potential Role and Delivery Approaches of Nitric Oxide in Chronic Wound Healing Management. *Curr. Pharm. Des.* **2021**, *27*, 1999–2014. [CrossRef] [PubMed]
11. Soneja, A.; Drews, M.; Malinski, T. Role of nitric oxide, nitroxidative and oxidative stress in wound healing. *Pharmacol. Rep.* **2005**, *57*, 108–119. [PubMed]
12. Isenberg, J.S.; Ridnour, L.A.; Espey, M.G.; Wink, D.A.; Roberts, D.A. Nitric oxide in wound-healing. *Microsurgery* **2005**, *25*, 442–451. [CrossRef]
13. Isbary, G.; Morfill, G.; Schmidt, H.; Georgi, M.; Ramrath, K.; Heinlin, J.; Karrer, S.; Landthaler, M.; Shimizu, T.; Steffes, B.; et al. A first prospective randomized controlled trial to decrease bacterial load using cold atmospheric argon plasma on chronic wounds in patients. *Br. J. Dermatol.* **2010**, *163*, 78–82. [CrossRef]
14. Suschek, C.V.; Opländer, C. The application of cold atmospheric plasma in medicine: The potential role of nitric oxide in plasma-induced effects. *Clin. Plasma Med.* **2016**, *4*, 1–8. [CrossRef]
15. HHeuer, K.; Hoffmanns, M.A.; Demir, E.; Baldus, S.; Volkmar, C.M.; Röhle, M.; Fuchs, P.C.; Awakowicz, P.; Suschek, C.V.; Opländer, C. The topical use of non-thermal dielectric barrier discharge (DBD): Nitric oxide related effects on human skin. *Nitric Oxide* **2015**, *44*, 52–60. [CrossRef] [PubMed]

16. Rajasekaran, P.; Opländer, C.; Hoffmeister, D.; Bibinov, N.; Suschek, C.V.; Wandke, D.; Awakowicz, P. Characterization of Dielectric Barrier Discharge (DBD) on Mouse and Histological Evaluation of the Plasma-Treated Tissue. *Plasma Process. Polym.* **2011**, *8*, 246–255. [CrossRef]

17. Park, K.-H.; Cho, O.H.; Jung, M.; Suk, K.-S.; Lee, J.H.; Park, J.S.; Ryu, K.N.; Kim, S.-H.; Lee, S.-O.; Choi, S.-H.; et al. Clinical characteristics and outcomes of hematogenous vertebral osteomyelitis caused by gram-negative bacteria. *J. Infect.* **2014**, *69*, 42–50. [CrossRef]

18. Masters, E.; Trombetta, R.; de Mesy Bentley, K.; Boyce, B.; Gill, A.; Gill, S.; Nishitani, K.; Ishikawa, M.; Morita, Y.; Ito, H.; et al. Evolving concepts in bone infection: Redefining "biofilm", "acute vs. chronic osteomyelitis", "the immune proteome" and "local antibiotic therapy". *Bone Res.* **2019**, *7*, 20. [CrossRef]

19. Arora, D.P.; Hossain, S.; Xu, Y.; Boon, E.M. Nitric Oxide Regulation of Bacterial Biofilms. *Biochemistry* **2015**, *54*, 3717–3728. [CrossRef]

20. Barraud, N.; Kelso, M.; Rice, S.; Kjelleberg, S. Nitric oxide: A key mediator of biofilm dispersal with applications in infectious diseases. *Curr. Pharm. Des.* **2015**, *21*, 31–42. [CrossRef] [PubMed]

21. MacMicking, J.; Xie, Q.; Nathan, C. Nitric oxide and macrophage function. *Annu. Rev. Immunol.* **1997**, *15*, 323–350. [CrossRef]

22. Saura, M.; Tarin, C.; Zaragoza, C. Recent insights into the implication of nitric oxide in osteoblast differentiation and proliferation during bone development. *Sci. World J.* **2010**, *10*, 624–632. [CrossRef] [PubMed]

23. Teixeira, C.; Agoston, H.; Beier, F. Nitric oxide, C-type natriuretic peptide and cGMP as regulators of endochondral ossification. *Dev. Biol.* **2008**, *319*, 171–178. [CrossRef] [PubMed]

24. Suschek, C.V.; Paunel, A.; Kolb-Bachofen, V. Nonenzymatic nitric oxide formation during UVA irradiation of human skin: Experimental setups and ways to measure. *Method Enzymol.* **2005**, *396*, 568–578.

25. Yang, F.; Troncy, E.; Francoeur, M.; Vinet, B.; Vinay, P.; Czaika, G.; Blaise, G. Effects of reducing reagents and temperature on conversion of nitrite and nitrate to nitric oxide and detection of NO by chemiluminescence. *Clin. Chem.* **1997**, *43*, 657–662. [CrossRef]

26. Campoccia, D.; Montanaro, L.; Arciola, C.R. The significance of infection related to orthopedic devices and issues of antibiotic resistance. *Biomaterials* **2006**, *27*, 2331–2339. [CrossRef] [PubMed]

27. Wright, J.A.; Nair, S.P. Interaction of staphylococci with bone. *Int. J. Med. Microbiol.* **2010**, *300*, 193–204. [CrossRef]

28. Costerton, J.W.; Montanaro, L.; Arciola, C.R. Biofilm in implant infections: Its production and regulation. *Int. J. Artif. Organs* **2005**, *28*, 1062–1068. [CrossRef]

29. Stewart, P.S.; Costerton, J.W. Antibiotic resistance of bacteria in biofilms. *Lancet* **2001**, *358*, 135–138. [CrossRef] [PubMed]

30. Buckingham-Meyer, K.; Goeres, D.M.; Hamilton, M.A. Comparative evaluation of biofilm disinfectant efficacy tests. *J. Microbiol. Methods* **2007**, *70*, 236–244. [CrossRef]

31. Mah, T.F.; O'Toole, G.A. Mechanisms of biofilm resistance to antimicrobial agents. *Trends Microbiol.* **2001**, *9*, 34–39. [CrossRef]

32. An, Y.H.; Friedman, R.J. Concise review of mechanisms of bacterial adhesion to biomaterial surfaces. *J. Biomed. Mater. Res.* **1998**, *43*, 338–348. [CrossRef]

33. Hudson, M.C.; Ramp, W.K.; Frankenburg, K.P. Staphylococcus aureus adhesion to bone matrix and bone-associated biomaterials. *FEMS Microbiol. Lett.* **1999**, *173*, 279–284. [CrossRef]

34. Hotchen, A.J.; McNally, M.A.; Sendi, P. The Classification of Long Bone Osteomyelitis: A Systemic Review of the Literature. *J. Bone Jt. Infect.* **2017**, *2*, 167–174. [CrossRef]

35. Brooks, B.D.; Sinclair, K.D.; Grainger, D.W.; Brooks, A.E. A resorbable antibiotic-eluting polymer composite bone void filler for perioperative infection prevention in a rabbit radial defect model. *PLoS ONE* **2015**, *10*, e0118696. [CrossRef]

36. Geurts, J.; Arts, J.J.C.; Walenkamp, G.H. Bone graft substitutes in active or suspected infection. Contra-indicated or not? *Injury* **2011**, *42* (Suppl. 2), S82–S86. [CrossRef]

37. Barraud, N.; Hassett, D.J.; Hwang, S.H.; Rice, S.A.; Kjelleberg, S.; Webb, J.S. Involvement of nitric oxide in biofilm dispersal of Pseudomonas aeruginosa. *J. Bacteriol.* **2006**, *188*, 7344–7353. [CrossRef]

38. Barraud, N.; Storey, M.V.; Moore, Z.P.; Webb, J.S.; Rice, S.A.; Kjelleberg, S. Nitric oxide-mediated dispersal in single- and multi-species biofilms of clinically and industrially relevant microorganisms. *Microb. Biotechnol.* **2009**, *2*, 370–378. [CrossRef]

39. Barraud, N.; Schleheck, D.; Klebensberger, J.; Webb, J.S.; Hassett, D.J.; Rice, S.A.; Kjelleberg, S. Nitric oxide signaling in Pseudomonas aeruginosa biofilms mediates phosphodiesterase activity, decreased cyclic di-GMP levels, and enhanced dispersal. *J. Bacteriol.* **2009**, *191*, 7333–7342. [CrossRef]

40. Ridnour, L.A.; Thomas, D.D.; Mancardi, D.; Espey, M.G.; Miranda, K.M.; Paolocci, N.; Feelisch, M.; Fukuto, J.; Wink, D.A. The chemistry of nitrosative stress induced by nitric oxide and reactive nitrogen oxide species. Putting perspective on stressful biological situations. *Biol. Chem.* **2004**, *385*, 1–10. [CrossRef]

41. Wallace, J.L. Nitric oxide as a regulator of inflammatory processes. *Mem. Inst. Oswaldo Cruz.* **2005**, *100* (Suppl. 1), 5–9. [CrossRef] [PubMed]

42. Malone-Povolny, M.J.; Maloney, S.E.; Schoenfisch, M.H. Nitric Oxide Therapy for Diabetic Wound Healing. *Adv. Healthc. Mater.* **2019**, *8*, e1801210. [CrossRef] [PubMed]

43. Yamamoto, T.; Bing, R.J. Nitric oxide donors. *Proc. Soc. Exp. Biol. Med.* **2000**, *225*, 200–206. [CrossRef] [PubMed]

44. Jennison, T.; McNally, M.; Pandit, H. Prevention of infection in external fixator pin sites. *Acta Biomater.* **2014**, *10*, 595–603. [CrossRef]

45. Xu, J.; Xu, X.; Verstraete, W. The bactericidal effect and chemical reactions of acidified nitrite under conditions simulating the stomach. *J. Appl. Microbiol.* **2001**, *90*, 523–529. [CrossRef]
46. Xia, D.S.; Liu, Y.; Zhang, C.M.; Yang, S.H.; Wang, S.L. Antimicrobial effect of acidified nitrate and nitrite on six common oral pathogens In Vitro. *Chin. Med. J.* **2006**, *119*, 1904–1909. [CrossRef]
47. Szabo, J.G.; Adcock, N.J.; Rice, E.W. Disinfection of Bacillus spores with acidified nitrite. *Chemosphere* **2014**, *113*, 171–174. [CrossRef] [PubMed]
48. Heaselgrave, W.; Andrew, P.W.; Kilvington, S. Acidified nitrite enhances hydrogen peroxide disinfection of Acanthamoeba, bacteria and fungi. *J. Antimicrob. Chemother.* **2010**, *65*, 1207–1214. [CrossRef]
49. Balzer, J.; Heuer, K.; Demir, E.; Hoffmanns, M.A.; Baldus, S.; Fuchs, P.C.; Awakowicz, P.; Suschek, C.V.; Oplander, C. Non-Thermal Dielectric Barrier Discharge (DBD) Effects on Proliferation and Differentiation of Human Fibroblasts Are Primary Mediated by Hydrogen Peroxide. *PLoS ONE* **2015**, *10*, e0144968. [CrossRef]
50. Naitali, M.; Kamgang-Youbi, G.; Herry, J.M.; Bellon-Fontaine, M.N.; Brisset, J.L. Combined effects of long-living chemical species during microbial inactivation using atmospheric plasma-treated water. *Appl. Environ. Microbiol.* **2010**, *76*, 7662–7664. [CrossRef]
51. Saha, A.; Goldstein, S.; Cabelli, D.; Czapski, G. Determination of optimal conditions for synthesis of peroxynitrite by mixing acidified hydrogen peroxide with nitrite. *Free Radic. Biol. Med.* **1998**, *24*, 653–659. [CrossRef] [PubMed]
52. Aryal, S.; Bisht, G. New paradigm for a targeted cancer therapeutic approach: A short review on Potential Synergy of Gold nanoparticles and cold atmospheric plasma. *Biomedicines* **2017**, *5*, 38. [CrossRef] [PubMed]
53. Boehm, D.; Bourke, P. Safety implications of plasma-induced effects in living cells—A review of In Vitro and In Vivo findings. *Biol. Chem.* **2018**, *400*, 3–17. [CrossRef] [PubMed]

MDPI AG
Grosspeteranlage 5
4052 Basel
Switzerland
Tel.: +41 61 683 77 34

Biomedicines Editorial Office
E-mail: biomedicines@mdpi.com
www.mdpi.com/journal/biomedicines